Order this book online at www.trafford.com
or email orders@trafford.com

Most Trafford titles are also available at major online book retailers.

Print information available on the last page.

ISBN: 978-1-4120-2093-0 (sc)
ISBN: 978-1-4122-2096-5 (e)

Trafford rev. 11/11/2015

www.trafford.com
North America & international
toll-free: 1 888 232 4444 (USA & Canada)
fax: 812 355 4082

MECHANICS MADE EASY
How To Solve Mechanics Problems

⇒ Features over 300 fully-solved problems
⇒ Has over 500 line diagrams
⇒ Supplements standard introductory-level texts
⇒ Covers twelve topics:
   1. Linear Velocity and Acceleration
   2. Relative Motion
   3. Projectiles
   4. Circular Motion
   5. Collisions
   6. Laws of Motion
   7. Jointed Rods
   8. Equilibrium
   9. Motion of a Rigid Body
   10. Hydrostatics
   11. Differential Methods
   12. Simple Harmonic Motion

By
David G. Reynolds

Introduction

The fascinating subject of mechanics provides an insight and the inter-relationships between mass, time, distance, velocity, momentum, acceleration, force, energy and power. In turn this improves our understanding of the workings of our everyday world.

An effective way to learn about mechanics is to solve mechanics problems. "Mechanics Made Easy (How To Solve Mechanics Problems)" is designed to supplement standard introductory-level school, college and university texts on this subject. The book consists of over 300 mechanics problems and step-by-step worked solutions in twelve topics:
1. Linear Velocity and Acceleration
2. Relative Motion
3. Projectiles
4. Circular motion
5. Collisions
6. Laws of Motion
7. Jointed Rods
8. Equilibrium
9. Motion of a Rigid Body
10. Hydrostatics
11. Differentiation and Integration
12. Simple Harmonic Motion

Over 500 clear, concise diagrams are provided to assist understanding of both problems and solutions. Working through these problems can help the reader improve problem-solving skills and gain the confidence to tackle similar questions.

CONTENTS

Chapter 1   Linear Velocity and Acceleration

1. Define the terms displacement, velocity and acceleration

Displacement: A displacement is a change is position.
Velocity: Velocity is the rate of change of displacement with respect to time.
Acceleration: Acceleration is the rate of change of velocity with respect to time.

2. Equations of motion
For a particle: t = time in seconds, u = initial velocity in m/s (at time t = 0), v = velocity in m/s after t seconds, s = distance travelled in metres in t seconds, a = acceleration in m/s². Prove that: (a) $v = u + at$ (b) $s = ut + \frac{1}{2}at^2$ (c) $v^2 = u^2 + 2as$

(a) Show: $v = u + at$
With initial velocity = u m/s and a constant acceleration of a m/s² the velocity after 1, 2 and 3 seconds is: $u + (a)(1) = u + a$ m/s, $u + 2a$ m/s, $u + 3a$ m/s, etc.
In general, velocity after t seconds = $u + at \Rightarrow v = u + at$ (i)
This is represented graphically on a velocity-time graph (See Fig. 1).

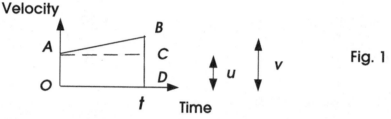

Velocity

Fig. 1

OA represents initial velocity u at time t = 0. Note: $|OA| = |CD|$
BD represents velocity v after t seconds. But from Fig. 1: $|BD| = |BC| + |CD|$
$\Rightarrow v = |BC| + u \Rightarrow |BC| = a \times t = at$ = increase in velocity over time t

(b) Show: $s = ut + \frac{1}{2}at^2$

s = distance travelled = average velocity × time (= Area OABD)
From Fig. 1: Average velocity = ½ $(u + v)$

$$\Rightarrow s = \left(\frac{u+v}{2}\right)t \quad \text{(ii)}\quad \text{But, from equation (i) above :} \quad v = u + at$$

$$\Rightarrow s = \frac{(u+(u+at))}{2}t = \frac{2ut}{2} + \frac{at^2}{2} = ut + \frac{1}{2}at^2 \Rightarrow s = ut + \frac{1}{2}at^2$$

(c) Show: $v^2 = u^2 + 2as$

From equation (i): $v - u = at$ From equation (ii): $s = \frac{(u+v)}{2}t \Rightarrow \frac{2s}{t} = u + v$

$$\Rightarrow (v-u)(v+u) = (at)\frac{(2s)}{t} \Rightarrow v^2 - u^2 = 2as \Rightarrow v^2 = u^2 + 2as$$

3. Particle passing three fixed points in a line
A particle moves with constant acceleration and passes points A, B and C in a straight line. If $|AB|$ = 100 metres and $|BC|$ = 100 metres and the particle takes 12 seconds to travel $|AB|$ and 6 seconds to travel $|BC|$ find the initial velocity (at A), acceleration and final velocity (at B).

Solution

See Velocity – Time Graph (Fig. 1). Let the velocity at B be v_1

From above : $s = \left(\dfrac{u+v}{2}\right)t \Rightarrow 200 = \left(\dfrac{u+v}{2}\right)18 \Rightarrow u+v = 22.22$

But, from equation : $v = u + at \Rightarrow v_1 = u + 12a$ after 12 seconds

Using : $|AB| = \left(\dfrac{u+v_1}{2}\right)t \Rightarrow 100 = \left(\dfrac{u+u+12a}{2}\right)12 \Rightarrow 8.333 = u + 6a$

$200 = \left(\dfrac{u+u+18a}{2}\right)18 \Rightarrow 11.111 = u + 9a \Rightarrow a = 0.926 \text{ m/s}^2 \Rightarrow u = 2.777 \text{ m/s} \Rightarrow v = 19.44 \text{ m/s}$

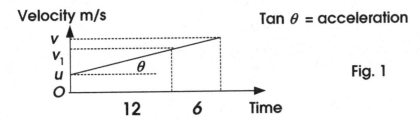

Velocity m/s

Tan θ = acceleration

Fig. 1

---

4. Car accelerates from rest

A car accelerates from rest at 4 m/s². Find its speed after it has travelled 50 m.

Solution using standard equations

Given: $u = 0$, $a = 4$, $s = 50$ $\Rightarrow$ use: $v^2 = u^2 + 2as \Rightarrow v^2 = 0 + 2(4)(50) = 400 \Rightarrow v = 20$ m/s

5. Car accelerates to a higher speed

A racing car accelerates from 40 m/s to 80 m/s over a distance of 400 m. Find the acceleration and the time taken.

Solution using Velocity-Time Graph

Distance covered in time t = area ABCD = $(80+40)/2 \times t = 400$

$\Rightarrow 60\,t = 400 \Rightarrow t = 20/3$ seconds $\Rightarrow$ acceleration = $40\ /\ (20/3) = 6$ m/s²

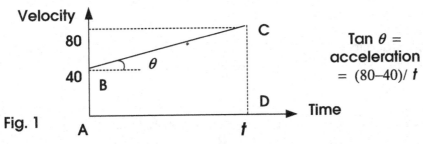

Tan θ = acceleration = $(80–40)/\,t$

Fig. 1

6. Car accelerating from rest

A car, starting from rest, travels 300 metres in 10 seconds. What is its acceleration?

Solution using standard equations

Information given: $u = 0$, $s = 300$, $t = 10$

Use equation : $s = ut + \dfrac{1}{2}at^2 \Rightarrow 300 = (0)(10) + \dfrac{1}{2}a(10)^2 = 50a \Rightarrow a = 6$ m/s²

7. Car accelerating to a higher speed

A car increases its speed from 2.5 m/s to 10 m/s over a distance of 35 metres. What is its acceleration?

Solution using standard equations

Information given: $u = 2.5$, $v = 10$, $s = 35$ $\Rightarrow$ use: $v^2 = u^2 + 2as$

Use equation : $v^2 = u^2 + 2as \Rightarrow (10)^2 = (2.5)^2 + 2a(35) \Rightarrow a = 1.34$ m/s^2

8. Car accelerating from rest
Find the velocity of a car which starts from rest and moves with acceleration of 0.6 m/s^2 over a distance of 6 metres. What is the time taken?

Solution using standard equations
Information given: $u = 0$, $s = 6$, $a = 0.6 \Rightarrow$ use: $v^2 = u^2 + 2as$
Find the velocity: Use equation : $v^2 = u^2 + 2as \Rightarrow v^2 = (0)^2 + 2(0.6)(6) \Rightarrow v = 2.68$ m/s
Time taken: Use equation : $s = ut + \dfrac{1}{2}at^2 \Rightarrow 6 = (0)t + \dfrac{1}{2}(0.6)t^2 \Rightarrow t = 2\sqrt{5} = 4.47$ seconds

Solution using the Velocity – Time graph: (See Fig. 1)
Distance travelled = area under $uv = \left(\dfrac{u+v}{2}\right)t \Rightarrow 6 = \dfrac{(0+v)}{2}(t) \Rightarrow v = \dfrac{12}{t}$

Acceleration $= \dfrac{v-u}{t} \Rightarrow 0.6 = \dfrac{\dfrac{12}{t} - 0}{t} \Rightarrow t = \sqrt{20} = 4.47$ seconds

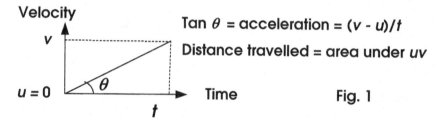

Velocity

Tan θ = acceleration = (v – u)/t

Distance travelled = area under uv

v

$u = 0$

θ

Time

t

Fig. 1

9. Car accelerates from rest and the decelerates to rest
A car starting from rest travels in a straight line, first with acceleration 2 m/s^2 to a maximum speed and then with deceleration 5 m/s^2 to rest. If the total time taken is $t = 21$ seconds and total distance covered = s, find: Time spent accelerating, Time spent decelerating, Maximum speed, Distance covered

Solution using standard equations
Consider the journey in two Stages. Use equations: (i) $s = ut + \frac{1}{2}at^2$ (ii) $v = u + at$

Stage 1: Acceleration:
Initial velocity = 0 m/s, Acceleration = 2 m/s^2, Time = t_1 seconds, Distance = s_1 metres.
From equations (i) and (ii) we have: $s_1 = t_1^2$ and $v = 2t_1$

Stage 2: Deceleration:
Initial velocity = $2t_1$ m/s, Final velocity = 0 m/s, Deceleration = -5 m/s^2, Time = t_2 seconds, Distance = s_2 metres
From equations (i), (ii) we have: $0 = v - 5t_2 \Rightarrow v = 5t_2 \Rightarrow 2t_1 = 5t_2$ (iii)$\Rightarrow t_2 = 0.4\,t_1$
But: $t_1 + t_2 = t = 21 = 1.4\,t_1 \Rightarrow t_1 = 15$ seconds (i) $\Rightarrow t_2 = 6$ seconds (ii)
$\Rightarrow v = 30$ m/s (iii) Also: $s = s_1 + s_2$ Using equation (i): $s = ut + \frac{1}{2}at^2$ gives:
$s_1 = (0)15 + 0.5(2)(225) = 225$, $s_2 = 30(6) - 0.5(5)(36) = 90 \Rightarrow s = 225 + 90 = 315$ metres

10. Descending lift

A lift starts from rest, descends part of its journey with acceleration a m/s² and the remainder with deceleration $4a$ m/s² until it comes to rest. Show that if h is the total depth of the lift shaft and t is the total time taken, then: $h = \frac{2}{5}at^2$.

Solution using Velocity-Time Graph

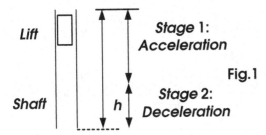

Lift

Shaft

h

Stage 1:
Acceleration

Fig.1

Stage 2:
Deceleration

From Fig. 2: $t = t_1 + t_2 = v/a + v/4a = 5v/4a \Rightarrow v = 4at/5$
Distance covered $= h = $ Area of triangle OAB + Area of triangle ABC
$\Rightarrow h = \frac{1}{2}vt_1 + \frac{1}{2}vt_2 = \frac{1}{2}vt = \frac{1}{2}(4at/5)t = \frac{2}{5}at^2$.

Velocity

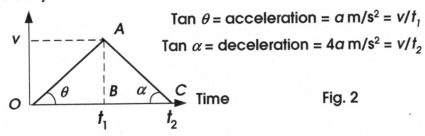

Tan θ = acceleration = a m/s² = v/t_1

Tan α = deceleration = $4a$ m/s² = v/t_2

Fig. 2

11. Particle projected vertically upwards
A particle is projected vertically upwards from the top of a 50-metre high cliff with a velocity of 40 m/s. Find the maximum height reached, time taken to reach this height, time taken to reach the foot of the cliff, and the particle speed at the foot of the cliff

Solution
(a) The maximum height reached
Use equation : $v^2 = u^2 + 2as \Rightarrow 0 = u^2 - 2gs \Rightarrow 0 = 1,600 - 2gs \Rightarrow s = 81.55$ metres
(b) The time taken to reach the maximum height
Use equation: $v = u + at$
At maximum height $v = 0 \Rightarrow 0 = u - gt \Rightarrow 0 = 40 - gt \Rightarrow t = 4.078$ seconds
(c) The time taken to reach the foot of the cliff
Use equation : $s = ut + \frac{1}{2}at^2$ i.e. $s = ut - \frac{1}{2}gt^2$ where : $s = -50$

$\Rightarrow -50 = 40t - 4.905t^2 \Rightarrow t^2 - 8.155t - 10.194 = 0 \Rightarrow t = \dfrac{8.155 \pm 10.358}{2} = 9.26$ s

(d) The particle's speed at the foot of the cliff
Use equation : $v^2 = u^2 + 2as$ i.e. $v^2 = u^2 - 2gs$ where : $s = -50$
$\Rightarrow v^2 = 1,600 - 2g(-50) \Rightarrow v = 50.8$ m/s

12. Stone falls from ascending balloon

A balloon at rest on the ground ascends from rest with uniform acceleration. A stone is dropped from the balloon after 7 seconds and reaches the ground 5 seconds later. Find the balloon's acceleration and the height from which the stone fell.

Solution

Use general equations as above: (i) $v = u + at$ (ii) $s = ut + \frac{1}{2}at^2$

Let: a = Acceleration of balloon in m/s², u = Initial velocity of balloon = 0 m/s

Assume that there are two Stages in the stone's motion:
Stage 1: the stone is accelerating upwards with the balloon (See Figs 1 and 2)
Stage 2: the stone falls from the balloon, rises against gravity for a short time until it reaches its greatest height and then falls to the ground. (See Fig. 3)

Stage 1:
Using equations (i) and (ii) it can be shown that, after 7 seconds:
Velocity = $v_1 = 0 + 7a = 7a$ m/s and

Height $= s = (0)(7) + \frac{1}{2}(a)(7)^2 = 0 + \frac{1}{2}(a)(49) = \frac{49}{2}a$ metres   Equation (iii)

Stage 2:
When the stone is dropped, it has a velocity of $7a$ m/s upwards and an acceleration of g m/s² downwards.

Thus, using the equation : $s = ut + \frac{1}{2}at^2$: Distance to ground $= s = (7a)t - \frac{1}{2}(g)(t)^2$

But : $s = -\frac{49}{2}a$,   (Given): $t = 5$ seconds $\Rightarrow \left(-\frac{49}{2}a\right) = (7a)5 - \frac{1}{2}(g)(5)^2$

$\Rightarrow a = 2.061$ m/s² $\Rightarrow s = 50.49 \Rightarrow$ Stone fell from a height of 50.49 metres

13. Stones dropped from tower

Stone A is dropped from the top of a tower of height H metres, and one second later stone B is dropped from a point on the tower which is 15 metres below the first point. If both stones reach the ground at the same instant find the height of the tower.

Solution

The height of the tower is (See Fig. 1): $H = 15 + h_2$ Also, from Fig. 2: $H = s_1 + s_2$
Use general equations as before: (i) $s = ut + \frac{1}{2}at^2$ (ii) $v = u + at$
Assume Stone A is falling from rest:
In the first second it covers the distance s_1: $s_1 = 0(1) + \frac{1}{2}g(1)^2 = \frac{1}{2}g$ (iii)
After 1 second it is travelling with a velocity of: $v = 0 + g(1) = g$ m/s
In t seconds it covers the distance s_2: $s_2 = gt + \frac{1}{2}gt^2$ (iv)

Using equations (iii) and (iv): $\Rightarrow H = s_1 + s_2 = \frac{1}{2}g + gt + \frac{1}{2}gt^2$ (v)

Stone B is falling from rest for t seconds
$\Rightarrow h_2 = (0)(t) + \frac{1}{2}gt^2 = \frac{1}{2}gt^2 \Rightarrow H = h_1 + h_2 = 15 + \frac{1}{2}gt^2$ (vi)

But : From equations (v) and (vi) :

$H = 15 + \frac{1}{2}gt^2 = \frac{1}{2}g + gt + \frac{1}{2}gt^2 \Rightarrow 15 = \frac{1}{2}g + gt \Rightarrow t = \dfrac{15 - \frac{1}{2}g}{g} = 1.029$ seconds

$\Rightarrow h_2 = 5.194$ metres $\Rightarrow H = h_1 + h_2 = 20.194$ metres

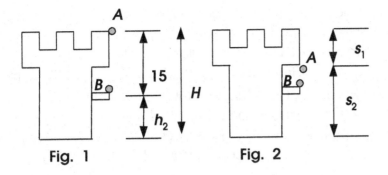

Fig. 1   Fig. 2

14. Ball projected vertically upwards
A ball is projected vertically upwards with a velocity of 19.62 m/s from a point A on a tower (see Fig. 1). (assume $g = 9.81$ m/s²). Find the:
(a) Time taken for the ball to reach the greatest height above A.
(b) Greatest height reached above A.
(c) Position of the ball when $t = 4$ seconds and $t = 6$ seconds.
(d) The time taken to reach a point at a distance of 9.81 metres above A

Solution
Use general equations as before: (i) $s = ut + \frac{1}{2}at^2$ (ii) $v = u + at$

Fig. 1 shows that the motion of the ball can be considered in two Stages:
Stage 1: AB, Ball rising against gravity to its greatest height.
Stage 2: BC, Ball falling from greatest height.

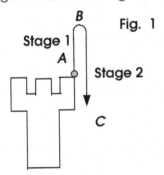

Fig. 1

Stage 1

Stage 2

(a) The time taken for the ball to reach the greatest height.
At its greatest height the velocity of the ball = 0 thus from equation (ii): Time taken to reach greatest height is t where: $0 = 19.62 - gt \Rightarrow t = 2$ seconds

(b) The greatest height reached.
Let the maximum height = h. Using equation (i) and letting: $s = h$ and $t = 2$ gives:
$$h = (19.62)(2) - \frac{1}{2}g(2)^2 = 19.62 \text{ metres}$$

(c) The position of the ball when $t = 4$ seconds and $t = 6$ seconds
The ball reaches the maximum height in 2 seconds. When $t = 4$ seconds the ball has been falling from this point for 2 seconds. The distance travelled by the ball (when falling from rest) in 2 seconds is: $s = 0 + \frac{1}{2}g(2)^2 = \frac{1}{2}(9.81)(2)^2 = 19.62$ metres

Thus at $t = 4$ seconds the ball is level with point A. Its velocity is (from equation (ii)):
$v = 0 + gt = 2g$ m/s $= 19.62$ m/s
When $t = 6$ seconds the ball has been falling for 2 seconds from a point level with the point of projection. The distance travelled by the ball is:

9

$s = (19.62)(2) + \frac{1}{2}g(2)^2 = (9.81)(6) = 58.86$ metres (below the point of projection).

NOTE: the positions of the ball after $t = 4$ seconds and $t = 6$ seconds relative to the point of projection can also be found by putting these values of t into equation (i):

$t = 4 \Rightarrow s = (19.62)(4) - \frac{1}{2}g(4)^2 = 0$ metres; $t = 6 \Rightarrow s = (19.62)(6) - \frac{1}{2}g(6)^2 = -58.86$ metres

(d) The time taken to reach a point at a distance of 9.81 metres above A:

Using equation (i): $9.81 = 19.62\,t - \frac{1}{2}gt^2 \Rightarrow \frac{1}{2}gt^2 - 19.62\,t + 9.81 = 0$

$\Rightarrow t = 2 \pm \sqrt{2}$ seconds $\Rightarrow$ ball reaches a height of 9.81 metres after $2 - \sqrt{2}$ seconds of flight while ascending and again after $2 + \sqrt{2}$ seconds of flight while descending.

15. Particle projected downwards

A particle, projected vertically down from the top of a 100-metre high cliff, reaches the foot of the cliff in 4 seconds. How long did it take to descend the final 30 metres?

Solution

Given: $a =$ acceleration $= g$ m/s²; time $= t = 4$ seconds; $s = 100$ metres

Use: $s = ut + \frac{1}{2}at^2 \quad s = ut + \frac{1}{2}gt^2 = 100 = 4u + 8g \Rightarrow u = 5.38$ m/s

Time for particle to descend 70 metres:

Use: $s = ut + \frac{1}{2}at^2 \quad s = ut + \frac{1}{2}gt^2 = 70 = 5.38t + \frac{1}{2}gt^2 \Rightarrow \frac{1}{2}gt^2 + 5.38t - 70 = 0$

$\Rightarrow t^2 + 1.097t - 14.27 = 0 \Rightarrow t = \dfrac{-1.097 \pm 7.634}{2} = 3.268$ seconds (t must be > 0)

$\Rightarrow$ Time to descend the last 30 metres $= 4 - 3.268 = 0.73$ seconds

16. Particle falls from the top of a tower

A particle falls from rest from the top of a tower of height H metres. If the particle falls through the final 9/25 ths of its journey in 1 second find H.

Solution using standard equations

The tower arrangement is shown in Fig. 1. Let the stone fall 16/25 ths of its journey in time t_1 seconds. (See Fig. 2).

$$\Rightarrow \frac{\frac{16}{25}H}{H} = \frac{16}{25} = \frac{0.(t_1) + \frac{1}{2}gt_1^2}{0.(1 + t_1) + \frac{1}{2}g(1 + t_1)^2} = \frac{\frac{1}{2}gt_1^2}{\frac{1}{2}g(1 + t_1)^2} = \frac{t_1^2}{(1 + t_1)^2}$$

$$\Rightarrow \frac{t_1}{1 + t_1} = \frac{4}{5} \Rightarrow 5t_1 = 4 + 4t_1 \Rightarrow t_1 = 4 \Rightarrow H = \frac{1}{2}g(1 + t_1)^2 = \frac{1}{2}g(5)^2 = 122.6 \text{ metres}$$

Fig. 1   **Fig. 2**

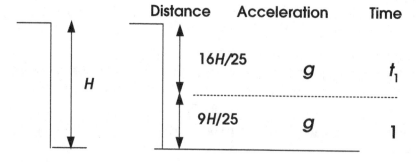

Distance	Acceleration	Time
16H/25	g	t_1
9H/25	g	1

17. Particle projected vertically upwards

A particle is projected vertically upwards with initial velocity = u m/s to a maximum height H. find the total time taken and the final velocity if the particle hits the ground a distance $2H$ vertically below the point of projection. See Fig. 1:

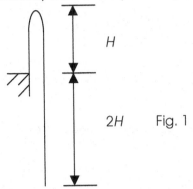

H

$2H$ Fig. 1

Solution using standard equations & a single stage:

Use equation: $s = ut - \frac{1}{2}gt^2$ Let: $s = -2H = -2\,u^2/2g = -u^2/g$

$\Rightarrow (-u^2/g) = ut - \frac{1}{2}gt^2 \Rightarrow \frac{1}{2}gt^2 - ut - u^2/g = 0 \Rightarrow t^2 - (2u/g)t - 2u^2/g^2 = 0$

$$\Rightarrow t = \frac{\left(\frac{2u}{g}\right) \pm \sqrt{\left(\frac{2u}{g}\right)^2 + 4\left(\frac{2u^2}{g^2}\right)}}{2} = \frac{\left(\frac{2u}{g}\right) \pm \sqrt{\frac{12u^2}{g^2}}}{2} = \frac{\left(\frac{2u}{g}\right) \pm \left(\frac{2u}{g}\right)\sqrt{3}}{2} = \left(\frac{u}{g}\right)\left(1 + \sqrt{3}\right)$$

Also: $v^2 = u^2 - 2gs$ where: $s = -2H = -2\,u^2/2g = -u^2/g$

$\Rightarrow v^2 = u^2 - 2g(-u^2/g) = u^2 + 2u^2 = 3\,u^2 \Rightarrow v = u\sqrt{3}$ m/s

18. Car travelling from rest to rest

A racing car covers a journey of 5,000 metres from rest to rest in three stages. Firstly, it accelerates uniformly in the first minute to reach its maximum speed of 50 m/s. It holds this speed for a certain time and then slows uniformly to rest with a retardation of four times that of the acceleration.

(a) Find the distances travelled in each three stage and the total time taken.
(b) If the maximum speed over the final 500 metres is 25 m/s show that the time taken from rest to rest would have been almost 8.2 seconds longer that before, assuming the same rates of acceleration and deceleration as before.

Solution

(a) Three Stages of travel in the journey (See Fig. 1):

1: Acceleration for t_1 (= 60 seconds) from rest to 50 m/s, over a distance of s_1 metres
2: Constant speed of 50 m/s for t_2 seconds covering a distance of s_2 metres
3: Deceleration for t_3 seconds from 50 m/s to rest, over a distance of s_3 metres
Note: (Given): $s_1 + s_2 + s_3 = 5,000$ metres. Total time = $60 + t_2 + t_3$
Use general equations as before: (i) $s = ut + \frac{1}{2}at^2$ (ii) $v = u + at$

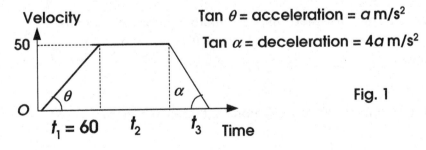

Velocity

50

O

$t_1 = 60$ t_2 t_3 **Time**

Tan θ = acceleration = a m/s^2

Tan α = deceleration = $4a$ m/s^2

Fig. 1

Stage 1: Acceleration from rest to 50 m/s:

Initial velocity = $u = 0$ m/s, Acceleration = a m/s^2, Final velocity = $v = 50$ m/s,
Time = $t_1 = 60$ seconds. But, from equation (ii): $50 = 0 + 60a \Rightarrow a = \frac{5}{6}$ m/s^2

From equation (i) the distance covered in 60 seconds:
$s_1 = ut + \frac{1}{2} at^2 = 0 \times 60 + \frac{1}{2} a(60)^2 = \frac{1}{2}\frac{5}{6}(3,600) = 1,500$ metres

Stage 3: Deceleration from 50 m/s to rest:

Initial velocity = 50 m/s, Acceleration = $-4a = -\frac{10}{3}$ m/s^2, Final velocity = 0 m/s

Thus from equation (ii): $0 = 50 - \frac{10}{3}t_3 \Rightarrow t_3 = 15$ seconds

Also, from equation (ii), $s_3 = ut_3 + \frac{1}{2} at_3^2 = (50)(15) - \frac{1}{2}\frac{10}{3}(15)^2 = 375$ metres

Stage 2: Constant speed of 50 m/s:

Thus, the distance travelled at constant speed is: $5,000 - 1,500 - 375 = 3,125$ metres

The time taken to cover this distance is: $t_2 = \dfrac{3,125}{50} = 62.5$ seconds

$\Rightarrow$ total time taken to complete the journey is $60 + 62.5 + 15 = 137.5$ seconds

(b) Maximum speed over the final 500 metres restricted to 25 m/s

If the maximum speed over the final 500 metres of the journey is restricted to 25 m/s the journey can be considered in five stages (See Fig. 2):
Stage 1: Acceleration for t_1 (= 60 seconds) from 0 to 50 m/s, over s_1 metres
Stage 2: Constant speed of 50 m/s for t_2 seconds covering a distance of s_2 metres
Stage 3: Deceleration for t_3 seconds from 50 m/s to 25 m/s, over a distance of s_3 metres
Stage 4: Constant speed of 25 m/s for t_4 seconds covering a distance of s_4 metres
Stage 5: Deceleration for t_5 seconds from 25 m/s to rest, covering s_5 metres
Note: $s_4 + s_5 = 500$ metres. Total time = $60 + t_2 + t_3 + t_4 + t_5$

Stage 1: Acceleration from 0 to 50 m/s

This will be unchanged: Initial speed = 0 m/s   Final speed = 50 m/s
Time taken = $t_1 = 60$ seconds   Distance travelled = 1,500 metres

Stage 3: Time taken and distance travelled while decelerating from 50 m/s to 25 m/s:

Initial speed = 50 m/s, Final speed = 25 m/s, Acceleration = $-\frac{10}{3}$ m/s^2

Using equations (i) and (ii): $25 = 50 - \frac{10}{3}(t) \Rightarrow t_3 = 7.5$ seconds

Distance travelled = $(50)(7.5) - \frac{1}{2}\frac{10}{3}(7.5)^2 = 281.25$ metres

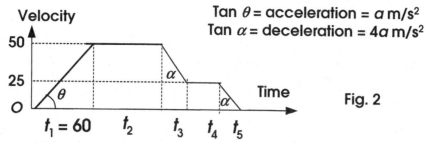

Tan θ = acceleration = a m/s^2
Tan α = deceleration = $4a$ m/s^2

Fig. 2

Stage 5: Time taken and distance travelled while decelerating from 25 to 0 m/s:

Initial speed = 25 m/s   Final speed = 0 m/s   Acceleration = $-\frac{10}{3}$ m/s^2

Using equations (i) and (ii): $0 = 25 - \frac{10}{3}t \Rightarrow t_5 = 7.5$ seconds

Distance travelled = $(25)(7.5) - \frac{1}{2}\frac{10}{3}(7.5)^2 = 93.75$ metres

Stage 4: Time taken and distance travelled at a constant speed at 25 m/s:

The car cannot exceed 25 m/s for the last 500 metres, and it only requires 93.75 metres stopping distance at this speed (see Stage 5 above), then the car can travel at 25 m/s for 500 – 93.75 = 406.25 metres. This takes: $406.25/25 = t_4 = 16.25$ seconds

Stage 2: Time taken and distance travelled at a constant speed at 50 m/s:
The 5 Stages are: 1,500 + Distance in Stage 2 + 281.25 + 406.25 + 93.75 = 5,000 metres
$\Rightarrow$ distance travelled in Stage 2 = 2,718.75 metres
This will take a time of $2,718.75/50 = t_2 = 54.375$ seconds.

Overall:
Thus the total time taken = 60 + 54.375 + 7.5 + 16.25 + 7.5 = 145.63 seconds, i.e.
145.63 – 137.5 = 8.13 seconds more than when no speed restriction applies.

19. Acceleration and constant speed stages
Starting from rest an athlete accelerates uniformly to a speed of 10 m/s, and then continues at that speed, covering 200 metres in 22 seconds. Find the acceleration.

Solution using Velocity-Time Graph
Given: Time: $t_1 + t_2 = 22$ seconds Eq. (i) Distance: $s_1 + s_2 = 200$ metres Eq. (ii), where:
s_1 = distance travelled during acceleration = Area $OAC = \frac{1}{2}(10)(t_1) = 5 t_1$
s_2 = distance travelled at constant velocity = Area $ABCD = (10)(t_2) = 10 t_2$
$\Rightarrow$ From equation (ii): $5 t_1 + 10 t_2 = 200 \Rightarrow t_1 + 2 t_2 = 40$
Using equations (i), (ii): $t_2 = 18$, $t_1 = 4$ seconds
But: From Fig. 1 above, acceleration = $a = 10/ t_1 = 10/ 4 = 2.5$ m/s^2

Velocity $\text{Tan } \theta = \text{acceleration} = a \text{ m/s}^2 = 10/t_1$

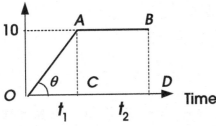

Fig. 1

20. Cyclist travelling from rest to rest
A cyclist has a maximum acceleration of 2.5 m/s^2, maximum speed of 15 m/s and maximum deceleration of 5 m/s^2. Find the shortest time in which the cyclist can travel distances of 100 and 50 metres respectively from rest to rest.

Solution
(a) $s = 100$ metres:
Fig. 1 shows the Velocity-Time Graph for the motion. Information given:
Time: $t_1 + t_2 + t_3$ = total time (to be a minimum)
Distance: $s_1 + s_2 + s_3 = 100$ metres   Equation (i), where:
s_1 = distance travelled during acceleration = Area $OAC = \frac{1}{2}(15)(t_1) = 7.5 t_1$
s_2 = distance travelled at constant velocity = Area $ABCD = (15)(t_2) = 15 t_2$
s_3 = distance travelled during deceleration = Area $BDE = \frac{1}{2}(15)(t_3) = 7.5 t_3$

But: From Fig. 1:
Acceleration = 2.5 m/s^2 = 15/ t_1 $\Rightarrow$ t_1 = 6 seconds $\Rightarrow$ s_1 = 45 metres   and
Deceleration = 5 m/s^2 = 15/ t_3 $\Rightarrow$ t_3 = 3 seconds $\Rightarrow$ s_3 = 22.5 metres
$\Rightarrow$ s_2 = 100 – 45 – 22.5 = 32.5 metres = 15 t_2 $\Rightarrow$ t_2 = 2.167 seconds
$\Rightarrow$ Total time taken = $t_1 + t_2 + t_3$ = 6 + 3 + 2.167 = 11.167 seconds

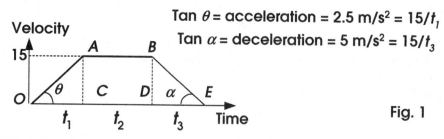

$$\text{Tan } \theta = \text{acceleration} = 2.5 \text{ m/s}^2 = 15/t_1$$
$$\text{Tan } \alpha = \text{deceleration} = 5 \text{ m/s}^2 = 15/t_3$$

Fig. 1

(b) $s = 50$ metres

From (a) above the acceleration and deceleration Stages alone will take over 50 metres. Therefore, to travel 50 metres in the shortest time involves only acceleration and deceleration stages (See Fig. 2).

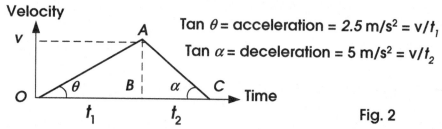

$$\text{Tan } \theta = \text{acceleration} = 2.5 \text{ m/s}^2 = v/t_1$$
$$\text{Tan } \alpha = \text{deceleration} = 5 \text{ m/s}^2 = v/t_2$$

Fig. 2

Fig. 2 shows the Velocity-Time Graph for the motion. Information given:
Time: $t_1 + t_2$ = total time (to be a minimum)
Distance: $s_1 + s_2 = 50$ metres   Equation (i), where:
s_1 = distance travelled during acceleration = Area $OAB = \frac{1}{2}(v)(t_1) = \frac{1}{2} v t_1$
s_2 = distance travelled during deceleration = Area $ABC = \frac{1}{2} (v)(t_2) = \frac{1}{2} v t_2$
$\Rightarrow \frac{1}{2} v t_1 + \frac{1}{2} v t_2 = 50 \Rightarrow v(t_1 + t_2) = 100$ Equation (i)

But: From Fig. 2:
Acceleration = 2.5 m/s^2 = $v/ t_1 \Rightarrow t_1 = v/2.5$ seconds
Deceleration = 5 m/s^2 = $v/ t_2 \Rightarrow t_2 = v/5$ seconds
$\Rightarrow t_1 + t_2 = 0.6 v$ Equation (ii). From Equations (i),(ii): $(t_1 + t_2)^2 = 60 \Rightarrow t_1 + t_2 = 7.745$ seconds

21. Sprinter accelerating
An athlete running with acceleration a m/s^2 takes 12 seconds to run 100 metres, 11 seconds to cover the next 100 metres. How long will it take to run the final 100 metres.

Solution using Velocity-Time Graph

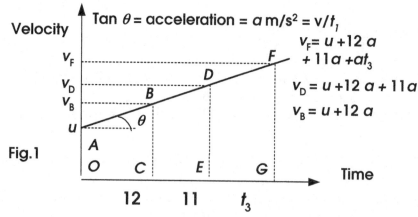

$$\text{Tan } \theta = \text{acceleration} = a \text{ m/s}^2 = v/t_1$$

$$v_F = u + 12 a$$
$$+ 11a + at_3$$

$$v_D = u + 12 a + 11a$$

$$v_B = u + 12 a$$

Fig.1

In Fig. 1 the Velocity-Time Graph shows the motion. (Given): Distances:

14

First and second 100 metres distances are represented by: Area $OABD$ and Area $BCDE$ respectively:

Area $OABD = 100 = \frac{1}{2}(u + (u + 12a))(12) \Rightarrow 25 = 3u + 18a$

Area $BCDE = 100 = \frac{1}{2}((u + 12a) + (u + 23a))(11) \Rightarrow 200 = 22u + 385a$

Use above equations to derive values of u, a: $u = 7.938$ m/s, $a = 0.06588$ m/s^2

Final 100 metres: Area $DEFG = 100 = \frac{1}{2}((u + 23a) + (u + 23a + at_3))(t_3)$

$\Rightarrow 100 = \frac{1}{2}(9.453\, t_3 + 9.453\, t_3 + 0.06588\, t_3^2)$

$\Rightarrow 0.06588\, t_3^2 + 18.906\, t_3 - 200 = 0 \Rightarrow t_3^2 + 286.984\, t_3 - 3035.82 = 0$

Solving for t_3 gives: $t_3 = (-286.984 \pm 307.41)/2 = 10.21$ seconds (Note: $t_3 > 0$)

22. Colliding particles

Ball A falls freely under gravity from rest at point C. One second after Ball A has fallen Ball B is projected vertically downwards from point C. (Fig. 1). If the particles collide two seconds after Ball A has fallen (See Fig. 2), find the distance from C at which the collision occurs and the velocity at which Ball B was initially projected downwards.

Solution

General equations: (i) $s = ut + \frac{1}{2}at^2$ (ii) $v = u + at$

Note: Collision occurs after Balls A, B have travelled for 2, 1 seconds respectively.

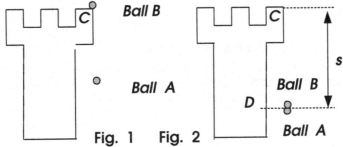

Fig. 1   Fig. 2

(1) Motion of Ball A:

Assume it falls from rest at time $t = 0$

At time $t = 2$ seconds it will have fallen a distance $s = 0(2) + \frac{1}{2}g(2)^2 = 2g$

$\Rightarrow$ Collision takes place after Ball A has fallen a distance of $2g$

(2) Motion of Ball B:

Assume it is projected from rest with initial velocity u m/s

After 1 second it will have traveled a distance: $s = u(1) + \frac{1}{2}g(1)^2$

$\Rightarrow$ Collision takes place after Ball B has fallen a distance of $u + \frac{1}{2}g$

$\Rightarrow 2g = u + \frac{1}{2}g \Rightarrow u = 1.5\,g$ m/s $\Rightarrow s = 2g = 19.62$ metres

Ball B is projected downwards with initial velocity $= 1.5g = 14.72$ m/s. The collision takes place two seconds after Ball A has been dropped, at 19.62 metres below C.

23. Lift and falling object

A lift starts from rest at point A and descends with acceleration of 0.6 m/s^2. When the lift is at B, 25 metres below A, a loose bolt falls from the shaft structure at A. Find the time taken for the bolt to strike the roof of the lift (at point C), the depth below A at which this impact occurs and the speed of the bolt at impact

Solution
Solution using the standard equations

a) The time taken for the bolt to strike the roof of the lift

Time for lift to travel 25 metres : Use equation : $s = ut + \frac{1}{2}at^2 \Rightarrow 25 = 0 + \frac{1}{2}(0.6)t^2$

$\Rightarrow t = 9.129$ seconds.

Assume an additional time of T seconds elapses before impact.

When the impact occurs the lift and the particle will have travelled for :

Lift : $T + 9.129$ to reach a distance of h metres below A

Bolt : T seconds to reach a distance of h metres below A

$\Rightarrow h = (0)(T + 9.129) + \frac{1}{2}(0.6)(T + 9.129)^2 = (0)T + \frac{1}{2}(g)T^2$

$\Rightarrow 0.3(T + 9.129)^2 = 4.905T^2 \Rightarrow 4.605T^2 - 5.477T - 25 = 0 \Rightarrow T = \dfrac{5.477 \pm 22.147}{9.21} = 3$ seconds

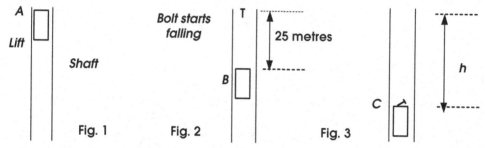

A		Bolt starts falling	T	25 metres			h
Lift							
Shaft							
		B				C	
Fig. 1	Fig. 2			Fig. 3			

(b) The depth below A at which this impact occurs

From above: $h = \frac{1}{2}gT^2 = 4.905(3)^2 = 44.145$ metres

(c) The speed of the bolt at impact

The velocity of the bolt on impact is: $v = 0 + g(3) = 29.43$ m/s

24. Two cars

Car A starts from rest at a point P and moves with constant acceleration a m/s². Five seconds later Car B passes through P in the same direction with constant velocity u m/s.

(a)   Show that Car B will just catch up on Car A if $u = 10a$.

(b)   If $u > 10a$ find the greatest distance Car B will be ahead of Car A.

Solution

(a) Show that Car B will just catch up on Car A if $u = 10a$.

Use general equations as before: (i) $s = ut + \frac{1}{2}at^2$ (ii) $v = u + at$

Assume Car B just catches up on Car A after Car A has been travelling for T seconds and Car B for T – 5 seconds.

The distance travelled by Car A in T seconds is, from equation (i):

$s_A = 0(T) + \frac{1}{2}a(T)^2 = \frac{1}{2}aT^2$

The distance travelled by Car B in T - 5 seconds is, from equation (i): $s_B = u(T - 5)$.

If Car B catches up with Car A: $\Rightarrow s_A = s_B \Rightarrow \frac{1}{2}aT^2 = u(T - 5) \Rightarrow aT^2 - 2uT + 10u = 0$

Solving this quadratic equation for T gives : $T = -\dfrac{2u \pm \sqrt{4u^2 - 40au}}{2}$

But, if the Car B just catches up with the first, there is only one value of T:

$\Rightarrow 4u^2 - 40au = 0 \Rightarrow (4u)(u - 10a) = 0 \Rightarrow u = 0, 10a$ (Note : The question states that $u > 0$ so that $u = 0$ is not a valid solution) $\Rightarrow u = 10a$

(b) If $u > 10a$ find the greatest distance Car B will be ahead of Car A.

Let D = greatest distance Car B will be ahead of the Car A = $D = s_B - s_A$

But, from above: $D = u(T-5) - \frac{1}{2}aT^2$ (iii)

Differentiating D with respect to T and giving the resulting function a value of 0 will determine the time at which D is a maximum: $\dfrac{dD}{dT} = u - aT = 0 \Rightarrow T = \dfrac{u}{a}$

Put this value into equation (iii): $D = u\left(\dfrac{u-5a}{a}\right) - \dfrac{1}{2}a\left(\dfrac{u}{a}\right)^2 \Rightarrow D = \dfrac{u(u-10a)}{2a}$

$\Rightarrow$ Greatest distance Car B will be ahead of Car A is: $\dfrac{u(u-10a)}{2a}$

25. Train travel

A train makes a three-stage journey from rest to rest with an average speed of $\frac{3}{4}v$ m/s.

It travels a distance s_1 with constant acceleration a m/s², a distance s_2 with constant speed v m/s and a distance s_3 with constant retardation d m/s². Find expressions for the distance travelled at constant speed and the time spent at constant speed.

Solution

Total distance travelled = $s = s_1 + s_2 + s_3$ (Distance travelled at constant speed = s_2)
Total time travelling = $T = t_1 + t_2 + t_3$ (Time spent at constant speed = t_2)

Given: average speed of $\frac{3}{4}v \Rightarrow \dfrac{\text{total distance}}{\text{total time}} = \dfrac{s}{T} = \dfrac{s_1 + s_2 + s_3}{t_1 + t_2 + t_3}$

Distance travelled at constant speed

The Velocity – Time graph is shown in Fig. 1.
Use general equations as before: (i) $s = ut + \frac{1}{2}at^2$　　(ii) $v = u + at$　(iii) $s = (u + v)t/2$

$\Rightarrow s_1 = \dfrac{(0+v)}{2}t_1 \Rightarrow t_1 = \dfrac{2s_1}{v}$, 　$s_3 = \dfrac{(v+0)}{2}t_3 \Rightarrow t_3 = \dfrac{2s_3}{v}$, 　$s_2 = vt_2 \Rightarrow t_2 = \dfrac{s_2}{v}$

$\Rightarrow \dfrac{s_1 + s_2 + s_3}{t_1 + t_2 + t_3} = \dfrac{s_1 + s_2 + s_3}{\dfrac{2s_1}{v} + \dfrac{s_2}{v} + \dfrac{2s_3}{v}} = \dfrac{3}{4}v \Rightarrow \dfrac{s_1 + s_2 + s_3}{2s_1 + s_2 + 2s_3} = \dfrac{3}{4} \Rightarrow s_2 = 2(s_1 + s_3)$

$\Rightarrow s_1 + s_2 + s_3 = s_2 + \dfrac{1}{2}s_2 = \dfrac{3}{2}s_2 \Rightarrow \dfrac{s_2}{s_1 + s_2 + s_3} = \dfrac{2}{3}$

Time spent at constant speed

Find : $\dfrac{t_2}{T} : \dfrac{t_2}{T} = \dfrac{t_2}{t_1 + t_2 + t_3} = \dfrac{\dfrac{s_2}{v}}{\dfrac{2s_1}{v} + \dfrac{s_2}{v} + \dfrac{2s_3}{v}} = \dfrac{s_2}{2s_1 + s_2 + 2s_3} = \dfrac{s_2}{2s_2} = \dfrac{1}{2}$

Two thirds of the total journey and one half of the total time is spent travelling at constant speed.

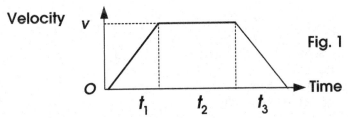

Velocity

Fig. 1

26. Journey from rest to rest

17

A body starting from rest travels with acceleration a m/s^2 and then with deceleration b m/s^2. If it comes to rest when it has covered a total distance of s metres show that the overall time for the journey, t seconds, is given by $t = \sqrt{2s\left(\dfrac{1}{a} + \dfrac{1}{b}\right)}$

Solution

Consider the journey as having two Stages (Fig. 1).
Stage 1: Acceleration from 0 to v m/s over a distance s_1 metres taking t_1 seconds
Stage 2: Deceleration from v to 0 m/s over a distance s_2 metres taking t_2 seconds
Total distance travelled $= s_1 + s_2 = s$, Total time taken $= t_1 + t_2 = t$
Use the general equations: $v^2 = u^2 + 2as$, $v = u + at$

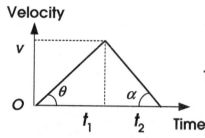

Velocity

$Tan\ \theta$ = acceleration = a m/s^2

$Tan\ \alpha$ = deceleration = b m/s^2

Fig. 1

$v^2 = u^2 + 2as$: $v^2 = 0^2 + 2as_1$, $0^2 = v^2 - 2bs_2$

$\Rightarrow s = s_1 + s_2 = \dfrac{v^2}{2a} + \dfrac{v^2}{2b} = \dfrac{v^2}{2}\left(\dfrac{1}{a} + \dfrac{1}{b}\right) \Rightarrow v = \sqrt{\dfrac{2s}{\left(\dfrac{1}{a} + \dfrac{1}{b}\right)}}$ Equation (i)

$v = u + at \Rightarrow v = 0 + at_1$, $0 = v - bt_2 \Rightarrow t = t_1 + t_2 = \dfrac{v}{a} + \dfrac{v}{b} = v\left(\dfrac{1}{a} + \dfrac{1}{b}\right)$ Equation (ii)

Putting the value for v into Equation (ii): $t = \sqrt{\dfrac{2s}{\left(\dfrac{1}{a} + \dfrac{1}{b}\right)}\left(\dfrac{1}{a} + \dfrac{1}{b}\right)} = \sqrt{2s\left(\dfrac{1}{a} + \dfrac{1}{b}\right)}$

27. Body travelling from rest to rest

A body starts from rest at A, travels for 900 metres in a straight line and then comes to rest at B. The time taken is 60 seconds. For the first 10 seconds it has a uniform acceleration a_1 m/s^2. It then travels at constant speed and is finally brought to rest by uniform deceleration a_2 m/s^2 acting for 20 seconds.
(a) Find a_1 and a_2
(b) If the journey from rest at A to rest at B had been travelled with no interval of constant speed, but subject to acceleration a_1 m/s^2 for t_1 seconds, followed by deceleration a_2 m/s^2 for t_2 seconds show that the journey will take $30\sqrt{3}$ seconds.

Solution

Use the general equations: (i) $s = ut + \frac{1}{2}at^2$ (ii) $v = u + at$ (iii) $v^2 = u^2 + 2as$

(a) Find a_1 and a_2

Consider the journey in three Stages. See Velocity-Time graph (Fig. 1):
Stage 1: Distance s_1 travelled in time t_1 while accelerating.
Stage 2: Distance s_2 travelled in time t_2 while at constant speed.
Stage 3: Distance s_3 travelled in time t_3 while decelerating.
(Given): $s_1 + s_2 + s_3 = 900$ metres and $t_1 + t_2 + t_3 = 60$ seconds
Since $t_1 = 10$ seconds, $t_3 = 20$ seconds $\Rightarrow t_2 = 60 - 10 - 20 = 30$ seconds

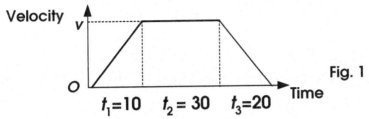

Fig. 1

Distance travelled = 900 metres = Area under lines in Fig. 1:

$$\Rightarrow 900 = \left(\frac{0+v}{2}\right)(10) + 30v + \left(\frac{v+0}{2}\right)(20) = 45v \Rightarrow v = 20 \quad \text{m/s}$$

But : In acceleration stage : $v = u + a_1 t_1 \Rightarrow 20 = 0 + 10a_1 \Rightarrow a_1 = 2 \quad \text{m/s}^2$

And : In deceleration stage : $0 = v - a_2 t_3 \Rightarrow 0 = 20 + 20a_2 \Rightarrow a_2 = -1 \quad \text{m/s}^2$

(b) When there is no interval of constant speed:

Fig. 2 shows the Velocity – Time graph and the two Stages of travel:

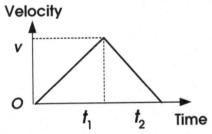

Fig. 2

Using equation $v^2 = u^2 + 2as$: $v^2 = 0^2 + 2(2)s_1 \Rightarrow s_1 = 0.25 \ v^2$; $0^2 = v^2 - 2(1)s_2 \Rightarrow s_2 = 0.5 \ v^2$

$\Rightarrow s_1 + s_2 = 900 = 0.75v^2 \Rightarrow v = 20\sqrt{3}$ Using equation $v = u + at$: $v = (2) \ t_1, 0 = v - (1)t_2$

Total time taken = $t = t_1 + t_2 = 0.5v + v = 1.5v = 30\sqrt{3}$

28. Two cars

Cars A and B are travelling along a straight road in the same direction. At point C, A and B have velocities v, $1.5v$ m/s respectively and accelerations $2a$, a m/s² respectively.

The cars are side by side at two different points C and D. Show that $|CD| = 2\dfrac{v^2}{a}$

Solution

Use general equations: (i) $s = ut + \frac{1}{2} at^2$ (ii) $v = u + at$

For Car A let: $s_1 = CD$, Initial velocity = v, Final velocity = v_1

Thus from equations (i) and (ii): $s_1 = vt + \frac{1}{2} 2at^2$ (iii) and $v_1 = v + 2at$

For Car B let: $s_2 = CD$, Initial velocity = $1.5v$, Final velocity = v_2

Thus from equations (i) and (ii): $s_2 = 1.5vt + \frac{1}{2} at^2$ (iv)

Assume the cars are at C when $t = 0$ and at D when $t = t$

$\Rightarrow s_1 = CD = s_2 \Rightarrow$ From equations (iii) and (iv): $vt + \frac{1}{2}(2a)t^2 = 1.5vt + \frac{1}{2} at^2$

$\Rightarrow \frac{1}{2} at^2 = 0.5vt \Rightarrow at = v \Rightarrow t = \dfrac{v}{a}$

Putting this into equation (i): $s_1 = |CD| = vt + \frac{1}{2}(2a)t^2 = v\left(\dfrac{v}{a}\right) + \frac{1}{2}(2a)\left(\dfrac{v}{a}\right)^2 = 2\dfrac{v^2}{a}$

Chapter 2   Relative Motion

1. Cars crossing at an intersection

Car A is travelling North at 20 m/s while car B is travelling East at 25 m/s. What is the velocity of B relative to A?

Solution

Let the velocities of cars A and B be represented by vectors (See Fig. 1):
$v_A = 0\,i + 20\,j,\ v_B = 25\,i + 0\,j$

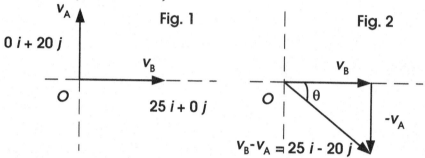

The velocity of B relative to A (See Fig. 2) $= v_{BA} = v_B - v_A = (25\,i + 0\,j) - (0\,i + 20\,j)$
$= (25 - 0)\,i - (0 + 20)\,j = 25\,i - 20\,j \Rightarrow$ magnitude $= \sqrt{25^2 + 20^2} = 32.02$ m/s

and this vector makes an angle of: $\theta = Tan^{-1}\dfrac{20}{25} = 38.66°$ south of east

2. Apparent direction of wind

A boat is moving East at 6 m/s. A passenger feels a wind which appears to blow from the North at 4 m/s. What is the true velocity of the wind?

Solution

Let the velocities of the passenger and the wind be represented by the vectors:
$v_{PASSENGER} = v_{BOAT} = 6\,i + 0\,j$
Let the true velocity of the wind $= v_{WIND} = x\,i + y\,j\,(x, y$ not known). But, (given): the wind appears to blow from the North at 4 m/s:
Let the velocity of the wind relative to the boat (and passenger) $= v_{(WIND)(BOAT)}$
$\Rightarrow v_{(WIND)(BOAT)} = v_{WIND} - v_{BOAT} = (x\,i + y\,j) - (6\,i + 0\,j) = (x - 6)\,i - (y)j = 0\,i - 4\,j$
$\Rightarrow x - 6 = 0 \Rightarrow x = 6$ and $y = -4 \Rightarrow v_{WIND} = 6\,i - 4\,j$
i.e. the wind is blowing at a speed of $\sqrt{6^2 + 4^2} = 7.211$ m/s in a direction making an

angle $\theta = Tan^{-1}\dfrac{4}{6} = 33.69°$ South of East (See Fig. 1)

Fig. 1

3. Two cars

The velocity of Car A is 10 m/s North, and the velocity of Car B relative to Car A is 15 m/s due East, find the velocity of Car B.

Solution

Velocity of Car A = $V_A = 0\,i + 10\,j$ Velocity of Car B relative to Car A = $V_{BA} = 15\,i + 0\,j$

But : $V_{BA} = V_B - V_A \Rightarrow V_B = V_{BA} + V_A$

$\Rightarrow$ Velocity of Car B = $V_B = V_{BA} + V_A = (15\,i + 0\,j) + (0\,i + 10\,j = 15\,i + 10\,j$

$\Rightarrow$ Car B travels with speed $\sqrt{15^2 + 10^2} = 5\sqrt{13}$ m/s in a direction θ degrees North of East

where $\theta = Tan^{-1}\dfrac{10}{15} = Tan^{-1}\dfrac{2}{3}$

4. Man travelling in windy conditions

A man walks due East at 5 km/hr and the wind appears to come from North. He then cycles in a North-Easterly direction at 20 km/hr and the wind now appears to come from 30° East of North. Find the true velocity of the wind.

Solution

Consider the man's travel in two different stages:

(a) Man travelling east:

Velocity of the man = $V_m = 5\,i + 0\,j$ (i)

Let the true velocity of the wind = $V_w = x\,i + y\,j$ (ii)

But velocity of the wind relative to man = $V_{wm} = V_w - V_m$

As the wind appears to come from the north it will have a j component of velocity only: assume that $V_{wm} = 0\,i - P\,j$ where P = magnitude of the velocity of the

wind $\Rightarrow 0\,i - P\,j = (x - 5)i + (y - 0)j \Rightarrow x = 5,\ y = -P$

(b) The man cycles in a north easterly direction at 20 km/hr:

$V_m = 10\sqrt{2}\,i + 10\sqrt{2}\,j$ (iii) $V_w = x\,i + y\,j$ (iv)

The wind appears to come from a direction 30° East of North: Assume that V_{wm} has magnitude of Q:

$$\Rightarrow V_{wm} = V_w - V_m = -Q\,Sin30°\,i - Q\,Cos30°\,j = -Q\left(\frac{1}{2}i + \frac{\sqrt{3}}{2}j\right)\ \ (v)$$

From equations (iii),(iv) : $V_{wm} = V_w - V_m = \left(x - 10\sqrt{2}\right)i + \left(y - 10\sqrt{2}\right)j$ (vi)

$\Rightarrow$ Using equations (v), (vi) : $-Q\left(\dfrac{1}{2}i + \dfrac{\sqrt{3}}{2}j\right) = \left(x - 10\sqrt{2}\right)i + \left(y - 10\sqrt{2}\right)j$

$\Rightarrow -\frac{1}{2}Q = x - 10\sqrt{2},\ \ \Rightarrow Q = 20\sqrt{2} - 2x$ and $\dfrac{-\sqrt{3}}{2}Q = y - 10\sqrt{2},\ \ \Rightarrow Q\sqrt{3} = 20\sqrt{2} - 2y$

But the true velocity of the wind is constant and from (a): $x = 5$

$\Rightarrow Q = 20\sqrt{2} - 2(5) = 20\sqrt{2} - 10$

But from abve : $2y = 20\sqrt{2} - Q\sqrt{3} \Rightarrow y = \frac{1}{2}\left(20\sqrt{2} - \left(20\sqrt{2} - 10\right)\sqrt{3}\right)$

Therefore, the true velocity of the wind is given by,

$V_w = x\,i + y\,j = 5\,i + (10\sqrt{2} - 10\sqrt{2}\sqrt{3} + 5\sqrt{3})\,j = 5\,i - 1.69\,j$

i.e. 5.278 m/s in a direction 18.68° South of East.

5. Two cars approaching crossroads

21

Two roads intersect at right angles at O. Car A heads South towards O at 60 km/hr and car B heads West towards O at 100 km/hr. When car A is 0.5 km from O, car B is 1 km from O. Find the minimum distance between the cars during the subsequent motion.

Solution

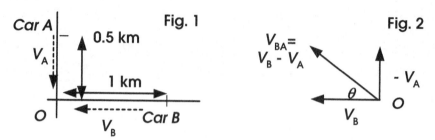

The space diagram in Fig. 1 shows the initial car locations (and indicates the velocities of cars A and B, V_A and V_B respectively). Fig. 2 shows the velocity diagram.

Given: $V_A = 0\,i - 60\,j$, $V_B = -100\,i + 0\,j$

Subtracting V_A from both velocities gives:

Velocity of car A = $V_A - V_A = 0$. Thus car A is stopped at co-ordinates (0, 0.5).

Velocity of car B relative to A = $V_{BA} = V_B - V_A = -100\,i + 60\,j$ along BD

Car B is closest to car A when at point C, i.e. where $AC \perp BD \Rightarrow$ closest distance = AC.

Car B travels along BD, containing points BCED. From Figs. 3, 4:

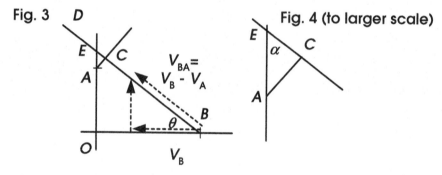

$$Tan\theta = \frac{60}{100} = 0.6 \Rightarrow \theta = 30.964° \Rightarrow \alpha = 59.036°$$

$$Tan\theta = \frac{OE}{OB} = \frac{OA + AE}{OB} = \frac{0.5 + AE}{1} = 0.6 \Rightarrow AE = 0.1$$

$$Sin\alpha = \frac{AC}{AE} = \frac{AC}{0.1} = 0.8575 \Rightarrow AC = 0.08575\,km = 85.75 \text{ metres}$$

6. Closest approach of two cars

At time $t = 0$, Car A is at O and drives off at 20 m/s in a direction 30° East of North. Car B is 300 metres East of O and heads towards O at 25 m/s. Find:

(a)   Their distances from O when the cars are closest to one another.

(b)   The distance between the cars when they are closest to one another

Solution

(a) Their distances from O when the cars are closest to one another.

Fig. 1 shows the initial location of the vehicles and their velocities.

Car A: Velocity = V_A = 20 $Cos60°$ i + 20 $Sin60°$ j = 10 i + $10\sqrt{3}$ j
Car B: Velocity = V_B = - 25 i

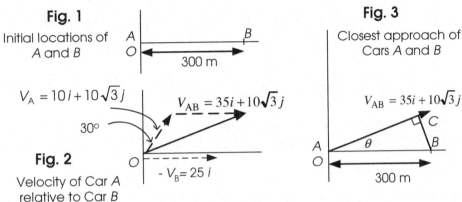

Fig. 1
Initial locations of A and B

300 m

$V_A = 10i + 10\sqrt{3}j$

30°

$V_{AB} = 35i + 10\sqrt{3}j$

Fig. 2
Velocity of Car A relative to Car B

- V_B= 25 i

Fig. 3
Closest approach of Cars A and B

$V_{AB} = 35i + 10\sqrt{3}j$

θ

300 m

Velocity of A relative to B = V_{AB} = V_A – V_B = 10 i + $10\sqrt{3}$ j + 25 i = 35 i + $10\sqrt{3}$ j
(See Fig. 2). The magnitude of this velocity is: 39.05 m/s

The closest approach occurs at point C (See Fig. 3) where $BC \perp AC$

$Tan\theta = \dfrac{10\sqrt{3}}{35} \Rightarrow \theta = 26.33° \Rightarrow Cos\theta = 0.896 = \dfrac{AC}{AB} = \dfrac{AC}{300} \Rightarrow AC = 268.88$ metres

Time taken to cover this distance = $\dfrac{268.88}{39.05}$ = 6.885 seconds

In 6.885 seconds Car A travels 20 × 6.885 = 137.7 metres from O and Car B travels 25 × 6.885 = 172.125 metres towards O and therefore will be 127.875 metres from O.

(b) The distance between the cars when they are closest to one another

$Sin\theta = \dfrac{BC}{300} = 0.4435 \Rightarrow BC = 133.06$ metres

7. Closest approach of two cars

At time $t = 0$, Car A is at point O (at an intersection) and drives off with a velocity v m/s in a Northerly direction. Car B is at a distance D metres East of O and heading towards O with a velocity of u m/s. Find:
(a)   The time taken for the cars to reach their closest approach
(b)   Their distances from O when the cars are closest to one another.

Solution

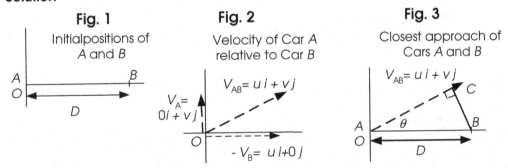

Fig. 1
Initialpositions of A and B

D

Fig. 2
Velocity of Car A relative to Car B

$V_{AB}= ui + vj$

$V_A= 0i + vj$

- V_B= $ui+0j$

Fig. 3
Closest approach of Cars A and B

$V_{AB}= ui + vj$

θ

D

(a) The time taken for the cars to reach their closest approach
Fig. 1 shows the initial positions of the vehicles. Fig. 2 shows the velocity diagram: Car A:
Velocity = V_A = 0 i + vj Car B: Velocity = V_B = - ui + $0j$

Velocity of A relative to $B = V_{AB} = V_A - V_B$

This resulting velocity has a magnitude $\sqrt{u^2 + v^2}$ and is directed along line AC (which makes an angle θ with the x-axis). To find the shortest distance between the cars, draw a line from B to a point C such that $BC \perp AC$. $|BC|$ = shortest distance. (See Fig. 3).

$$Tan\theta = \frac{v}{u} \Rightarrow Cos\theta = \frac{u}{\sqrt{u^2 + v^2}} = \frac{AC}{D} \Rightarrow AC = \frac{uD}{\sqrt{u^2 + v^2}} \Rightarrow \text{Time required to}$$

cover distance AC at speed $\sqrt{u^2 + v^2} = t = \dfrac{AC}{\sqrt{u^2 + v^2}} = \dfrac{uD}{u^2 + v^2}$ seconds

(b) Their distances from O when the cars are closest to one another.

Cars A and B are at the following distances from O after time t seconds:

Car A: $(v)\left(\dfrac{uD}{u^2 + v^2}\right) = \dfrac{uvD}{u^2 + v^2}$ metres, Car B: $D - (u)\left(\dfrac{uD}{u^2 + v^2}\right) = \dfrac{v^2 D}{u^2 + v^2}$ metres

8. Swimmer crossing a river

A river is flowing with speed u m/s and a swimmer who has a maximum speed of v m/s sets out from O to cross the river.

(a) What is the minimum value of v (v_{MIN}) which will permit the swimmer to reach the opposite bank at a point A directly opposite O (See Fig. 1)

(b) If $v < v_{MIN}$, and the swimmer reaches a point B on the other side of the river, find the minimum ratio of AB to OA.

Solution

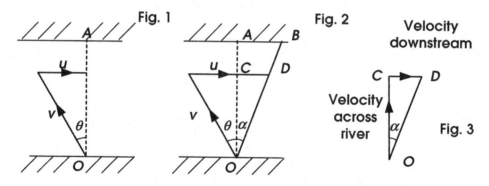

(a) Reaching point A from point O:

To reach A the swimmer must head upstream at an angle θ to OA, such that the resultant velocity of the swimmer is in the direction of OA. (See Fig. 1). Furthermore, for the swimmer to reach A, then $0° < \theta < 90°$.

But: $Sin\theta = \dfrac{u}{v} = \dfrac{u}{v_{MIN}}$ and $0 < Sin\,\theta < 1$ from the limits placed on θ

$$\Rightarrow 0 < \frac{u}{v_{MIN}} < 1 \Rightarrow u < v_{MIN} \Rightarrow v_{MIN} > u$$

(b) Reaching point B from point O:

If $v < u$ then the resultant velocity of the swimmer makes an angle α on the downstream of OA (See Fig. 2).

The resultant velocity component of the swimmer across the river is $v\,Cos\theta$

Therefore, the total time spent in crossing the river is:

$$t = \text{time} = \frac{\text{distance}}{\text{speed}} = \frac{OA}{vCos\theta} \quad \text{(i)}$$

But the resultant <u>downstream</u> component of the swimmer's speed is given by $u - vSin\,\theta$ in the CD direction (See Fig. 3):

$\Rightarrow$ distance downstream $= AB =$ (resultant downstream speed) $\times t$

$$\Rightarrow AB = (u - vsin\theta)\frac{OA}{vCos\theta} \qquad \text{But } \frac{AB}{OA} = \text{ratio}: \frac{\text{downstream distance}}{\text{width of the river}}$$

$$\Rightarrow \frac{AB}{OA} = \frac{u - vSin\theta}{vCos\theta} \quad \text{(ii)}$$

The value of θ which gives a minimum value of the ratio is found by differentiation: and

$$\frac{d(\text{Ratio})}{d\theta} = \frac{(vCos\theta)(-vCos\theta) - (u - vSin\theta)(-vSin\theta)}{v^2 Cos^2\theta} = 0$$

$$\Rightarrow -v^2 Cos^2\theta + uvSin\theta - v^2 Sin^2\theta = 0 \Rightarrow uvSin\theta = v^2(Sin^2\theta + Cos^2\theta) = v^2(1) = v^2$$

$$\Rightarrow Sin\theta = \frac{v^2}{uv} = \frac{v}{u} \text{ for minimum value of the ratio}$$

If $\quad Sin\theta = \frac{v}{u} \Rightarrow Cos\,\theta = \frac{\sqrt{u^2 - v^2}}{u} \qquad$ Substituting into equation (ii):

$$\Rightarrow \text{Minimum value of Ratio,} \frac{AB}{OA} = \frac{u - v\left(\frac{v}{u}\right)}{v\frac{\sqrt{u^2-v^2}}{u}} = \frac{\frac{u^2-v^2}{u}}{v\sqrt{u^2-v^2}} = \frac{\sqrt{u^2-v^2}}{v}$$

9. Crossing a river: Shortest path and shortest time

A river of width L metres is flowing with speed nu m/s and a swimmer who has a maximum speed of u m/s sets out to cross from O to cross the river. If $n < 1$ in what direction must the swimmer head and how much time is required to cross the river: (a) By the shortest path   (b) In the shortest time

Solution

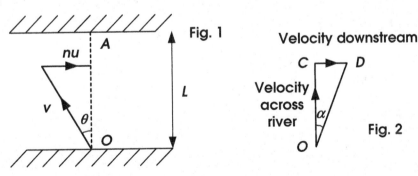

Fig. 1

Velocity downstream

Fig. 2

(a) By the shortest path.

The shortest path is OA. To reach A the swimmer must head upstream at an angle θ to OA, and have a resultant velocity in the direction OA. (See Fig. 1). But:

$$Sin\theta = \frac{nu}{u} = n(<1) \Rightarrow \theta = Sin^{-1}n \qquad \text{Thus, the time taken}$$

$$= \frac{L}{\text{Speed component along } OA} = \frac{L}{uCos\theta} = \frac{L}{u\left(\frac{\sqrt{u^2-n^2u^2}}{u}\right)} = \frac{L}{u\sqrt{1-n^2}} \qquad \text{seconds}$$

(b) In the shortest time

To cross by the shortest time the swimmer must minimise the time taken (see above). The value of θ which gives a minimum value for time taken found by differentiation:

$$\text{Time taken} = \frac{L}{uCos\theta} \Rightarrow \frac{d(\text{Time taken})}{d\theta} = \frac{(uCos\theta)(0)-(L)(-uSin\theta)}{(uCos\theta)^2} = \frac{uLSin\theta}{(uCos\theta)^2} = 0$$

$\Rightarrow \theta = 0° \Rightarrow$ Swimmer swims in the direction OA. His resultant velocity is in the direction

OD (making an angle α with OA) (See Fig. 2). $\Rightarrow$ Time taken $= \frac{L}{uCos(0)} = \frac{L}{u}$ seconds

10. Crossing a river: Shortest path and shortest time

A river of width L metres is flowing with speed nu m/s and a swimmer who has a maximum speed of u m/s sets out to cross from O to cross the river. If $n > 1$ in what direction must the swimmer head to cross the river and how much time is required to cross: (a) By the shortest path (b) In the shortest time

Solution
(a) By the shortest path.

The shortest path is OA. But, $u < nu$ so the resultant velocity of the swimmer makes an angle α on the downstream of OA (See Fig. 1).

The resultant velocity component of the swimmer along OA is $u\,Cos\theta$.

$\Rightarrow$ the total time spent in crossing the river is: $t = \text{time} = \frac{\text{distance}}{\text{speed}} = \frac{L}{uCos\theta}$ (i)

But the resultant <u>downstream</u> component of the swimmer's speed is given by $nu - uSin\theta$ in the CD direction (See Fig. 2):

$\Rightarrow$ distance downstream $= AB = (\text{resultant downstream speed}) \times t$

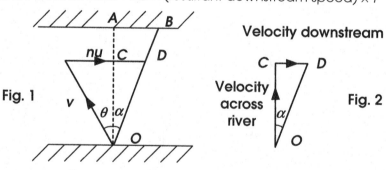

Fig. 1

Velocity downstream

Velocity across river

Fig. 2

$$\Rightarrow AB = (nu - uSin\theta)\left(\frac{L}{u\,Cos\theta}\right) \quad \text{(i)}$$

The value of θ which gives a minimum value of distance downstream will give the

shortest path across the river: $\dfrac{d(AB)}{d\theta} = \dfrac{(uCos\theta)(-uCos\theta)-(nu-uSin\theta)(-uSin\theta)}{u^2Cos^2\theta} = 0.$

$$\Rightarrow -u^2 Cos^2\theta + nu^2 Sin\theta - u^2 Sin^2\theta = 0 \Rightarrow nu^2 Sin\theta = u^2\left(Sin^2\theta + Cos^2\theta\right) = u^2(1) = u^2$$

$$\Rightarrow Sin\theta = \frac{1}{n} \text{ for minimum value of } |AB| \Rightarrow \theta = Sin^{-1}\left(\frac{1}{n}\right) = \text{direction of travel}$$

If $Sin\theta = \frac{1}{n} \Rightarrow Cos\theta = \frac{\sqrt{n^2-1}}{n}$ From equation (i): Time taken $= \frac{L}{uCos\theta} = \frac{nL}{u\sqrt{n^2-1}}$

(b) In the shortest time

To cross by the shortest time the swimmer must aim to minimise the time taken (see above). The value of θ which gives a minimum value for time taken is found by differentiation: Time taken $= \dfrac{L}{uCos\theta}$

$$\frac{d(Time\,taken)}{d\theta} = \frac{(uCos\theta)(0) - (L)(-uSin\theta)}{(uCos\theta)^2} = \frac{uLSin\theta}{(uCos\theta)^2} = 0$$

$\Rightarrow \theta = 0° \Rightarrow$ swimmer heads in direction OA. $\Rightarrow$ time taken $= \dfrac{L}{u}$ seconds.

11. Swimmer crossing river

A swimmer capable of swimming at a maximum speed of 4 m/s must cross a 100 metres wide river flowing of 5 m/s. Find the time taken to cross the river by the shortest path

Solution

Assume the swimmer heads at an angle α to the upstream bank and has a resultant velocity v m/s at an angle β to the downstream (See Fig. 1). The shortest time to cross will occur when the angle β is a maximum. The equations of motion are:

Parallel to river: $5 - 4\,Cos\,\alpha = v\,Cos\,\beta$ (i)   Perpendicular to river: $4\,Sin\,\alpha = v\,Sin\,\beta$ (ii).

Using equations (i) and (ii) $\Rightarrow \dfrac{vSin\beta}{vCos\beta} = Tan\,\beta = \dfrac{4Sin\alpha}{5 - 4Cos\alpha}$ (iii)

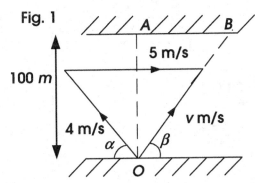

Fig. 1

100 m

5 m/s

4 m/s

v m/s

α β

O

A   B

When the value of $Tan\,\beta$ is a maximum, the angle β will be a maximum. To find the maximum value of $Tan\,\beta$, differentiate it with respect to α:

$$\frac{d(Tan\beta)}{d\alpha} = \frac{(5 - 4\,Cos\alpha)(4\,Cos\alpha) - (4\,Sin\alpha)(4\,Sin\alpha)}{(5 - 4\,Cos\alpha)^2} = 0$$

$$\Rightarrow (5 - 4\,Cos\alpha)(4\,Cos\alpha) = 16Sin^2\alpha = 16\left(1 - Cos^2\alpha\right)$$

$$\Rightarrow 20\,Cos\alpha - 16Cos^2\alpha = 16\left(1 - Cos^2\alpha\right) \Rightarrow 20\,Cos\alpha = 16$$

$\Rightarrow Cos\alpha = \frac{4}{5} \Rightarrow Tan\beta$ is a maximum when: $Cos\alpha = \frac{4}{5}$

$\Rightarrow$ Angle β is a maximum when: $Cos\alpha = \frac{4}{5}$

But, when: $Cos\alpha = \frac{4}{5} \Rightarrow Sin\alpha = \frac{3}{5}$

$\Rightarrow$ From equation (iii): maximum value of $Tan\beta = \dfrac{4\left(\frac{3}{5}\right)}{5 - 4\left(\frac{4}{5}\right)} = \dfrac{4}{3}$

$\Rightarrow Sin\beta = \frac{4}{5}$ and $Cos\beta = \frac{3}{5}$

From equation (ii): $4\ Sin\ \alpha = v\ Sin\ \beta \Rightarrow v = 3$ m/s

The actual path followed by the swimmer across the river is OB.

But: $OB\ Sin\ \beta = OA = 100$ metres.

$\Rightarrow OB\ Sin\ \beta = 100$ gives : $OB\,\frac{4}{5} = 100 \Rightarrow OB = \dfrac{500}{4} = 125$ metres

$\Rightarrow$ Time to cross by shortest path $= \dfrac{\text{distance}}{\text{velocity}} = \dfrac{125}{3} = 41.67$ seconds

12. Motorcycle and car at intersection

Two straight roads intersect at A. At time $t = 0$, a car is at point A and is travelling due North at 25 m/s while a motorcycle is 200 metres West of A and is travelling East at 20 m/s.

(a) Find the velocity of the motorcycle relative to the car.

(b) Calculate the least distance between the vehicles

(c) Show that the vehicles are closest in $t < 4$ seconds

Solution

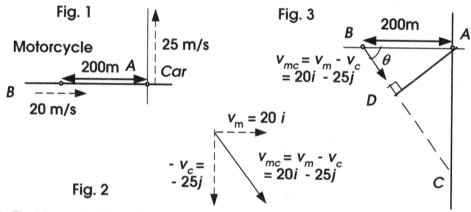

(a) Find the velocity of the motorcycle relative to the car

Fig. 1 shows the space diagram. Fig. 2 shows the velocity diagram.

Let velocity of car $= Vc = 0\ i + 25\ j$

Let velocity of motorcycle $= Vm = 20\ i + 0\ j \Rightarrow$ velocity of the motorcycle relative to the car $= Vmc = Vm - Vc = 20\ i - 25\ j$ Equation (i)

Thus, to the car driver, the motion of the motorcycle relative to him is identical to that he would see if his car was stopped at A and the motorcycle was travelling with speed

$\sqrt{(20)^2 + (25)^2} = 32.016$ m/s along BC i.e. in a direction whose angle to the horizontal is

$\theta = Tan^{-1}\left(-\dfrac{25}{20}\right) = Tan^{-1}\left(-\dfrac{5}{4}\right)$

(b) Calculate the least distance between the vehicles

The motorcycle is nearest A when the line joining A to the position of the motorcycle (at D, say) i.e. AD is such that: $AD \perp BC$. (See Fig. 3)

$$|Tan\theta| = \frac{5}{4} \Rightarrow Sin\theta = \frac{5}{\sqrt{5^2 + 4^2}} = \frac{5}{\sqrt{41}} = \frac{AD}{AB} = \frac{AD}{200} \Rightarrow AD = 156.17 \text{ metres}$$

(c) Show that the vehicles are closest in $t < 4$ seconds

The time taken to reach the least distance = the time taken for the motorcycle to travel from B to D. The distance BD is

$AB \; Cos \; \theta = 200(0.6247) = 124.94$ metres

Since $Vmc = 32.016$ m/s, then the time taken for the motorcycle to reach its least distance from the car is: $124.94/32.016 = t = 3.9$ seconds

13. Interception of a ship

At time $t = 0$, a ship at point O is steaming at 8 m/s in a direction 30° North of East and a patrol boat capable of moving at 15 m/s is at point Y, 5 km East of O (See Fig. 1).
(a) Find the minimum speed with which the patrol boat can intercept the ship
(b) Find the minimum time in which the patrol boat can intercept the ship

Solution

(a) Find the minimum speed with which the patrol boat can intercept the ship

At $t = 0$:
The ship is at O; the patrol boat is at Y. If the patrol boat is to intercept the ship with minimum speed it must travel the shortest distance possible i.e. along line

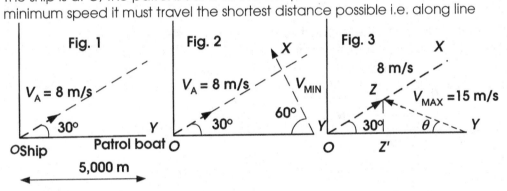

XY where $XY \perp OX$ to intercept at X. Since OX is at 30° to the horizontal, XY must lie at an angle of 60° with the horizontal, (see Fig. 2).

Let the ship move with velocity $V_S = 8$ m/s to X and the patrol boat travel with velocity V_{MIN} to X.

Thus, when interception takes place at X: The time taken, t, for the ship to travel from O to X = the time taken for the patrol boat to travel from Y to X. Thus: $t = \dfrac{OX}{V_S} = \dfrac{YX}{V_{MIN}}$ (i)

From geometry (See Fig. 2): $OY = 5,000$ metres, $XY = 2,500$ metres, $OX = 4,330$ metres. But the ship can travel at a speed of 8 m/s $\Rightarrow$ from equation (i) the minimum speed with which the patrol boat can intercept the ship = $\dfrac{4,330}{8} = \dfrac{2,500}{V_{MIN}} \Rightarrow V_{MIN} = 4.62$ m/s

(b) Find the minimum time in which the patrol boat can intercept the ship

For interception in minimum time the patrol boat must travel at its maximum speed of 15 m/s. Assume that, travelling at 15 m/s, it can intercept the ship at Z (See Fig. 3):

To find the direction of travel:
The ship travels from O along OX at 8 m/s. The patrol boat travels from Y along YZ at 15 m/s. Assume that the interception occurs at Z after time t, then the vertical distance ZZ' can be expressed as a function of the speeds of both the ship and the patrol boat. Resolving horizontally and vertically gives:
Vertically :

Patrol boat : $ZZ' = 15\,Sin\theta\,(t)$ Ship : $ZZ' = 8\,Sin30°\,(t) = 4t$

$\Rightarrow 15\,Sin\theta(t) = 4t \Rightarrow 15\,Sin\theta = \dfrac{4}{15} \Rightarrow \theta = 15.47°$

Horizontally : $OZ' + Z'Y = 5{,}000$ metres   (given)
= combined distance travelled horizontally by both ships in time t
= $8\,Cos30°\,t + 15\,Cos15.47°\,(t) \Rightarrow 5{,}000 = t(6.928 + 14.457)$
$\Rightarrow$ Minimum interception time = $t = 233.8$ seconds

14. Plane travel
A jet capable of travelling at V m/s flies from airport A to airport B, x metres North of A. Just before landing the jet is diverted to airport C, x metres East of B. A constant wind, W m/s, is blowing towards the direction θ degrees East of North.
(a)   Find the directions (in terms of θ) in which the jet must fly from A to B and from B to C to reach these destinations in the shortest times.
(b)   Find the ratio of the times taken for flying the journeys AB and BC.

Solution
(a) Find the directions in which the jet must fly to arrive in the shortest times.
Let the jet depart from A in a direction at an angle of α degrees west of north, and set out from B in a direction at an angle of β degrees east of south (See Fig. 1).

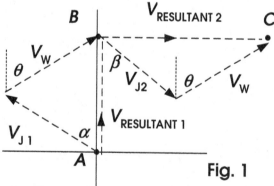

Fig. 1

Flying from A to B:
The velocities of the wind and the plane can be represented by the following vectors:
V_W = $(W\,Sin\,\theta)\,i + (W\,Cos\,\theta)\,j$ (i) $V_{J1} = -(V\,Sin\,\alpha)\,i + (V\,Cos\,\alpha)\,j$ (ii)
To make this journey in the minimum time, the resultant velocity of the jet must lie along the line AB. During this flight the resulting velocity of the jet ($V_{RESULTANT1}$) is: $V_{RESULTANT1} = V_W + V_{J1} = (W\,Sin\,\theta - V\,Sin\,\alpha)\,i + (W\,Cos\,\theta + V\,Cos\,\alpha)\,j$ (iii)

But $V_{RESULTANT1}$ must be in a Northerly direction only so it has only a j component $\Rightarrow$ From equation (iii): $(W\,Sin\,\theta - V\,Sin\,\alpha)\,i = 0 \,|\Rightarrow W\,Sin\,\theta = V\,Sin\,\alpha$

$\Rightarrow$ Direction of flight is at an angle α West of North where: $\alpha = Sin^{-1} \dfrac{W Sin\theta}{V}$ (iv)

<u>Flying from B to C.</u>
The velocity of the wind is unchanged: $V_w = (W Sin \theta) i + (W Cos \theta) j$
Let the velocity of the jet be: $V_{J2} = (V Sin \beta) i - (V Cos \beta) j$ (v)

The resulting velocity of the jet is:
$V_{RESULTANT2} = V_w + V_{J2} = (W sin \theta + V Sin \beta) i - (W Cos \theta - V Cos \beta) j$ (vi)
To make this journey in the minimum time, the resultant velocity of the plane must lie along the line $BC \Rightarrow V_{RESULTANT2}$ has an i component only:
$\Rightarrow$ From equation (vi): $(W Cos \theta - V Cos \beta) j = 0 j$
$\Rightarrow W Cos \theta = V Cos \beta \Rightarrow \beta = Cos^{-1} \dfrac{W Cos\theta}{V}$ (vii)

(b) Find the ratio of the times taken for flying the journeys AB and BC
Let $|AB| = |BC| = x$; But: (velocity)(time, t) = distance travelled = x. Thus:

For $|AB|$: $V_{RESULTANT1} = (W Cos \theta + V Cos \alpha) j \Rightarrow (W Cos \theta + V Cos \alpha) t_1 = x$
For $|BC|$: $V_{RESULTANT2} = (W Sin \theta + V Sin \beta) i \Rightarrow (W Sin \theta) + (V Sin \beta) t_2 = x$
But: $|AB| = |BC| \Rightarrow (W Cos \theta + V Cos \alpha) t_1 = (W Sin \theta + V Sin \beta) t_2$
$\Rightarrow \dfrac{t_1}{t_2} = \dfrac{W Sin\theta + V Sin \beta}{W Cos\theta + V Cos\alpha}$
But, from equations (iv) and (vii): $W Sin \theta = V Sin \alpha$ and $W Cos \theta = V Cos \beta$
$\Rightarrow \dfrac{t_1}{t_2} = \dfrac{V Sin\alpha + V Sin\beta}{V Cos\beta + V Cos\alpha} = \dfrac{Sin \alpha + Sin \beta}{Cos\alpha + Cos\beta}$

15. Plane on a return journey
B is located d metres North of A. A plane flies from A to B and back to A (without landing) with speed V m/s. During the flight there is a wind of W m/s blowing from a point $30°$ North of East. If the complete journey takes T seconds, find the distance d.

Solution

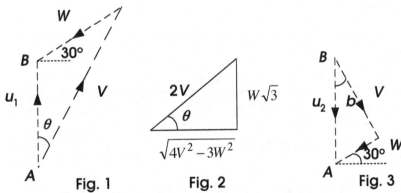

Fig. 1   Fig. 2   Fig. 3

Consider the journey as taking place in two stages:
(a) Journey from A to B:
To reach B the plane must travel in a direction θ degrees East of North so that (from Fig. 1):

$$W \cos 30° = V \sin\theta \Rightarrow \frac{W\sqrt{3}}{2} = V \sin\theta \Rightarrow \sin\theta = \frac{W\sqrt{3}}{2V}$$

Thus, see Fig. 2 : $\cos\theta = \dfrac{\sqrt{4V^2 - 3W^2}}{2V}$

If the plane travels at V m/s, θ degrees East of North, then its resulting velocity North, u_1, (i.e. along AB directly towards B) can be expressed as:

$$u_1 = V\cos\theta - W\cos 60° = V\frac{\sqrt{4V^2 - 3W^2}}{2V} - \frac{W}{2} \Rightarrow u_1 = \frac{1}{2}\left(\sqrt{4V^2 - 3W^2} - W\right)$$

$\Rightarrow$ Time taken to travel to B:

$$= \frac{\text{Distance from } A \text{ to } B}{\text{Resultant speed along } AB} = \frac{d}{u_1} = \frac{2d}{\sqrt{4V^2 - 3W^2} - W} \qquad \text{seconds}$$

(b) Journey from B to A:

On the return journey to A, the plane will travel in a direction β degrees East of South (See Fig. 3) so the resultant velocity South can be found by:

$$V \sin\beta = W \cos 30° = \frac{W\sqrt{3}}{2} \Rightarrow \sin\beta = \frac{W\sqrt{3}}{2V} ; \text{Thus}: \cos\beta = \frac{\sqrt{4V^2 - 3W^2}}{2V}$$

If the plane travels at V m/s, θ degrees East of South, then the resultant speed directly towards A is u_2 where:

$$u_2 = V\cos\beta + W\cos 60° = \frac{V}{2V}\sqrt{4V^2 - 3W^2} + \frac{W}{2} \Rightarrow u_2 = \frac{\sqrt{4V^2 - 3W^2} + W}{2}$$

$\Rightarrow$ time to travel from : B to A ($|BA| = d$) is $\dfrac{d}{u_2} = \dfrac{2d}{\sqrt{4V^2 - 3W^2} + W} \qquad$ seconds

Total journey

Therefore the total time taken to fly from A to B and from B back to A is:

$$T = \frac{2d}{\sqrt{4V^2 - 3W^2} - W} + \frac{2d}{\sqrt{4V^2 - 3W^2} + W} = \frac{4d\sqrt{4V^2 - 3W^2}}{(4V^2 - 3W^2) - W^2} \qquad \text{seconds}$$

$$\Rightarrow d = \frac{T(4V^2 - 4W^2)}{4\sqrt{4V^2 - 3W^2}} = \frac{T(V^2 - W^2)}{\sqrt{4V^2 - 3W^2}} \qquad \text{metres}$$

16. Two ships moving at sea

A minesweeping ship, initially at point M, is steaming due South at 25 km/hr. An aircraft carrier capable of travelling at 40 km/h is located at point A, a distance D metres North-East of this position, and is ordered to take up and maintain a position 2 km North of the minesweeper. If the aircraft carrier arrives at a position 2 km North of the minesweeper in 20 minutes, find the original distance, D, separating the ships.

Solution

The initial position of both ships is shown in Fig. 1: the aircraft carrier is at A while the minesweeper is at M. The final positions are shown in Fig. 2: the aircraft carrier is at B while the minesweeper is at N.

Assume the co-ordinates of the initial position of minesweeper = $(x, y) = (0, 0)$

$\Rightarrow$ position after 20 minutes is $(0, -8.33)$

$\Rightarrow$ position of aircraft carrier in 20 minutes = $(0, -6.33)$

But, aircraft carrier travels at 40 km/hr ⇒ in 20 minutes it will travel 13.33 km.

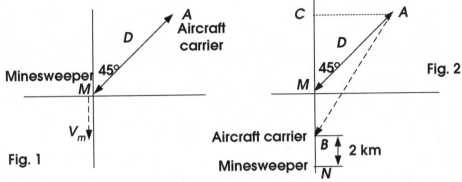

Fig. 1

Fig. 2

Initial position of aircraft carrier is at A, at coordinates $\left(\dfrac{D}{\sqrt{2}}, \dfrac{D}{\sqrt{2}}\right)$

$$\Rightarrow 13.33 = \sqrt{\left(\dfrac{D}{\sqrt{2}} - 0\right)^2 + \left(\dfrac{D}{\sqrt{2}} + 6.33\right)^2} \Rightarrow (13.33)^2 = \left(\dfrac{D}{\sqrt{2}} - 0\right)^2 + \left(\dfrac{D}{\sqrt{2}} + 6.33\right)^2$$

$\Rightarrow D^2 + 8.95D - 137.62 = 0$ Solving gives : $D = 8.08$ km

$\Rightarrow$ Original distance separating ships = 8.08 km

17. Closest approach of two planes

Plane A is travelling North East at 200 km/hr when, at noon, its radar detects Plane B, 16 km to the South East and travelling at 400 km/hr in a direction 22.5° East of North. Find:
(a) The closest distance between the planes (b) The time taken to reach this distance

Solution
(a) The closest distance between the planes
Fig. 1 shows the initial positions at noon:

<u>Plane A</u>
Velocity: Plane A is at O and is travelling at 200 km/hr in a North-East direction. Its direction of travel is 45° East of North. The velocity of A can be expressed as a vector V_A where: $V_A = 141.42\,i + 141.42\,j$
Plane A's position: at O, i.e. (0,0)

<u>Plane B</u>
Velocity: Plane B is at X, a distance of 16 km South-East of O. B is travelling at 400 km/hr in a direction 22.5° East of North. The velocity of B can be expressed as a vector V_B

where: $V_B = 153.07\,i + 369.55\,j$

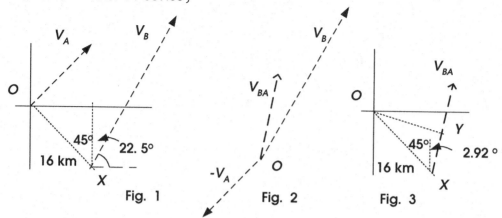

Fig. 1   Fig. 2   Fig. 3

Plane B's position (in x, y co-ordinates) at $X = (16\,Sin\,45°, -16\,Cos\,45°) = (11.314, -11.314)$
Fig. 2 shows the velocity diagram. The relative velocity of B to $A = V_{BA} = V_B - V_A$ giving
$V_{BA} = 11.65\,i + 228.13\,j$
This resultant velocity is of magnitude 228.43 km/hr and is in a direction 2.92° East of
North. See Fig. 3.
Fig. 3 shows the approach to finding the time when the planes are closest and the
distance then separating them.

V_{BA} indicates the direction of travel along line XY. The shortest distance between Planes
A and B occurs at point Y where the line joining O and Y is perpendicular to the line
joining X and Y.
The shortest distance = $|OY| = OX\,Sin\,(45°+2.92°) = 16\,Sin\,47.92° = 11.876$ km
(b) The time taken to reach this distance
Time taken to reach point Y = time taken by Plane B to travel from X to Y at a velocity

$$V_{BA} = \frac{XY}{V_{BA}} = \frac{OXCos47.92°}{228.43} = \frac{10.723}{228.43} = 0.0469\,hours = 168.9\,seconds$$

18. Mid-air interception
In a light patrol plane travelling at 192 km/hr the radar detects a helicopter at a
distance of 184 km in a direction 30° North of East which is travelling due North at 128
km/hr. How quickly can the plane intercept the helicopter?

Solution
See Fig. 1 for initial locations.
Plane
Velocity: The plane is at A and is travelling at $v_P = 192$ km/hr. The direction in which the
plane must fly to intercept the helicopter as soon as possible must be found.
Helicopter
Velocity: The helicopter is at O. Its velocity in km/hr is: $v_H = 0\,i + 128\,j$
Add $-v_H$ to the velocities of the plane and helicopter.
Helicopter: resulting velocity = $v_H - v_H = 0 \Rightarrow$ Helicopter is "stopped" at O.
Plane: The relative velocity of the plane to the helicopter = $v_{PH} = v_P - v_H$
= resulting velocity.

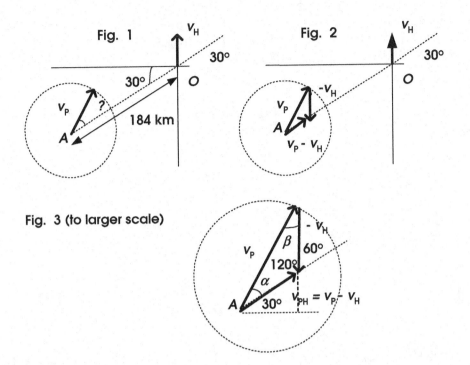

Fig. 1

Fig. 2

Fig. 3 (to larger scale)

But, If the plane is to reach the helicopter in the shortest possible time its resultant velocity should lie along the straight line joining A to O (where the helicopter is "stopped". But, (from Fig. 3) using the Sine Rule gives:

$$\frac{|v_P|}{Sin\,120°} = \frac{|-v_H|}{Sin\,\alpha} \Rightarrow \frac{192}{0.866} = \frac{128}{Sin\,\alpha} \Rightarrow Sin\,\alpha = 0.5773 \Rightarrow \alpha = 35.26° \Rightarrow \beta = 24.74°$$

$$\frac{192}{Sin\,120°} = \frac{|v_{PH}|}{Sin\,24.74°} \Rightarrow |v_{PH}| = 92.79\ \text{km/hr} \Rightarrow \text{Minimum time} = \frac{184}{92.79} = 1.98\ \text{hours}$$

19. Mid-air interception

Two ships can only communicate by signal lamps which have a maximum range of 15 km. Ship A is 50 km due North of Ship B. Ship A is travelling South West at 25 km/hr and ship B is travelling due West at 15 km/hr. Find the total amount of time during which the ships can communicate by signal lamp.

Solution

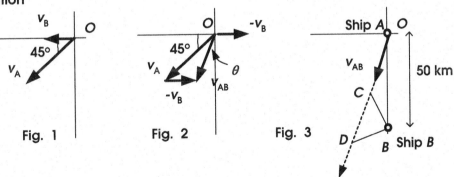

Fig. 1   Fig. 2   Fig. 3

The velocities are shown in Fig. 1.

$v_A = -25\,Cos\,45°\,i - 25\,Sin\,45°\,j = 17.678\,i - 17.678\,j$ $v_B = -15\,i + 0\,j$

35

Add a velocity " - v_B " (i.e. 15 i + 0 j) to the velocity of each ship (See Fig. 2). The result is: The velocity of ship A relative to ship B = v_A - v_B = v_{AB} = - 2.678 i – 17.678 j. It can be seen that ship A will proceed on the line OD. Note that this line makes an angle of θ = Tan^{-1}(17.678/2.678) = 8.61° with the North-South axis (See Figs. 1 and 2). Also, magnitude of v_{AB} = 17.88 km/hr.

v_B - v_B = 0 i + 0 j (i.e. ship B is "stopped" at point B)

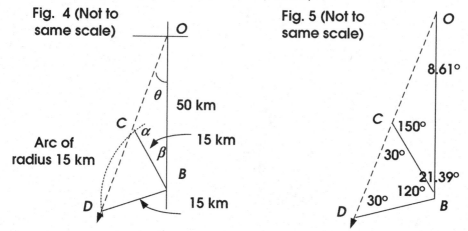

Fig. 4 (Not to same scale)

Fig. 5 (Not to same scale)

Draw an arc of radius 15 km centred on B (See Fig. 3). The two ships can communicate by lamp when Ship A passes between points C and D as C, D and all points between them are within a distance of 15 km from B. Using the Sine Rule in triangle OBC:

$$\frac{|15|}{Sin\, 8.61°} = \frac{|50|}{Sin\alpha} \Rightarrow Sin\alpha = 0.5 \Rightarrow \alpha = 30°, 150° \Rightarrow \beta = 180° - 8.61° - 150° = 21.39°$$

From the Sine Rule in triangle CDB: $\dfrac{|15|}{Sin\, 30°} = \dfrac{|CD|}{Sin\, 120°} \Rightarrow |CD| = 25.98$ km

$\Rightarrow$ Time over which signalling can occur = $\dfrac{25.98 \ km}{17.88 \ km/hr}$ = 1.45 hours

Chapter 3   Projectiles

1. Projectile on horizontal plane

A particle is projected at an angle θ to a horizontal plane (See Fig. 1). If u m/s is the initial velocity of projection and there is no air resistance, find expressions for:

(i)   Horizontal and vertical components of velocity after time t
(ii)   Horizontal and vertical components of displacement after time t
(iii)   Maximum height reached
(iv)   Total time of flight
(v)   Horizontal range
(vi)   Maximum range possible

Solution

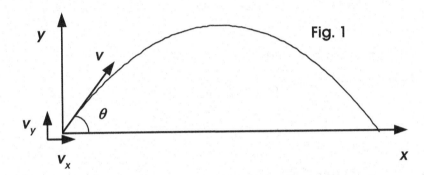

Fig. 1

The components of velocity are shown in Fig. 1 where the x-axis represents the horizontal plane and the y-axis, the vertical plane:

(i) Horizontal component of velocity = $v_x = u\,Cos\,\theta$
  Vertical component of velocity = $v_y = u\,Sin\,\theta - gt$

(ii) Horizontal component of displacement = $x = v_x\,t = ut\,Cos\,\theta$
  Vertical component of displacement = $y = ut\,Sin\,\theta - \frac{1}{2}gt^2$

(iii) Maximum height reached:
At maximum height, $v_y = 0 \Rightarrow u\,Sin\,\theta - gt = 0$

$\Rightarrow$ Time elapsed to reach maximum height $= \dfrac{u\,Sin\theta}{g}$

$\Rightarrow$ To calculate maximum height, put this expression for t into the expression for y in (ii) above.

$\Rightarrow$ maximum height $= u\left(\dfrac{u\,Sin\theta}{g}\right)Sin\theta - \dfrac{1}{2}g\left(\dfrac{u\,Sin\theta}{g}\right)^2 = \dfrac{u^2\,Sin^2\theta}{2g}$

(iv) Total time of flight
At maximum height the particle has travelled half its total time of flight. Thus, the total time of flight can be expressed as: $2 \times \left(\dfrac{u\,Sin\theta}{g}\right) = \dfrac{2u\,Sin\theta}{g}$

(v) Horizontal range

Horizontal range = (horizontal component of velocity) × (total time of flight)

$$= (u\,Cos\theta)\left(\frac{2u\,Sin\theta}{g}\right) = \frac{u^2\,Sin2\theta}{g}$$

(vi) Maximum range possible

This is achieved when the horizontal range has a maximum value i.e. when $\frac{u^2\,Sin2\theta}{g}$ is a maximum. This occurs when $Sin\,2\theta$ is a maximum, i.e. when $Sin\,2\theta =$

$$1 \Rightarrow \theta = 45° \Rightarrow \quad \text{maximum range} = \frac{u^2}{g}$$

2. Projectile on horizontal plane

Using the same symbols as above, derive an expression for y in terms of x and θ only.

Solution

After time t in flight the displacements are:

$$y = ut\,Sin\theta - \tfrac{1}{2}gt^2, \quad x = ut\,Cos\theta \Rightarrow t = \frac{x}{u\,Cos\theta}$$

$$\Rightarrow y = u\,Sin\theta\left(\frac{x}{u\,Cos\theta}\right) - \tfrac{1}{2}g\left(\frac{x}{u\,Cos\theta}\right)^2 = x\,Tan\theta - \frac{gx^2}{2u^2\,Cos^2\theta}$$

3. Horizontal range of a projectile

Show that, in general, a particle can be projected at two different angles to reach a given horizontal range, R.

Solution

Where θ = angle of projection and u = initial velocity then the range, R is given by

$$R = \frac{u^2\,Sin2\theta}{g} \Rightarrow Sin2\theta = \frac{gR}{u^2} \quad \text{But: } Sin2\theta = Sin(180° - 2\theta) = Sin2(90° - \theta)$$

$\Rightarrow$ Two possible angles of projection are: $\theta, 90° - \theta$

When $\theta = 45°$ the range is a maximum and there is only one value of θ possible.

4. Greatest Height reached

Express the greatest height, H, reached by a projectile in terms of the time T required to reach H.

Solution

When the projectile reaches its greatest height, its vertical component of velocity = 0 $\Rightarrow$
$v_y = 0 = u\,Sin\,\theta - gt \Rightarrow t = T\,(\text{say}) = (u\,Sin\,\theta)/g$
But, Maximum height is as follows: $H = ut - \tfrac{1}{2}gt^2$

$$\Rightarrow H = \text{Maximum height} = u\left(\frac{u\,Sin\theta}{g}\right)Sin\theta - \frac{1}{2}g\left(\frac{u\,Sin\theta}{g}\right)^2 = \frac{u^2\,Sin^2\theta}{2g}$$

$$\Rightarrow H = (gT)^2/2g = \tfrac{1}{2}gT^2$$

5. Maximum range achieved

Express the maximum range reached by a projectile in terms of its maximum height, H during its flight.

Solution

$\text{Range} = \dfrac{u^2 Sin\, 2\theta}{g}$ and Maximum Range occurs when : $Sin\, 2\theta$ is a maximum.

This occurs when $Sin\, 2\theta = 1$ i.e when : $\theta = 45^O \Rightarrow$ Maximum range $= \dfrac{u^2}{g}$

$\theta = 45^O \Rightarrow$ maximum height $= H = \dfrac{u^2 Sin^2\theta}{2g} = \dfrac{u^2}{4g} \Rightarrow$ Maximum Range $= 4H$

6. Displacement of a projectile

A particle is projected from O at an angle θ with initial velocity u m/s. After time t seconds the particle is at point A, a horizontal distance d metres and vertical height h metres from the point of projection.

(a)   Find the two possible values of t for this displacement (say, t_1 and t_2).
(b)   If $d = h\sqrt{3}$, show that: $t_1 t_2 = 4h/g$

Solution
(a) Find the two possible values of t for this displacement (say, t_1 and t_2).
From Fig. 1 the following equations can be written:

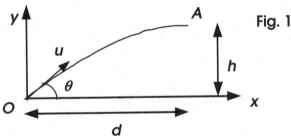

Fig. 1

$d = ut\, Cos\theta$ and $h = ut\, Sin\theta - \tfrac{1}{2}gt^2 \Rightarrow ut\, Sin\theta = h + \tfrac{1}{2}gt^2$

$\Rightarrow d^2 + \left(h + \tfrac{1}{2}gt^2\right)^2 = (ut\, Cos\theta)^2 + (ut\, Sin\theta)^2$

$= (ut)^2\left(Cos^2\theta + Sin^2\theta\right) = (ut)^2 = u^2 t^2 \Rightarrow d^2 + \left(h + \tfrac{1}{2}gt^2\right)^2 = u^2 t^2$

$d^2 + h^2 + ght^2 + \dfrac{g^2 t^4}{4} = u^2 t^2$ which can be expressed as :

$\Rightarrow \left(\dfrac{g^2}{4}\right) t^4 + \left(gh - u^2\right) t^2 + \left(d^2 + h^2\right) = 0$

Solving for t^2 gives : $t^2 = \dfrac{\left(u^2 - gh\right) \pm \sqrt{\left(gh - u^2\right)^2 - g^2\left(d^2 + h^2\right)}}{\tfrac{1}{2}g^2}$

$\Rightarrow t^2$ has two values $\Rightarrow t$ has two values, t_1 and t_2, say

$$\Rightarrow t_1^2 = \frac{2(u^2 - gh) + 2\sqrt{(gh - u^2)^2 - g^2(d^2 + h^2)}}{g^2}$$

and $\quad t_2^2 = \dfrac{2(u^2 - gh) - 2\sqrt{(gh - u^2)^2 - g^2(d^2 + h^2)}}{g^2}$

(b) If $d = h\sqrt{3}$ $\quad$ show : $t_1 t_2 = \dfrac{4h}{g}$

NOTE : Finding $t_1^2 t_2^2$ will allow a simpler expression to be derived :

$$\Rightarrow t_1^2 t_2^2 = \frac{2^2(u^2 - gh)^2 - 2^2\left[(gh - u^2)^2 - g^2(d^2 + h^2)\right]}{g^4}$$

$$= \frac{4}{g^4}\left[u^4 + g^2 h^2 - 2ghu^2 - g^2 h^2 - u^4 + 2ghu^2 + g^2(d^2 + h^2)\right] = \frac{4(d^2 + h^2)}{g^2}$$

$$\Rightarrow t_1 t_2 = \frac{2\sqrt{d^2 + h^2}}{g}$$

But (given) : $d = h\sqrt{3} \Rightarrow t_1 t_2 = \dfrac{4h}{g}$

7. Particle thrown from a cliff

A ball is thrown out to sea from a point O on the edge of a cliff which is 30 metres high. The ball reaches a maximum height of 80 metres above the sea level before hitting the water at a distance 100 metres from the foot of the cliff (See Fig. 1). Find the angle at which the ball was thrown.

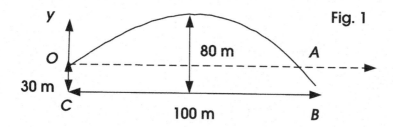

Fig. 1

Solution

Consider the curve OA. The maximum height above the horizontal through O attained by the ball is $(80 - 30) = 50$ metres. If u = initial velocity of the ball, then the maximum height is (see Question 1):

$\dfrac{u^2 Sin^2\theta}{2g} = 50 \Rightarrow u^2 = \dfrac{100g}{Sin^2\theta}$ $\qquad$ From Question 2 above :

$y = x Tan\theta - \dfrac{gx^2}{2u^2 Cos^2\theta} = x Tan\theta - \dfrac{gx^2}{2\left(\dfrac{100g}{Sin^2\theta}\right) Cos^2\theta}$

$\Rightarrow y = x Tan\theta - \dfrac{x^2 Sin^2\theta}{200 Cos^2\theta} = x Tan\theta - \dfrac{x^2 Tan^2\theta}{200}$

But, if $x = 100$ metres, then $y = -30$ metres (given)

$\Rightarrow -30 = 100\,Tan\theta - \dfrac{10,000\,Tan^2\,\theta}{200} \Rightarrow -30 = 100\,Tan\theta - 50\,Tan^2\theta$

$\Rightarrow 50\,Tan^2\,\theta - 100\,Tan\theta - 30 = 0$ i.e. $Tan^2\,\theta - 2Tan\theta - 0.6 = 0$

Solving for $Tan\theta$:

$\Rightarrow Tan\theta = \dfrac{2 \pm \sqrt{4 + 2.4}}{2} = \dfrac{2 \pm 2.53}{2} = 2.265,\ -0.265 \Rightarrow \theta = 66.17°,\ -14.84°$

But : $\theta > 0,\ \Rightarrow \theta = 66.17°$

8. Stone projected over vertical wall

A stone is projected with velocity u m/s at an angle $45°$ from a point on the ground a distance d metres from the foot of the vertical wall of height h metres. If the stone just clears the wall, show that $h = d - \dfrac{d^2}{4H}$ where H is the greatest height reached.

Solution
Assume the stone passes over the top of the wall when $t =$ time elapsed after initial projection (in seconds). Then:

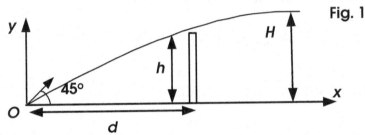

Fig. 1

$x = d = ut\,Cos\theta$ $\qquad y = ut\,Sin\theta - \frac{1}{2}gt^2;$ Substituting : $t = \dfrac{d}{u\,Cos\theta}$ into the equation for y

gives: $y = dTan\theta - \dfrac{gd^2}{2u^2\,Cos^2\theta}$

As the stone just passes over the wall, of height h, then:

$y = h = d\,Tan\theta - \dfrac{gd^2}{2u^2\,Cos^2\theta}$ $\qquad$ (i)

But $\theta = 45°$ (given) $\Rightarrow h = d - \dfrac{gd^2}{u^2}$

But (from Question 1) $H =$ greatest height reached $\Rightarrow$

$H = \dfrac{u^2\,Sin^2\theta}{2g} \Rightarrow H = \dfrac{u^2}{4g}$ i.e. $u^2 = 4gH \Rightarrow h = d - \dfrac{d^2}{4H}$

9. Projecting a ball over the wall

A ball is projected with velocity u m/s at an angle θ from a point O on the ground so as just to clear a wall of height a metres, at a distance of a metres from O. Find the possible angles of projection, maximum height and maximum range.

Solution

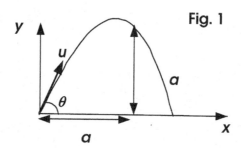

Fig. 1

Writing $y = x\,Tan\theta - \dfrac{gx^2}{2u^2 Cos^2\theta}$ where $y = a$, $x = a$, gives :

$a = a\,Tan\theta - \dfrac{ga^2}{2u^2 Cos^2\theta} \Rightarrow 2u^2 Cos^2\theta(a) = 2u^2 Cos^2\theta(a\,Tan\theta) - ga^2$

$\Rightarrow a = \dfrac{2u^2}{g}\left(Sin\theta\,Cos\theta - Cos^2\theta\right) \Rightarrow$ value of a is a maximum when $\dfrac{d(a)}{d\theta} = 0$

$\Rightarrow \dfrac{d(a)}{d\theta} = \dfrac{2u^2}{g}\left(Cos^2\theta - Sin^2\theta + 2Sin\theta\,Cos\theta\right) = 0$

$\Rightarrow Cos^2\theta - Sin^2\theta + 2Sin\theta\,Cos\theta = 0$

Letting $Cos\theta = x$ and $Sin\theta = \sqrt{1 - x^2}$

$\Rightarrow x^2 - (1 - x^2) + 2x\sqrt{1 - x^2} = 0 \Rightarrow 2x^2 - 1 = \left(-2x\sqrt{1 - x^2}\right)^2$

$\Rightarrow 4x^4 - 4x^2 + 1 = 4x^2 - 4x^4 \Rightarrow 8x^4 - 8x^2 + 1 = 0 \Rightarrow x^2 = \dfrac{2 \pm \sqrt{2}}{4}$

$\Rightarrow x = 0.9238,\ 0.3827 \Rightarrow$ Possible angles of projection $= \theta = 22.5°,\ 67.5°$

Maximum height $= \dfrac{u^2 Sin^2\theta}{2g} = \dfrac{0.427u^2}{g}$ (at $\theta = 67.5°$)

Maximum range $= \dfrac{u^2 Sin2\theta}{g} = \dfrac{u^2}{g\sqrt{2}}$ for either value of θ

10. Particle projected from cliff

A particle is projected at initial speed u m/s from the top of a cliff of height h metres. The particle strikes the sea at a distance d metres from the foot of the cliff (See Fig. 1). Show that if $u = \sqrt{gh}$ the maximum value of d is $h\sqrt{3}$

Solution

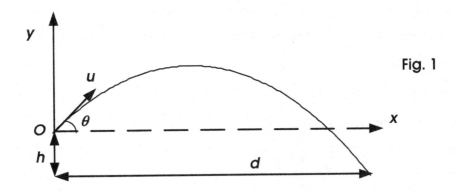

Fig. 1

Using the fact that when $y = -h$, $x = d$ it can be written that:

$d = ut\,Cos\theta$ and $-h = ut\,Sin\theta - \frac{1}{2}gt^2$

$\Rightarrow$ From above : $d^2 = u^2t^2Cos^2\theta$, $\left(\frac{1}{2}gt^2 - h\right)^2 = u^2t^2Sin^2\theta$

$\Rightarrow d^2 + \left(\frac{1}{2}gt^2 - h\right)^2 = u^2t^2Sin^2\theta + u^2t^2Cos^2\theta = u^2t^2$

$\Rightarrow d^2 + \frac{1}{4}g^2t^4 - \left(u^2 + gh\right)t^2 + h^2 = 0 \Rightarrow d^2 = \left(u^2 + gh\right)t^2 - \frac{g^2}{4}t^4 - h^2$

The value of d is a maximum when d^2 has a maximum value. The maximum value of d can be found by first determining the maximum value of d^2. This value occurs when:

$\frac{d(d^2)}{dt} = 2\left(u^2 + gh\right)t - g^2t^3 = 0 \Rightarrow 2\left(u^2 + gh\right) - g^2t^2 = 0$

$\Rightarrow t^2 = \frac{2\left(u^2 + gh\right)}{g^2}$ gives the flight time for maximum value of d

$\Rightarrow$ Maximum value of $d^2 = \left(u^2 + gh\right)\frac{2\left(u^2 + gh\right)}{g^2} - \frac{g^2}{4}\left(\frac{2\left(u^2 + gh\right)}{g^2}\right)^2 - h^2$

$= \frac{2}{g^2}\left(u^2 + gh\right)^2 - \frac{g^2}{g^4}\left(u^2 + gh\right)^2 - h^2 = \frac{\left(u^2 + gh\right)^2}{g^2} - h^2 = \frac{u^4 + 2u^2gh}{g^2}$

$\Rightarrow$ Maximum value of $d = \frac{u\sqrt{u^2 + 2gh}}{g}$ metres

But, (given): $u = \sqrt{gh} \Rightarrow$ Maximum value of $d = h\sqrt{3}$ metres

11. Direction of projectile

A particle of mass m kg is projected with initial velocity u m/s at an angle $\theta = Sec^{-1}\sqrt{a}$ to the horizontal. Find its minimum kinetic energy during its flight.

Solution

43

Initial value of kinetic energy $= \frac{1}{2}mu^2$ Joules

If velocity at time t seconds is v m/s then the velocity components are:

v_x = horizontal component of velocity in metres = $u\,Cos\theta$

v_y = vertical component of velocity in metres = $u\,Sin\theta - gt$

$\Rightarrow v^2 = v_x^2 + v_y^2 \Rightarrow$

$v^2 = u^2 Cos^2\theta + u^2 Sin^2\theta + g^2 t^2 - 2gtu\,Sin\theta$

$\Rightarrow v^2 = u^2 + g^2 t^2 - 2gtu\,Sin\theta$　(i)

$\Rightarrow$ The value of kinetic energy $= \frac{1}{2}mv^2$ Joules

But: $\theta = Sec^{-1}\sqrt{a} \Rightarrow Cos\theta = \dfrac{1}{\sqrt{a}}$ (given):

The minimum value of kinetic energy of the particle occurs when v^2 is a minimum. This occurs when the value of v^2 (as in equation (i)) is a minimum:

$$\frac{d(v^2)}{dt} = 0 + 2g^2 t - 2gu\,Sin\theta = 0$$

$\Rightarrow$ minimum kinetic energy occurs after time, $t = \dfrac{u\,Sin\theta}{g}$　seconds

$$\Rightarrow v^2 = u^2 + g^2\left(\frac{u\,Sin\theta}{g}\right)^2 - 2ug\,Sin\,\theta\left(\frac{u\,Sin\theta}{g}\right)$$

$$= u^2 + u^2 Sin^2\theta - 2u^2 Sin^2\theta = u^2 - u^2 Sin^2\theta = u^2 Cos^2\theta = \frac{u^2}{a}$$

$\Rightarrow$ Minimum kinetic energy $= \frac{1}{2}mv^2 = \frac{1}{2}m\dfrac{u^2}{a} = \frac{1}{2}mu^2\left(\dfrac{1}{a}\right)$　Joules

$\Rightarrow$ Minimum kinetic energy $= \left(\dfrac{1}{a}\right) \times$ (Initial kinetic energy)

12. Angle of projection

A particle projected at u m/s at an angle of $30°$ to the horizontal lands 5 metres short of the target. When projected at u m/s at $45°$ it lands 10 metres beyond the target (See Fig. 1). Find the correct angle of projection to hit the target.

Solution

Assume the correct angle of projection = θ. Then the desired horizontal range is: $R = \dfrac{u^2 Sin2\theta}{g}$　(i)

$\theta = 30° \Rightarrow$ horizontal range $= \dfrac{u^2 Sin60°}{g} = \dfrac{u^2\sqrt{3}}{2g} = R - 5$　(given) (ii)

$\theta = 45° \Rightarrow$ horizontal range $= \dfrac{u^2 Sin90°}{g} = \dfrac{u^2}{g} = R + 10$　(given) (iii)

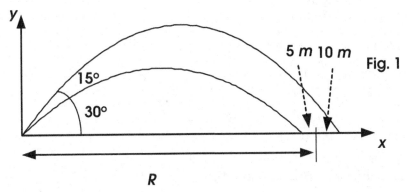

Fig. 1

Equations (ii) and (iii) give :

$$u^2 = \frac{2g(R-5)}{\sqrt{3}} = g(R+10) \Rightarrow 2gR - 10g = g\sqrt{3}R + 10g\sqrt{3} \Rightarrow R = \frac{10+10\sqrt{3}}{2-\sqrt{3}} = 101.96 \quad \text{metres}$$

Putting this value into equation (ii) gives: $u = 33.14$ m/s
Putting these values for R and u into equation (i) gives the correct angle of projection =
$\theta = 32.8°$

13. Two particles projected in perpendicular directions

Two particles, A and B, are projected simultaneously from a single point at the same speed at angles θ and $\theta + 90°$ respectively (See Fig. 1). A strikes the ground first.
Show that while both particles are in motion:
(a)   The slope of the line joining A, B remains constant during the flight
(b)   The distance $|AB|$ increases as a function of time, t
(c)   Find B's horizontal displacement when A strikes the ground.

Solution
(a)   The slope of the line joining A, B remains constant during the flight

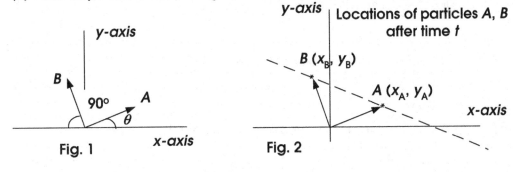

Fig. 1

Locations of particles A, B after time t

Fig. 2

See Fig. 2 for details of how slope of line is derived.
Let the co-ordinates of A and B at any time t be (x_A, y_A) and (x_B, y_B) respectively, where:

$x_A = utCos\theta, \quad y_A = utSin\theta - \frac{1}{2}gt^2$ and

$x_B = utCos(\theta + 90°), \quad y_B = utSin(\theta + 90°) - \frac{1}{2}gt^2$

The slope of the line joining A and B is : $\dfrac{y_A - y_B}{x_A - x_B}$

$$= \frac{\left(utSin\theta - \frac{1}{2}gt^2\right) - \left(utSin(\theta + 90°) - \frac{1}{2}gt^2\right)}{(utCos\theta) - (utCos(\theta + 90°))} = \frac{Sin\theta - Sin(\theta + 90°)}{Cos\theta - Cos(\theta + 90°)}$$

$$= \frac{Sin\theta - (Sin\theta\, Cos90° + Cos\theta\, Sin90°)}{Cos\theta - (Cos\theta\, Cos90° - Sin\theta\, Sin90°)} = \frac{Sin\theta - Cos\theta}{Cos\theta + Sin\theta}$$

$\Rightarrow$ The slope remains constant for any value of θ.

(b) The distance $|AB|$ increases as a function of time, t

Distance $|AB| = \sqrt{(x_A - x_B)^2 + (y_A - y_B)^2}$ From equation (i):

$= ut\sqrt{(Cos\theta + Sin\theta)^2 + (Sin\theta - Cos\theta)^2} = ut\sqrt{2(Sin^2\theta + Cos^2\theta)} = ut\sqrt{2}$ (ii)

$\Rightarrow |AB|$ increases as a function of t.

(c) Find B's horizontal displacement when A strikes the ground.

Assume A strikes the ground after time T_A then:

Time of flight of $A = T_A = \dfrac{2uSin\theta}{g}$

But horizontal distance travelled by B in time T_A is x_B where :

$x_B = utCos(\theta + 90°) = u\left(\dfrac{2uSin\theta}{g}\right)Cos(\theta + 90°) = -\dfrac{2u^2 Sin^2\theta}{g}$

Note : minus sign as B travels in the opposite direction to A

14. Two particles projected simultaneously

Two particles, A and B, are projected simultaneously from a point at angles θ and α to the horizontal respectively. Show that the line joining A, B:

(a)   Is unaltered in its direction throughout the motion

(b)   Makes an angle of $\frac{1}{2}(\theta + \alpha)$ with the vertical

Solution

Let the positions of A and B after a time t be as follows (See Fig. 1):

A: $(x_A, y_A) = (utCos\,\theta, utSin\,\theta - \frac{1}{2}gt^2)$

B: $(x_B, y_B) = (utCos\,\alpha, utSin\,\alpha - \frac{1}{2}gt^2)$

(a)   The line joining A, B is unaltered in its direction throughout the motion

The slope of the line joining A and B is given by:

Slope $= \dfrac{y_A - y_B}{x_A - x_B} = \dfrac{\left(utSin\theta - \frac{1}{2}gt^2\right) - \left(utSin\alpha - \frac{1}{2}gt^2\right)}{utCos\theta - utCos\alpha} = \dfrac{Sin\theta - Sin\alpha}{Cos\theta - Cos\alpha}$

$\Rightarrow$ for any given values of θ and α the slope remains constant throughout the motion.

(b) The line joining A, B makes an angle of $\frac{1}{2}(\theta + \alpha)$ with the vertical

From (a):

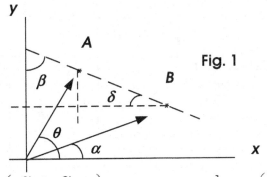

Fig. 1

$$Tan\ \delta = -\text{Slope} = -\left(\frac{Sin\theta - Sin\alpha}{Cos\theta - Cos\alpha}\right) \quad \text{But}: Tan\ \beta = \frac{1}{Tan\ \delta} = -\left(\frac{Cos\theta - Cos\alpha}{Sin\theta - Sin\alpha}\right)$$

Let $\theta = X + Y, \alpha = X - Y; \Rightarrow 2X = (\theta + \alpha) \Rightarrow X = \frac{1}{2}(\theta + \alpha)$

$\Rightarrow Cos\theta - Cos\alpha = Cos(X + Y) - Cos(X - Y) = -2 Sin X Sin Y$

$\Rightarrow Sin\theta - Sin\alpha = Sin(X + Y) - Sin(X - Y) = 2 Cos X Sin Y$

$$\Rightarrow Tan\ \beta = -\left(\frac{Cos\theta - Cos\alpha}{Sin\theta - Sin\alpha}\right) = \frac{2 Sin X Sin Y}{2 Cos X Sin Y} = Tan X = Tan\left(\frac{1}{2}(\theta + \alpha)\right)$$

$\Rightarrow$ Magnitude of angle made by line joining A, B and the vertical is : $\beta = \frac{1}{2}(\theta + \alpha)$

15. Collision of two particles

A mass M_1 of m kg is projected from point O with velocity u m/s at an angle θ to the horizontal. Simultaneously, a mass M_2 of $5m$ kg is dropped from a point A. The masses collide at M_1's greatest height, (point B), coalesce and fall to the ground at point D. (See Fig. 1)

(a)   Find $|AC|$
(b)   Find the time of flight between B and D
(c)   Find $|OD|$

Solution

(a) Find $|AC|$

See Fig. 1. Assume the masses are projected and dropped respectively at time $t = 0$; Assume the collision occurs at time $t = t$.

Given: As the collision occurs at greatest height reached by M_1 it occurs at B when M_1 has been in motion for half of its total flight time as a projectile:

i.e. $t = \dfrac{uSin\theta}{g}$ and $x = \dfrac{u^2 Sin2\theta}{2g} = \dfrac{u^2 Sin\theta\ Cos\theta}{g}$ (i)

$|BC|$ = greatest height reached by $M_1 = \dfrac{u^2 Sin^2\theta}{2g}$ (ii)

In time t, M_2 mass falls a distance $|AB|$.
Using the standard equation: $s = ut + \frac{1}{2}at^2$

$$\Rightarrow |AB| = 0 + \frac{1}{2}gt^2 = \frac{1}{2}g\left(\frac{uSin\theta}{g}\right)^2 = \frac{u^2 Sin^2\theta}{2g}$$

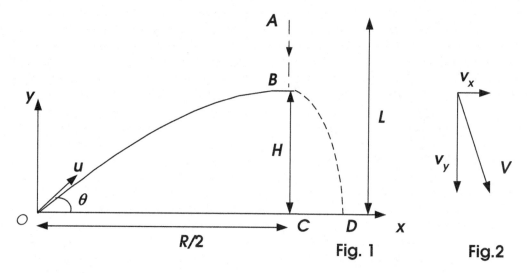

Fig. 1   Fig.2

If there is a collision, then both masses must be at the same height:

$$\Rightarrow \left|AC\right| = \left|AB\right| + \left|BC\right| = \frac{u^2 Sin^2\theta}{2g} + \frac{u^2 Sin^2\theta}{2g} = \frac{u^2 Sin^2\theta}{g}$$

(b) Find the time of flight between B and D
At the point of collision, the velocity components are:
M_1 has a velocity with horizontal component only, $v_x = u\,Cos\theta$

M_2 has a velocity with vertical component only: $v_y = u + gt = 0 + gt = g\dfrac{u\,Sin\theta}{g} = u\,Sin\theta$

To find the speed and direction of the new mass of 6m:
Let the 6m mass have a resultant velocity v with horizontal and vertical velocity
components v_{hor} and v_{ver} where $v = \sqrt{v_{hor}^2 + v_{ver}^2}$
The horizontal and vertical components of momentum will be conserved:

Horizontal component: $6m(v_{hor}) = mu\,Cos\theta \Rightarrow v_{hor} = \dfrac{u\,Cos\theta}{6}$ (iii)

Vertical component: $6m(v_{ver}) = 5mu\,Sin\theta \Rightarrow v_{ver} = \dfrac{5u\,Sin\theta}{6}$

Time taken for the 6m mass to hit the ground:
Vertical component of speed at instant of collision $= \dfrac{5u\,Sin\theta}{6}$

From equation (ii) above: Height above ground $= \left|BC\right| = \dfrac{u^2 Sin^2\theta}{2g}$

Using the general equation : $s = ut + \frac{1}{2}gt^2$ where $s = |BC|$

$$|BC| = \frac{u^2 Sin^2\theta}{2g} = ut + \frac{1}{2}gt^2 = \left(\frac{5u Sin\theta}{6}\right) t + \frac{1}{2}gt^2 = \quad \text{(from equation (ii))}$$

$$\Rightarrow t^2 + \left(\frac{5u Sin\theta}{3g}\right) t - \frac{u^2 Sin^2\theta}{g^2} = 0$$

$$\Rightarrow t = \frac{-\dfrac{5u Sin\theta}{3g} \pm \sqrt{\dfrac{25u^2 Sin^2\theta}{9g^2} + \dfrac{4u^2 Sin^2\theta}{g^2}}}{2}$$

$$\Rightarrow t = \frac{-5u Sin\theta}{3g} \pm \frac{u Sin\theta \sqrt{61}}{3g} = \frac{u Sin\theta\left(\sqrt{61} - 5\right)}{3g} \quad \text{(as } t > 0) \quad \text{(iv)}$$

(c) Find $|OD|$

Horizontal distance travelled after the collision: see equations (iii) and (iv):

$$= v_{hor} \times t = \left(\frac{u Cos\theta}{6}\right)\left(\frac{u Sin\theta\left(\sqrt{61} - 5\right)}{3g}\right) = \frac{u^2 Cos\theta \, Sin\theta\left(\sqrt{61} - 5\right)}{18g} \quad \text{(v)}$$

$\Rightarrow$ Total horizontal distance travelled is (from equations (i) and (v)):
"Distance travelled before collision" + " Distance travelled after collision"

$$= \frac{u^2 Sin\theta \, Cos\theta}{g} + \frac{u^2 Cos\theta \, Sin\theta\left(\sqrt{61} - 5\right)}{18g} = \frac{u^2 Cos\theta \, Sin\theta\left(13 + \sqrt{61}\right)}{18g}$$

16/17. Particle projected up an inclined plane:

A plane is inclined at an angle α. A particle is projected up the plane with velocity u m/s at angle β to the plane. See Fig. 1. Find:
(a) The total time of flight, T
(b) The range up the plane, R
(c) The maximum range up the plane, R_{MAX}
(d) The time to reach maximum perpendicular height above the plane, T_M
(e) The maximum perpendicular height above the plane, H
(f) Two angles of projection to obtain any given range on the inclined plane
(g) The angle which the direction of motion of particle makes with the plane when the particle strikes the plane

16. Approach 1 (using Horizontal and Vertical velocity and displacement components)

Solution

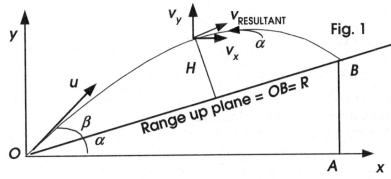

49

(a) The total time of flight, T

Horizontal velocity component at time $t = v_x = u\,Cos\,(\alpha + \beta)$

Vertical velocity component at time $t = v_y = uSin(\alpha + \beta) - gt$

Horizontal displacement component at time $t = x = ut\,Cos\,(\alpha + \beta)$

Vertical displacement component at time $t = y = utSin(\alpha + \beta) - \frac{1}{2}gt^2$ Eq. (i)

Fig. 1: At $t = T$: $Tan\alpha = \dfrac{AB}{OA} = \dfrac{y}{x} \Rightarrow y = xTan\alpha = uTCos(\alpha + \beta)Tan\alpha$ Eq. (ii)

Using equations (i), (ii): $uTSin(\alpha + \beta) - \frac{1}{2}gT^2 = (\,uT\,Cos\,(\alpha + \beta)\,)Tan\,\alpha$

$\Rightarrow uSin(\alpha + \beta) - \frac{1}{2}gT = (\,u\,Cos\,(\alpha + \beta)\,)Tan\,\alpha$

$\Rightarrow T = \dfrac{2u}{g}\left(\dfrac{Sin(\alpha + \beta)Cos\alpha - Cos(\alpha + \beta)Sin\alpha}{Cos\alpha}\right) = \dfrac{2u}{g}\dfrac{Sin(\alpha + \beta - \alpha)}{Cos\alpha} = \dfrac{2u}{g}\dfrac{Sin\beta}{Cos\alpha}$

(b) Find the range up the plane, R

See Fig. 1: $R = \dfrac{OA}{Cos\alpha} = \dfrac{(\text{Total time of flight})(\text{Horizontal velocity})}{Cos\alpha}$

$\Rightarrow R = \dfrac{\left(\dfrac{2uSin\beta}{gCos\alpha}\right)(uCos(\alpha + \beta))}{Cos\alpha} = \left(\dfrac{2u^2}{gCos^2\alpha}\right)(Cos(\alpha + \beta)Sin\beta)$

(c) Find the maximum range up the plane, R_{MAX}

$R = \left(\dfrac{2u^2}{gCos^2\alpha}\right)(Cos(\alpha + \beta)Sin\beta)$

But: $Cos(\alpha + \beta)Sin\beta = Cos\alpha\,Cos\beta\,Sin\beta - Sin\alpha\,Sin^2\beta$

$= \frac{1}{2}[(Sin\alpha Cos2\beta + Cos\alpha\,Sin2\beta) - Sin\alpha] = \frac{1}{2}[Sin(\alpha + 2\beta) - Sin\alpha]$

$\Rightarrow R = \left(\dfrac{u^2}{gCos^2\alpha}\right)[Sin(\alpha + 2\beta) - Sin\alpha]$

Usually: u, α are known $\Rightarrow R$ has a maximum value when $Sin\,(\alpha + 2\beta)$ has a maximum value i.e. when $Sin\,(\alpha + 2\beta) = 1$. It is then possible to re-write R:

$R_{MAX} = \left(\dfrac{u^2}{gCos^2\alpha}\right)[1 - Sin\alpha] = \left(\dfrac{u^2(1 - Sin\alpha)}{g(1 - Sin^2\alpha)}\right) = \left(\dfrac{u^2(1 - Sin\alpha)}{g(1 - Sin\alpha)(1 + Sin\alpha)}\right) = \dfrac{u^2}{g(1 + Sin\alpha)}$

(d) Time to reach maximum perpendicular height above the plane, T_M

When the particle is at the maximum perpendicular height above the plane, H, then its velocity component perpendicular to the plane = 0. Thus, the direction of its velocity ($v_{RESULTANT}$ in Fig. 1) is parallel to the inclined plane. Thus, $v_{RESULTANT}$ is inclined at an angle α to the horizontal plane.

$\Rightarrow Tan\alpha = \dfrac{v_Y}{v_X} = \dfrac{uSin(\alpha + \beta) - gT_M}{uCos(\alpha + \beta)} \Rightarrow T_M = \dfrac{uSin(\alpha + \beta) - uCos(\alpha + \beta)Tan\alpha}{g}$

$= \left(\dfrac{u}{gCos\alpha}\right)(Sin(\alpha + \beta)Cos\alpha - Cos(\alpha + \beta)Sin\alpha) = \left(\dfrac{u}{gCos\alpha}\right)Sin(\alpha + \beta - \alpha)$

$\Rightarrow T_M = \left(\dfrac{u}{gCos\alpha}\right)Sin\beta = \dfrac{uSin\beta}{gCos\alpha}$ Note: $T_M = \frac{1}{2}T$

(e) Maximum perpendicular height above the plane, H

See Fig. 2: Using the terminology of the previous section: $H = (H_{VERTICAL})Cos\alpha$

50

Fig. 2

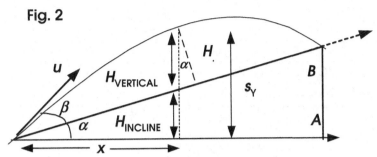

Vertical height of particle above horizontal plane at time $t = T_M = \dfrac{u\,Sin\beta}{g\,Cos\alpha}$

$s_Y = H_{VERTICAL} + H_{INCLINE} = u\,T_M\,Sin(\alpha + \beta) - \dfrac{1}{2}g\,T_M^2 = \dfrac{u^2\,Sin\beta Sin(\alpha + \beta)}{g\,Cos\alpha} - \dfrac{u^2\,Sin^2\beta}{2g\,Cos^2\alpha}$

But: $H_{INCLINE} = x\,Tan\,\alpha = \left(uCos(\alpha + \beta)\right)\left(\dfrac{uSin\beta}{g\,Cos\alpha}\right) = \dfrac{u^2\,Cos(\alpha + \beta)Sin\beta}{g\,Cos\alpha}$

But: $s_Y = H_{VERTICAL} + H_{INCLINE} \Rightarrow H_{VERTICAL} = s_Y - H_{INCLINE}$

$\Rightarrow H_{VERTICAL} = \dfrac{u^2\,Sin(\alpha + \beta)Sin\beta}{g\,Cos\alpha} - \dfrac{u^2\,Sin^2\beta}{2g\,Cos^2\alpha} - \dfrac{u^2\,Cos(\alpha + \beta)Sin\beta}{g\,Cos\alpha}$

$= \dfrac{2u^2\,Sin(\alpha + \beta)Sin\beta Cos\alpha}{2g\,Cos^2\alpha} - \dfrac{u^2\,Sin^2\beta}{2g\,Cos^2\alpha} - \dfrac{2u^2\,Cos(\alpha + \beta)Sin\beta Cos\alpha}{2g\,Cos^2\alpha}$

$= \dfrac{u^2\,Sin\beta[2Sin(\alpha + \beta)Cos\alpha - Sin\beta - 2Cos(\alpha + \beta)Cos\alpha]}{2g\,Cos^2\alpha} = \dfrac{u^2\,Sin\beta[2Sin\beta - Sin\beta]}{2g\,Cos^2\alpha}$

$= \dfrac{u^2\,Sin^2\beta}{2g\,Cos^2\alpha}$ But : $H = (H_{VERTICAL})Cos\alpha \Rightarrow H = \dfrac{u^2\,Sin^2\beta}{2g\,Cos\alpha}$

(f) Find two angles of projection to obtain any given range on the inclined plane

From above : $R = \left(\dfrac{u^2}{g\,Cos^2\alpha}\right)(Sin(\alpha + 2\beta) - Sin\alpha) \Rightarrow Sin(\alpha + 2\beta) = \dfrac{(g\,Cos^2\alpha)R}{u^2} + Sin\alpha$

In general, there will be two solutions for $(\alpha + 2\beta)$: $(\alpha + 2\beta), 180° - (\alpha + 2\beta)$

(g) Angle θ which the direction of motion of particle makes with the plane when the particle strikes the plane

Fig. 3

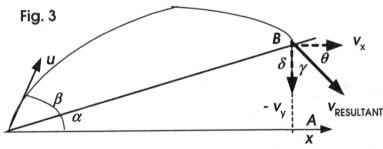

Time of flight $= t = T = \dfrac{2u\,Sin\beta}{g\,Cos\alpha}$ Direction of motion $= \theta = Tan^{-1}\dfrac{-V_y}{V_x}$

$$v_x = uCos(\alpha + \beta) \quad v_y = uSin(\alpha + \beta) - g\left(\frac{2uSin\beta}{gCos\alpha}\right) = uSin(\alpha + \beta) - \left(\frac{2uSin\beta}{Cos\alpha}\right)$$

$$\Rightarrow Tan\theta = \frac{-uSin(\alpha + \beta) + \left(\frac{2uSin\beta}{Cos\alpha}\right)}{uCos(\alpha + \beta)} = \frac{2Sin\beta - Sin(\alpha + \beta)Cos\alpha}{Cos(\alpha + \beta)Cos\alpha}$$

If particle strikes at 90° to the inclined plane:

From Fig. 3: If the particle strikes the plane at 90° then $\gamma = \alpha$. But: $\theta + \gamma = 90°$

$$\Rightarrow \theta + \alpha = 90° \Rightarrow Tan\theta = \frac{1}{Tan\alpha} \Rightarrow Tan\alpha = \frac{Cos(\alpha + \beta)Cos\alpha}{2Sin\beta - Sin(\alpha + \beta)Cos\alpha}$$

$$\Rightarrow 2Sin\alpha Sin\beta - Sin\alpha Cos\alpha Sin(\alpha + \beta) = Cos^2\alpha Cos(\alpha + \beta)$$

$$\Rightarrow 2Sin\alpha Sin\beta = Cos\alpha[Cos(\alpha + \beta)Cos\alpha + Sin(\alpha + \beta)Sin\alpha] = Cos\alpha[Cos(\alpha + \beta - \alpha)]$$

$$\Rightarrow 2Sin\alpha Sin\beta = Cos\alpha Cos\beta \Rightarrow 2Tan\alpha Tan\beta = 1$$

If particle strikes the plane horizontally (i.e 0° to the horizontal plane):

From Fig. 3: If the particle strikes the plane at 0° then $\theta = 0°$

$$\Rightarrow Tan\theta = \frac{2Sin\beta - Sin(\alpha + \beta)Cos\alpha}{Cos(\alpha + \beta)Cos\alpha} = 0 \Rightarrow 2Sin\beta - Sin(\alpha + \beta)Cos\alpha = 0$$

$$\Rightarrow 2Sin\beta = Sin\alpha Cos\beta Cos\alpha + Cos\alpha Sin\beta Cos\alpha$$

$$\Rightarrow 2Tan\beta = Sin\alpha Cos\alpha + Cos^2\alpha Tan\beta = \tfrac{1}{2}Sin2\alpha + Cos^2\alpha Tan\beta$$

$$\Rightarrow 2Tan\beta = \tfrac{1}{2}\left(\frac{2Tan\alpha}{1 + Tan^2\alpha}\right) + \left(\frac{1}{1 + Tan^2\alpha}\right)Tan\beta = \frac{Tan\alpha + Tan\beta}{1 + Tan^2\alpha}$$

$$\Rightarrow Tan\alpha + Tan\beta = 2Tan\beta + Tan^2\alpha Tan\beta \Rightarrow Tan\alpha - Tan\beta = 2Tan^2\alpha Tan\beta$$

17. Approach 2 (using parallel and perpendicular velocity, displacement components)

Solution

The velocity and displacement can be resolved into components as follows:

Parallel to the inclined plane:

The velocity at time t is: $v_{parallel} = uCos\beta - (gSin\alpha)\,t$ Equation (i)

The displacement at time t is: $x = (uCos\beta)t - \tfrac{1}{2}(g\,Sin\alpha)t^2$ Equation (ii)

Perpendicular to the inclined plane

The velocity at time t is: $v_y = uSin\beta - (g\,Cos\alpha)\,t$ Equation (iii)

The displacement at time t is: $y = (uSin\beta)t - \tfrac{1}{2}(g\,Cos\alpha)t^2$ Equation (iv)

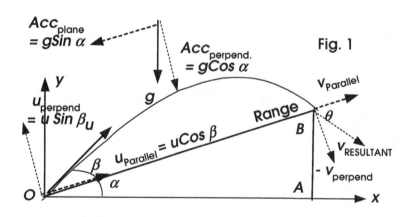

52

(a) The time of flight, T

The particle's flight ends on impact with the inclined plane: when $y = 0$. Thus, time of flight is: $0 = uSin\beta \, T - \frac{1}{2}(gCos\alpha) T^2 \Rightarrow T = \dfrac{2uSin\beta}{gCos\alpha}$ Equation (v)

(b) Find the range up the plane, R

R = displacement along plane at time $t = T$

$$R = u\left(\frac{2uSin\beta}{gCos\alpha}\right)Cos\beta - \frac{1}{2}(g\,Sin\alpha)\left(\frac{2uSin\beta}{gCos\alpha}\right)^2$$

$$= \frac{2u^2 gCos\alpha \, Sin\beta Cos\beta - 2u^2 gSin\alpha(Sin\beta)^2}{(gCos\alpha)^2}$$

$$= \frac{2u^2}{g}\frac{Cos\alpha \, Sin\beta Cos\beta - Sin\alpha \, Sin\beta \, Sin\beta}{(Cos\alpha)^2} = \frac{2u^2 Cos(\alpha + \beta)Sin\beta}{g(Cos\alpha)^2}$$

(c) Find the maximum range up the plane, R_{MAX} See Previous question

(d) Find two angles of projection to obtain any given range on the inclined plane See Previous question

(e) Time taken to reach its maximum perpendicular height above plane, T_M

The maximum height above the plane is achieved when $v_y = 0$ and $t = T_M$

$$\Rightarrow v_y = uSin\beta - (gCos\alpha)T_M = 0 \Rightarrow uSin\beta = (gCos\alpha)T_M \Rightarrow T_M = \frac{uSin\beta}{gCos\alpha} \quad \left(= \frac{T}{2}\right)$$

(f) Maximum perpendicular height above the plane, H

At time $t = T_M = \dfrac{uSin\beta}{gCos\alpha}$ $\Rightarrow y =$ Maximum perpendicular height above plane = H

(g)

$$\Rightarrow H = (uSin\beta)\left(\frac{uSin\beta}{gCos\alpha}\right) - \frac{1}{2}(gCos\alpha)\left(\frac{uSin\beta}{gCos\alpha}\right)^2 = \frac{u^2 Sin^2\beta}{gCos\alpha} - \frac{u^2 Sin^2\beta}{2gCos\alpha} = \frac{u^2 Sin^2\beta}{2gCos\alpha}$$

Angle which the direction of motion of particle makes with the plane when the particle strikes the plane

From Fig 1: $Tan\theta = \dfrac{-v_y}{v_x} = \dfrac{-(uSin\beta - (gCos\alpha)t)}{uCos\beta - (gSin\alpha)t}$ But : $t = T = \dfrac{2uSin\beta}{gCos\alpha}$

$$\Rightarrow Tan\theta = \frac{-\left(uSin\beta - (gCos\alpha)\left(\frac{2uSin\beta}{gCos\alpha}\right)\right)}{uCos\beta - (gSin\alpha)\left(\frac{2uSin\beta}{gCos\alpha}\right)} = \frac{Sin\beta}{Cos\beta - 2Tan\alpha Sin\beta}$$

If particle strikes at 90° to inclined plane:

$\Rightarrow Tan\theta = \infty \Rightarrow 2Tan\alpha Sin\beta - Cos\beta = 0 \Rightarrow 2Tan\alpha \, Tan\beta = 1$

If particle strikes the plane horizontally (i.e 0° to the horizontal plane):

$\Rightarrow \theta = \alpha \Rightarrow Tan\theta = Tan\alpha \Rightarrow Tan\alpha = \dfrac{Sin\beta}{Cos\beta - 2Tan\alpha Sin\beta}$

$\Rightarrow Cos\beta \, Tan\alpha - 2Tan\alpha \, Tan\alpha \, Sin\beta = Sin\beta$ Divide across by : $Cos\beta$

$\Rightarrow Tan\alpha - 2Tan^2\alpha \, Tan\beta = Tan\beta \Rightarrow Tan\alpha - Tan\beta = 2Tan^2\alpha \, Tan\beta$

18/19. Particle projected down an inclined plane

A particle is projected with velocity u m/s at an angle β down a plane which is inclined at an angle α to the horizontal (See Fig. 1). Find the:

(a)   Horizontal and vertical velocity and displacement components of the particle at
       time *t*.
(b)   Time of flight down the plane
(c)   Range, *R*, along the plane in time *T*.
(d)   Maximum range, R_{MAX}, along the plane
(e)   Two angles of projection to obtain any given range on the inclined plane
(f)   Time taken to reach its maximum perpendicular height above plane
(g)   Maximum perpendicular height above the plane

18. Approach 1: using Horizontal and Vertical velocity, displacement components
Solution

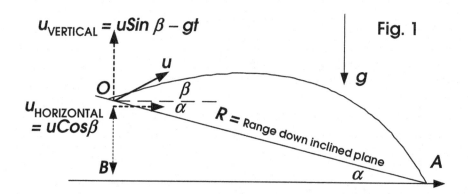

(a)   Horizontal and vertical velocity and displacement components of the particle at
**   time *t*.**
Consider the previous cases where there was no inclined plane. See Fig. 1 above. The
expressions derived for the <u>horizontal</u> and <u>vertical</u> components of velocity and
displacement were as follows:
Velocity
Horizontal component $= v_x = u\,Cos\,\beta$ Equation (i)
Vertical component $= v_y = u\,Sin\,\beta - gt$ Equation (ii)
Displacement
Horizontal component $= x = ut\,Cos\,\beta$ Equation (iii)
Vertical component $= y = ut\,Sin\,\beta - \frac{1}{2}\,gt^2$ Equation (iv)
(b) The time of flight down the plane, *T*

From Fig. 1 we can see that: $Tan\alpha = \dfrac{OB}{BA} = \dfrac{-y}{x}$ Equation   (v)

Using Equations (iii), (iv) and (v) we get:
$y = utSin\beta - \frac{1}{2}gt^2$ (Note: minus sign used here)

$- y = xTan\,\alpha = (ut\,Cos\,\beta)Tan\,\alpha \;\Rightarrow\; -utSin\beta + \frac{1}{2}gt^2 = (ut\,Cos\,\beta)Tan\,\alpha$

$\Rightarrow -uSin\,\beta + \frac{1}{2}gt = (u\,Cos\,\beta)Tan\,\alpha \Rightarrow +\frac{1}{2}gt = u\,Cos\,\beta\,Tan\,\alpha + uSin\,\beta$

$\Rightarrow t = T = \dfrac{2u}{g}\left(\dfrac{Sin\alpha\,Cos\beta + Cos\alpha\,Sin\beta}{Cos\alpha}\right) = \dfrac{2uSin(\alpha + \beta)}{gCos\alpha}$

(c)   Range, *R*, along the plane in time *T*.
From Equation (iii) the Horizontal displacement component $= x = ut\,Cos\,\beta$

54

In time $t = T$, $x = |BA| = |AB|$ From Fig. 1:

$$Cos\,\alpha = \frac{AB}{OA} = \frac{x \text{ travelled in time } T}{\text{Range, } R} \Rightarrow R = \frac{AB}{Cos\alpha} = \frac{uTCos\beta}{Cos\alpha}$$

$$\Rightarrow R = \frac{uCos\beta}{Cos\alpha}\left(\frac{2uSin(\alpha+\beta)}{gCos\alpha}\right) = \frac{2u^2 Sin(\alpha+\beta)Cos\beta}{gCos^2\alpha}$$

(d) Maximum range, R_{MAX}, along the plane

$$R = \frac{2u^2 Sin(\alpha+\beta)Cos\beta}{gCos^2\alpha} = \frac{u^2[Sin(\alpha+2\beta)+Sin\alpha]}{gCos^2\alpha}$$ This is a maximum value

when : $Sin(\alpha+2\beta)$ is a maximum $\Rightarrow Sin(\alpha+2\beta) = 1 \Rightarrow \alpha + 2\beta = \dfrac{\pi}{2} \Rightarrow \beta = \dfrac{1}{2}\left(\dfrac{\pi}{2} - \alpha\right)$

$$\Rightarrow R_{MAX} = \frac{u^2[1+Sin\alpha]}{gCos^2\alpha} = \frac{u^2[1+Sin\alpha]}{g(1-Sin^2\alpha)} = \frac{u^2[1+Sin\alpha]}{g(1-Sin\alpha)(1+Sin\alpha)} = \frac{u^2}{g(1-Sin\alpha)}$$

(e) Find two angles of projection to obtain any given range on the inclined plane

$$\Rightarrow R = \frac{2u^2\ Sin(\alpha+\beta)Cos\beta}{gCos^2\alpha} = \left(\frac{2u^2}{gCos^2\alpha}\right)(Sin(\alpha+\beta)Cos\beta)$$

$$= \left(\frac{2u^2}{gCos^2\alpha}\right)\frac{1}{2}(Sin(\alpha+2\beta)+Sin\alpha) = \left(\frac{u^2}{gCos^2\alpha}\right)(Sin(\alpha+2\beta)+Sin\alpha)$$

$$\Rightarrow Sin(\alpha+2\beta) = \frac{(gCos^2\alpha)R}{u^2} - Sin\alpha$$

In general, there will be two solutions for $(\alpha+2\beta)$: $(\alpha+2\beta), 180° -(\alpha+2\beta)$

(f) Time taken to reach its maximum perpendicular height above plane, T_M

When the particle is at its maximum perpendicular height above plane then its velocity component perpendicular to the plane = 0. Thus, the direction of its velocity is parallel to the inclined plane. Thus, $v_{RESULTANT}$ must be inclined at an angle "$-\alpha$" to the horizontal plane.

$$\Rightarrow Tan(-\alpha) = -Tan(\alpha) = \frac{-Sin\alpha}{Cos\alpha} = \frac{v_Y}{v_X} = \frac{uSin\beta - gt}{uCos\beta} \Rightarrow t = \frac{uCos\alpha Sin\beta + uSin\alpha Cos\beta}{gCos\alpha}$$

$$\Rightarrow t = \left(\frac{u}{gCos\alpha}\right)(Sin\beta Cos\alpha + Cos\beta Sin\alpha) = \left(\frac{u}{gCos\alpha}\right)Sin(\alpha+\beta) \Rightarrow t = \frac{uSin(\alpha+\beta)}{gCos\alpha}$$

(g) Maximum perpendicular height above the plane, H

See Fig. 2: $Cos\alpha = \dfrac{H}{H_{V1}+H_{V2}} \Rightarrow H = (H_{V1} + H_{V2})Cos\alpha$

Vertical height of particle above horizontal plane at time $t = \dfrac{uSin(\alpha+\beta)}{gCos\alpha}$

is: $H_{V1} = utSin\beta - \frac{1}{2}gt^2 = \dfrac{u^2\ Sin\beta Sin(\alpha+\beta)}{gCos\alpha} - \dfrac{u^2\ Sin^2(\alpha+\beta)}{2gCos^2\alpha}$

$H_{V2} = x\ Tan\ \alpha = (uCos\beta)\left(\dfrac{uSin(\alpha+\beta)}{gCos\alpha}\right)Tan\alpha = \dfrac{u^2Cos\beta Sin(\alpha+\beta)Sin\alpha}{gCos^2\alpha}$

$$\Rightarrow H_{V1} + H_{V2} = \frac{2u^2 Sin(\alpha+\beta)Sin\beta Cos\alpha}{2gCos^2\alpha} - \frac{u^2 Sin^2(\alpha+\beta)}{2gCos^2\alpha} + \frac{2u^2 Cos\beta Sin(\alpha+\beta)Sin\alpha}{2gCos^2\alpha}$$

$$= \frac{u^2 Sin(\alpha+\beta)[2Sin\beta Cos\alpha - Sin(\alpha+\beta) + 2Cos\beta Sin\alpha]}{2gCos^2\alpha}$$

$$= \frac{u^2 Sin(\alpha+\beta)[2Sin(\alpha+\beta) - Sin(\alpha+\beta)]}{2gCos^2\alpha} = \frac{u^2 Sin^2(\alpha+\beta)}{2gCos^2\alpha} \quad \text{But}: H = (H_{V1} + H_{V2})Cos\alpha$$

$$\Rightarrow H = \frac{u^2 Sin^2(\alpha+\beta)}{2gCos\alpha}$$

Fig. 2

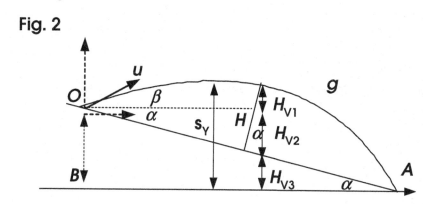

19. Approach 2: using Parallel and Perpendicular velocity and displacement components

Solution

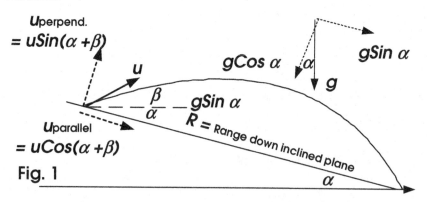

Fig. 1

(a) Displacement and velocity of the particle at time t

Parallel to the plane:

The velocity at time t is: $v_x = uCos(\alpha+\beta) + (gSin\alpha)t$ Equation (i)

The displacement at time t is: $x = utCos(\alpha+\beta) + \frac{1}{2}(gSin\alpha)t^2$ Equation (ii)

Perpendicular to the plane

The velocity at time t is: $v_y = uSin(\alpha+\beta) - (gCos\alpha)t$ Equation (iii)

The displacement is: $y = ut \, Sin(\alpha + \beta) - \frac{1}{2}(g \, Cos\alpha) \, t^2$ Equation (iv)

(b) Total time of flight, T

At $t = T$, $y = 0 \Rightarrow y = ut \, Sin(\alpha + \beta) - \frac{1}{2}(g \, Cos\alpha) \, t^2 = 0 \Rightarrow uSin(\alpha + \beta) = \frac{1}{2}(g \, Cos\alpha) \, t$

$$\Rightarrow T = \frac{2uSin(\alpha + \beta)}{gCos\alpha}$$

(c) Range, R, travelled down the plane at time, T

R = distance travelled parallel to plane in time, T. Substitute T in Equation (ii):

$$R = u\left(\frac{2uSin(\alpha + \beta)}{gCos\alpha}\right)Cos(\alpha + \beta) + \frac{1}{2}(g \, Sin\alpha)\left(\frac{2uSin(\alpha + \beta)}{gCos\alpha}\right)^2$$

$$= \left(\frac{2u^2 Sin(\alpha + \beta)}{gCos\alpha}\right)\left(Cos(\alpha + \beta) + (\, Sin\alpha)\left(\frac{Sin(\alpha + \beta)}{Cos\alpha}\right)\right)$$

$$= \left(\frac{2u^2 Sin(\alpha + \beta)}{gCos^2\alpha}\right)(Cos(\alpha + \beta)Cos\alpha + (\, Sin\alpha)Sin(\alpha + \beta) \,)$$

$$= \frac{2u^2 Sin(\alpha + \beta)Cos(\alpha + \beta - \alpha)}{gCos^2\alpha} = \frac{2u^2 Sin(\alpha + \beta)Cos\beta}{gCos^2\alpha} = \frac{u^2(Sin(\alpha + 2\beta) + Sin\alpha)}{gCos^2\alpha}$$

(d) Maximum range, R$_{MAX}$, along the plane

R is a max when: $Sin(\alpha + 2\beta)$ is a max i.e. when: $Sin(\alpha + 2\beta) = 1$

Note: Maximum range occurs when: $\alpha + 2\beta = \frac{\pi}{2} \Rightarrow \beta = \frac{1}{2}\left(\frac{\pi}{2} - \alpha\right)$

$$\Rightarrow R_{MAX} = \frac{u^2[1 + Sin\alpha]}{gCos^2\alpha} = \frac{u^2[1 + Sin\alpha]}{g(1 - Sin^2\alpha)} = \frac{u^2}{g(1 - Sin\alpha)}$$

(e) Find two angles of projection to obtain any given range on the inclined plane

From above: $R = \left(\frac{u^2}{gCos^2\alpha}\right)(Sin(\alpha + 2\beta) + Sin\alpha) \Rightarrow Sin(\alpha + 2\beta) = \frac{(gCos^2\alpha)R}{u^2} - Sin\alpha$

In general, there will be two solutions for $(\alpha + 2\beta)$: $(\alpha + 2\beta), 180° - (\alpha + 2\beta)$

(f) Time taken to reach its maximum perpendicular height above plane, T$_M$

The maximum height above the plane is achieved when $v_y = 0$ and $t = T_M$

$\Rightarrow v_y = uSin(\alpha + \beta) - (gCos\alpha)T_M = 0 \Rightarrow uSin(\alpha + \beta) = (gCos\alpha)T_M$

$\Rightarrow T_M = \frac{uSin(\alpha + \beta)}{gCos\alpha}$ $\Rightarrow T_M = \frac{1}{2}(\text{Total time of flight, } T) = \frac{1}{2}T$

(g) Maximum perpendicular height above plane, H

The displacement perpendicular to the plane is: $y = ut \, Sin(\alpha + \beta) - \frac{1}{2}(g \, Cos\alpha) \, t^2$

$\Rightarrow$ Maximum perpendicular height =

$$= u\left(\frac{uSin(\alpha + \beta)}{gCos\alpha}\right)Sin(\alpha + \beta) - \frac{1}{2}(g \, Cos\alpha)\left(\frac{uSin(\alpha + \beta)}{gCos\alpha}\right)^2$$

$$= \frac{u^2 Sin^2(\alpha + \beta)}{gCos\alpha} - \frac{u^2 Sin^2(\alpha + \beta)}{2gCos\alpha} = \frac{u^2 Sin^2(\alpha + \beta)}{2gCos\alpha}$$

20. Particle projected up inclined plane

A particle is projected from O up a plane which is inclined at an angle α to the horizontal. The direction of projection makes an angle of $60°$ with the inclined plane. See Fig. 1.
(a)   Find the value of α if the particle strikes the plane at right angles.
(b)   Show that the total energy at the end of the flight is the same as the initial energy

Solution

(a) Find the value of α if the particle strikes the plane at right angles
Consider the velocity of the particle perpendicular to this inclined plane. This is:
$Vy = u\,Sin60° - (g\,Cos\alpha)\,t$ (i)
Where $-g\,Cos\alpha$ is the acceleration perpendicular to the plane, the perpendicular distance of the particle from the plane is:
$Sy = ut\,Sin60° - \frac{1}{2}(g\,Cos\alpha)\,t^2$ (ii)

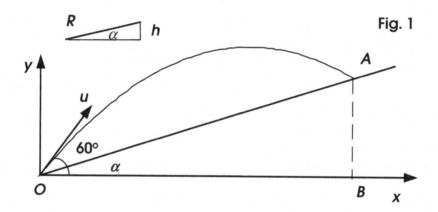

R α h   **Fig. 1**

When the particle strikes plane again (at A), $Sy = 0$. Thus, from equation (ii):
$\Rightarrow 0 = ut\,Sin60° - \frac{1}{2}(g\,Cos\alpha)t^2$
$\Rightarrow t = T\,(say) = \dfrac{2u\,Sin60°}{g\,Cos\alpha} = $ time of flight   (iii)
Consider the velocity of the particle parallel to the inclined plane, $Vplane$ say, at time T (i.e. when the particle strikes the plane). Given: the particle only has a velocity component perpendicular $\Rightarrow Vplane = 0$.
$$Vplane = u\,Cos60° - g\,TSin\alpha = u\,Cos60° - \dfrac{g\,Sin\alpha(2u\,Sin60°)}{g\,Cos\alpha} = 0$$
$\Rightarrow u\,Cos60° - 2u\,Tan\alpha\,Sin60° = 0$
$\Rightarrow \dfrac{1}{2}u - 2u\,Tan\alpha\,\dfrac{\sqrt{3}}{2} = 0 \Rightarrow Tan\alpha = \dfrac{1}{2\sqrt{3}} \Rightarrow \alpha = Tan^{-1}\dfrac{1}{2\sqrt{3}}$ (iv)

This is the value of α so the particle strikes the inclined plane at right angles.

(b) Show that the total energy at the end of the flight = initial energy
The energy of the particle at any point = its kinetic energy + its potential energy.
Initial energy (i.e. at O)

58

If initial velocity = $u \Rightarrow$ Initial Kinetic Energy = $\frac{1}{2}mu^2$ (where m = mass of particle). Assume initial potential energy (at point of projection) = 0.

$\Rightarrow$ Initial energy = $= \frac{1}{2}mu^2 + 0 = \frac{1}{2}mu^2$

Final energy (i.e. at A):
Final energy of particle = Kinetic Energy + Potential Energy
= $\frac{1}{2}mv^2$ + Potential Energy (gained by particle), (where v = final velocity)

Kinetic Energy at A
But, the final velocity has only a component perpendicular to the inclined plane:
From equation (i) the final velocity = $v = u\,Sin\,60° - (g\,Cos\,\alpha)\,t$ (v)

From equation (iii): $T = \dfrac{2u\,Sin60°}{g\,Cos\alpha}$ Putting this value into equation (v) gives :

$\Rightarrow$ Final velocity = $- u\,Sin60° = -u\dfrac{\sqrt{3}}{2} \Rightarrow$ Kinetic Energy $= \dfrac{1}{2}m\left(-u\dfrac{\sqrt{3}}{2}\right)^2 = \dfrac{3mu^2}{8}$ (vi)

Potential Energy at A
Potential Energy gained = mgh where h = perpendicular height reached = AB
From inset in Fig. 1:

$Sin\alpha = \dfrac{h}{R} \Rightarrow h = R\,Sin\alpha$

Range $= R = \dfrac{2u^2\,Sin60°\,Cos(60° + \alpha)}{g\,Cos^2\alpha} \Rightarrow h = \dfrac{2u^2\,Sin60°\,Cos(60° + \alpha)Sin\alpha}{g\,Cos^2\alpha}$

From equation (iv): $Cos\alpha = \dfrac{2\sqrt{3}}{\sqrt{13}}$, $Sin\alpha = \dfrac{1}{\sqrt{13}} \Rightarrow h = \dfrac{u^2}{8g}$

From equations (vi) and (vii): Final energy $= \dfrac{3mu^2}{8} + mg\dfrac{u^2}{8g} = \dfrac{4mu^2}{8} = \dfrac{1}{2}mu^2$

$\Rightarrow$ Initial energy of particle = Final energy of particle.

Chapter 4   Circular Motion

1. Car moving on circular track
Find the acceleration of a car moving at a constant speed of v m/s around a circular track of radius r metres.
Solution
As the speed is constant there is no tangential component of acceleration. The

acceleration towards the centre of the circular track is $\dfrac{v^2}{r}$ m/s²

2. Particle moving in a circle
Find the speed of a particle moving in a circle of radius r metres with an acceleration towards the centre of the circle of a m/s².
Solution
$$a = \frac{v^2}{r} \Rightarrow v = \sqrt{a\,r}\quad \text{m/s}$$

3. Particle moving in a circle
A particle of mass m kg is connected by a light inextensible string of length r metres to a fixed point. If the particle is performing X revolutions per minute about the fixed point, find the tension in the string, T.
Solution
Let: revolutions per minute = rpm and revolutions per second = rps. Then:

$$X \text{ rpm} = \frac{X}{60} \text{ rps} = \frac{2\pi X}{60} = \frac{\pi X}{30}\quad \text{rad/sec} = \omega, \quad T = \frac{mv^2}{r} = m\omega^2 r \Rightarrow T = m\left(\frac{\pi X}{30}\right)^2 r\quad \text{Newtons}$$

4. Motion in a horizontal circle
Two particles, A (on a smooth table) and B (beneath the table), each of mass m kg are connected by a string which passes through a hole in the table. (See Fig. 1). If particle A performs 59.8 revolutions per minute in a radius of r metres and particle B remains at rest, what is the value of r?
Solution
The forces in Newtons involved are shown in Fig. 2:
For particle A: $T = m\omega^2 r$ For particle B: $T = mg$
Therefore: $mg = m\omega^2 r$ for no movement of particle B to occur.

$$\Rightarrow g = \omega^2 r \Rightarrow \omega^2 = \frac{g}{r} \Rightarrow \omega = \sqrt{\frac{g}{r}} = \sqrt{\frac{9.81}{r}}\quad \text{radians/second}$$

But: $\omega = 2\pi f$ where f = frequency in revolutions per second
Given: $f = \omega/2\pi = 0.99667$ revolution per second (= 59.8 revolutions/minute).

$$\Rightarrow f = \frac{\omega}{2\pi} = 0.99667 = \frac{\sqrt{\dfrac{g}{r}}}{2\pi} \Rightarrow r = \frac{g}{4\pi^2(0.99667)^2} = 0.25 \text{ metres}$$

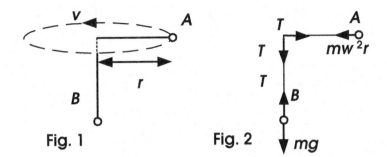

Fig. 1 **Fig. 2** *mg*

5. Motion of a conical pendulum

A mass *m* kg is suspended from a point *O* by a string of length *L*. If the particle is performing circular motion with constant speed with the string inclined at angle θ to the vertical (See Fig. 1). find:

(a)   The linear speed of the particle, *v* (in m/s)
(b)   The angular speed, ω, in radians/second
(c)   The angular speed in revolutions per minute

Solution

Let: *AB* = *a* metres, *OB* = string length = *L* metres, *m* = Particle mass at *B* in kg

(a) The linear speed of the particle (in m/s)

The forces in Newtons acting are shown in Fig. 2:

Vertically : $T\,Cos\,\theta = mg$ (i)   Horizontally: $T\,Sin\theta = \dfrac{mv^2}{a}$ (ii)

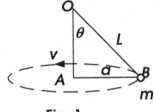   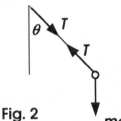

Fig. 1   **Fig. 2** *mg*

$$\frac{T\,Sin\theta}{T\,Cos\theta} = Tan\theta = \frac{\frac{mv^2}{a}}{mg} = \frac{v^2}{ag} \Rightarrow v^2 = ag\,Tan\theta \Rightarrow v = \sqrt{ag\,Tan\theta}\quad \text{m/s}$$

(b) The angular speed in radians/second

The relationship between linear and angular speed is given by: $v = \omega a$

$$\Rightarrow \omega = \frac{v}{a} = \frac{\sqrt{ag\,Tan\theta}}{a} = \sqrt{\frac{g\,Tan\theta}{a}}\quad \text{radians/second}$$

(c) The angular speed in revolutions per minute (rpm)

Let : *f* = revolutions/second $f = \dfrac{\omega}{2\pi} = \dfrac{1}{2\pi}\sqrt{\dfrac{g\,Tan\theta}{a}}$ rps $= \dfrac{30}{\pi}\sqrt{\dfrac{g\,Tan\theta}{a}}$ rpm

6. Motion of a conical pendulum

A mass *m* kg is fastened to a string of length *r* metres and is made to describe a horizontal circle with speed *v* m/s (See Fig. 1).

(a)   Find the tension of the string, *T*, and θ, its inclination to the vertical in terms of *v*, *r*.

(b)   If $r = 1$ metre for what value of v will the inclination to the vertical be 30°?

Solution

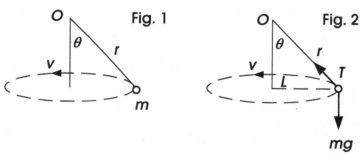

Fig. 1   Fig. 2

(a) Find the tension of the string, T (Newtons), and θ, its inclination to the vertical.

See Figs. 1 and 2.
Let θ be the inclination of the string to the vertical and T = Tension of the string
The forces acting are:

Horizontally: $T\,Sin\theta = \dfrac{mv^2}{L}$ (i)   Vertically: $T\,Cos\theta = mg$ (ii)

From the above equations:

$$\frac{T\,Sin\theta}{T\,Cos\theta} = Tan\theta = \frac{\dfrac{mv^2}{L}}{mg} = \frac{v^2}{gL}$$

But : $Sin\theta = \dfrac{L}{r} \Rightarrow L = r\,Sin\theta \Rightarrow Tan\theta = \dfrac{Sin\theta}{Cos\theta} = \dfrac{v^2}{gr\,Sin\theta} \Rightarrow Sin^2\theta = \dfrac{v^2}{gr}Cos\theta$

Since : $Sin^2\theta + Cos^2\theta = 1$, then $Sin^2\theta = 1 - Cos^2\theta$

$\Rightarrow 1 - Cos^2\theta = \dfrac{v^2}{gr}Cos\theta \Rightarrow Cos^2\theta + \dfrac{v^2}{gr}Cos\theta - 1 = 0$

Solving we get : $Cos\theta = \frac{1}{2}\left(-\dfrac{v^2}{gr} \pm \sqrt{\dfrac{v^4}{g^2r^2} + 4}\right) = \dfrac{-v^2 \pm \sqrt{v^4 + 4g^2r^2}}{2gr}$

But, from equation (ii) $T = \dfrac{mg}{Cos\theta} \Rightarrow T = \dfrac{mg}{\dfrac{-v^2 \pm \sqrt{v^4 + 4g^2r^2}}{2gr}} = \dfrac{2mg^2r}{-v^2 \pm \sqrt{v^4 + 4g^2r^2}}$ (iii)

and angle of inclination to the vertical $\theta = Cos^{-1}\left(\dfrac{-v^2 \pm \sqrt{v^4 + 4g^2r^2}}{2gr}\right)$ (iv)

(b) If $r = 1$ metre for what value of v will the inclination to the vertical be 30°?
If $r = 1$ metre and $\theta = 30°$ then

$Cos\theta = \dfrac{\sqrt{3}}{2} = \dfrac{-v^2 \pm \sqrt{v^4 + 4g^2}}{2g} \Rightarrow g\sqrt{3} = -v^2 \pm \sqrt{v^4 + 4g^2}$

$\Rightarrow \left(v^2 + g\sqrt{3}\right)^2 = v^4 + 4g^2$

$\Rightarrow v^4 + 2gv^2\sqrt{3} + 3g^2 = v^4 + 4g^2 \Rightarrow 2gv^2\sqrt{3} = g^2 \Rightarrow v = \sqrt{\dfrac{g}{2\sqrt{3}}}$ m/s

7. Motion of a conical pendulum

A mass m_1 kg rests on a rough horizontal table with coefficient of friction $\mu = 0.5$. A string, which passes through a smooth hole in the table at O links masses m_1 and m_2. (See Fig. 1).

(a) If m_2 describes a horizontal circle with a uniform velocity v and m_1 is just about to slip find the radius of the circle, r, and the length of the string below the table, L.

(b) If $m_1 = 3$ $m_2 = m$ kg and $v = 1$ m/s find r and L.

Solution

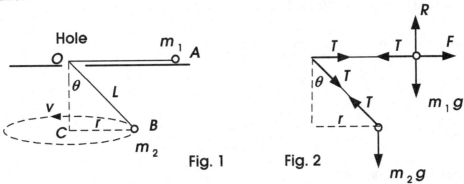

Fig. 1   **Fig. 2**

The forces in Newtons acting are shown in Fig. 2:
F = Friction between table and mass m_1, T = Tension of string, R = Reaction between mass m_1 and the table, m_1g, m_2g = Weights of m_1 and m_2 respectively

(a) Find the radius of the circle, r, and the length of the string below the table, L

Let $OB = L$, $BC = r$. Consider m_2 inclined at angle θ to the vertical OC

When the particle is on the point of slipping:

For mass m_1: resolving the forces gives: $R = m_1g$, $T = F = \mu R = \mu m_1 g = \dfrac{m_1 g}{2}$ (i)

For mass m_2: resolving the forces gives:

Vertically : $T \cos \theta = m_2 g$ (ii),   Horizontally: $T \sin\theta = \dfrac{m_2 v^2}{r}$ (iii)

Using equations (i) and (ii):

$$T\cos\theta = \frac{m_1 g}{2}\cos\theta = m_2 g \Rightarrow \cos\theta = \frac{2m_2}{m_1}$$

But : $\sin\theta = \sqrt{1 - \cos^2\theta} = \sqrt{1 - \dfrac{4m_2^2}{m_1^2}} = \sqrt{\dfrac{m_1^2 - 4m_2^2}{m_1^2}}$ (iv)   and

equations (i) and (iii) give :

$$T Sin\theta = \frac{m_1 g}{2}\sqrt{\frac{m_1^2 - 4m_2^2}{m_1^2}} = \frac{g}{2}\sqrt{m_1^2 - 4m_2^2} = \frac{m_2 v^2}{r} \Rightarrow r = \frac{2m_2 v^2}{g\sqrt{m_1^2 - 4m_2^2}} \quad (v)$$

But, from equations (iv) and (v): $Sin\theta = \frac{r}{L} \Rightarrow L = \frac{r}{Sin\theta} = \frac{\dfrac{2m_2 v^2}{g\sqrt{m_1^2 - 4m_2^2}}}{\sqrt{\dfrac{m_1^2 - 4m_2^2}{m_1^2}}} = \frac{2m_1 m_2 v^2}{g}$

(b) If $m_1 = 3\ m_2 = m$ kg and $v = 1$ m/s find r and L

From (a) above : $r = \dfrac{2m_2 v^2}{g\sqrt{m_1^2 - 4m_2^2}} = \dfrac{2\left(\dfrac{m}{3}\right)(1)^2}{g\sqrt{m^2 - \dfrac{4}{9}m^2}} = \dfrac{2}{g\sqrt{5}}$ metres

From (a): $L = \dfrac{2m_1 m_2 v^2}{g} = \dfrac{2m\left(\dfrac{m}{3}\right)(1)^2}{g} = \dfrac{2m^2}{3g}$ metres

8. Motion of a conical pendulum

A small ring of mass 2 kg is threaded on a smooth light flexible inelastic string of length 1.2 metres. The ends of the string are attached to A and B with B a distance of 0.8 metres vertically below A. The ring describes with constant speed a horizontal circle about B. Find the tension in the string and the speed of rotation (See Fig. 1).

Fig. 1 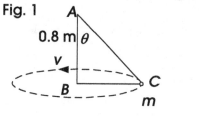   **Fig. 2**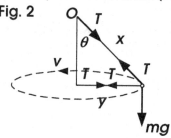

Solution

Assume the particle is travelling with speed v m/s in a horizontal circle. The forces acting on the particle are shown in Fig 2.

From geometry :

$|AB| = 0.8$ metres (given) and $|AC| + |BC| = 1.2$ metres (given)

$\Rightarrow (0.8)^2 + |BC|^2 = |AC|^2$ and $|AC| = 1.2 - |BC|$

$\Rightarrow (0.8)^2 + |BC|^2 = (1.2 - |BC|)^2 = (1.2)^2 + |BC|^2 - 2.4|BC|$

$\Rightarrow |BC| = \dfrac{(1.2)^2 - (0.8)^2}{2.4} = 0.333$ metres $\Rightarrow |AC| = 0.867$ metres

$\Rightarrow Sin\theta = \dfrac{0.333}{0.867} = 0.384,\quad Cos\theta = \dfrac{0.8}{0.867} = 0.923$

Resolving the forces vertically and horizontally:

Vertically: $T Cos\theta = mg$ (i) , Horizontally: $T + T Sin\theta = \dfrac{mv^2}{|BC|}$ (ii)

64

From equation (i): $T = \dfrac{mg}{Cos\theta} = \dfrac{2g}{0.923} = 2.167g$ Newtons

From equation (ii): $T + TSin\theta = \dfrac{mv^2}{|BC|}$

$$\Rightarrow v^2 = \frac{|BC|T}{m}(1 + Sin\theta) = \frac{0.333(2.167g)(1 + 0.384)}{2} = \frac{g}{2} = 4.9$$

$\Rightarrow$ constant speed of rotation: $v = 2.21$ m/s

9. Motion on internal surface of a sphere

The smooth inside surface of a bowl is a segment of a sphere of radius r metres, the height of the segment being x metres. See Fig. 1. The bowl rotates about its axis, OC. Find the greatest velocity of the bowl if a particle of mass m kg on the inside rim of the bowl (at A) can remain at rest relative to the bowl if: (a) $x = r/2$, (b) $x = r/3$

Solution

(a) $x = r/2$

Let the particle have mass m kg and be at A, a distance a metres from the axis, OB. (See Fig. 2: the dotted lines are provided to assist development of equations).

The forces in Newtons acting when the bowl is spinning are shown in Fig. 3:
R = Reaction between particle and bowl and R is directed towards O
mg = Weight of particle

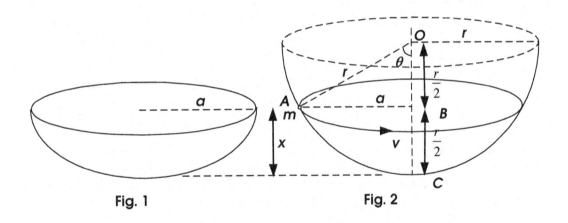

Fig. 1   Fig. 2

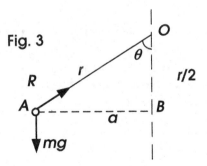

Fig. 3

Resolving these forces horizontally and vertically:

(i) Horizontally: $R\,Sin\theta = \dfrac{mv^2}{a}$ where v = velocity of particle and $a = |AB|$

(ii) Vertically: $R\,Cos\,\theta = mg$

$$\dfrac{R\,Sin\theta}{R\,Cos\theta} = Tan\,\theta = \dfrac{mv^2}{amg} = \dfrac{v^2}{ag} \quad \text{(iii)}$$

But, from Fig. 2 : $Cos\theta = \dfrac{\frac{r}{2}}{r} = 0.5 \Rightarrow \theta = 60° \Rightarrow Sin\,\theta = \dfrac{\sqrt{3}}{2}, \quad Cos\theta = \dfrac{1}{2}$

Also : $Sin\,\theta = Sin60° = \dfrac{\sqrt{3}}{2} = \dfrac{a}{r} \Rightarrow a = \dfrac{r\sqrt{3}}{2} \quad \text{(iv)}$

$\Rightarrow$ Using equations (iii) and (iv): $Tan\,\theta = Tan60° = \dfrac{v^2}{ag} = \dfrac{2v^2}{gr\sqrt{3}} = \sqrt{3} \Rightarrow v = \sqrt{\dfrac{3gr}{2}}$

$\Rightarrow$ Speed necessary to keep particle at rest relative to bowl just within rim is $\sqrt{\dfrac{3gr}{2}}$ m/s

(b) x = r/3

If $x = \dfrac{r}{3}$, $Sin\theta = \dfrac{a}{r} \Rightarrow a = r\,Sin\theta$ and $Cos\theta = \dfrac{r-x}{r} = \dfrac{r-\frac{r}{3}}{r} = \dfrac{2}{3}$

$\Rightarrow Sin\theta = \sqrt{1-\left(\dfrac{2}{3}\right)^2} = \dfrac{\sqrt{5}}{3} = \dfrac{a}{r} \Rightarrow a = \dfrac{r\sqrt{5}}{3} \Rightarrow Tan\theta = \dfrac{v^2}{ag} = \dfrac{\frac{a}{r}}{\frac{2}{3}} = \dfrac{3r\sqrt{5}}{6r} = \dfrac{\sqrt{5}}{2}$

$\Rightarrow \dfrac{v^2}{ag} = \dfrac{\sqrt{5}}{2} \Rightarrow v^2 = \dfrac{5gr}{6} \Rightarrow v = \sqrt{\dfrac{5gr}{6}}$

$\Rightarrow$ speed necessary to keep particle at rest relative to bowl just within rim is $\sqrt{\dfrac{5gr}{6}}$ m/s

10. Fixed Point and Sliding ring

A) A particle m_1 of mass 1 kg at C is attached by two light inelastic strings CA and CB, each of length 1 metre, to fixed points A and B. The line AB is vertical. The particle and strings rotate about AB with constant angular speed ω = 10 radians per second.
(a)   Show the forces acting on the particle
(b)   Calculate the tensions in the two strings.

B) Suppose the end of the string CB is attached to a ring of mass 0.4 kg at B which is free to slide on a smooth vertical wire AB. If the particle and strings rotate about AB with constant angular speed ω = 10 radians per second.
(c)   Show the forces acting on the particle and on the ring
(d)   Calculate the tensions in their two strings.

Solution
A)

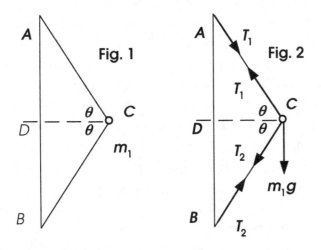

Fig. 1

Fig. 2

(a) Show the forces acting on the particle

Forces acting on Particle: (note: $m_1 = 1$ kg)

(i) Vertical forces: $T_1 \, Sin \, \theta = T_2 \, Sin \, \theta + m_1 g$

(ii) Horizontal forces: $T_1 \, Cos \, \theta + T_2 \, Cos \, \theta = m_1 \, \omega^2 \, |CD|$

Since $Cos\theta = \dfrac{|CD|}{|AC|} = \dfrac{|CD|}{1} = |CD|$

(b) Calculate the tensions in the two strings

$\Rightarrow$ equation (i) becomes : $(T_1 - T_2) \, Sin \, \theta = (1)(g) \Rightarrow (T_1 - T_2) = \dfrac{g}{Sin\theta}$

and equation (ii) becomes : $(T_1 + T_2)|CD| = (1)\omega^2|CD| \Rightarrow T_1 + T_2 = \omega^2 = 1 \times 10^2 = 100$

Equation (i)−(ii) gives : $-2T_2 = \dfrac{g}{Sin\theta} - 100 \Rightarrow T_2 = 50 - \dfrac{0.5g}{Sin\theta} \Rightarrow T_1 = 100 - T_2 = 50 + \dfrac{0.5g}{Sin\theta}$

B)

(c) Show the forces acting on the particle and ring

See Figs. 3 and 4.

Forces acting on Particle:

Vertical forces: $T_1 \, Sin \, \theta = T_2 \, Sin \, \theta + m_1 g$ (i)

Horizontal forces: $T_1 \, Cos \, \theta + T_2 \, Cos \, \theta = m_1 \, \omega^2 \, |CD|$ (ii)

Since $Cos\theta = \dfrac{|CD|}{|AC|} = \dfrac{|CD|}{1} = |CD|$

$\Rightarrow$ Equation (i) becomes : $(T_1 - T_2) Sin\theta = 1g \Rightarrow (T_1 - T_2) = \dfrac{g}{Sin\theta}$

and (ii) becomes : $(T_1 + T_2)|CD| = 1\omega^2|CD| = T_1 + T_2 = \omega^2 = 1 \times 10^2 = 100$

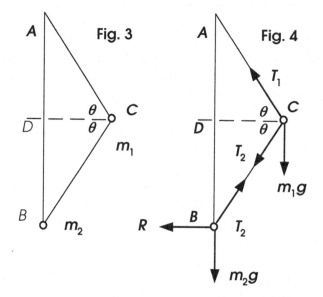

Fig. 3   Fig. 4

Equation (i)–Equation (ii): $-2T_2 = \dfrac{g}{Sin\theta} - 100 \Rightarrow T_2 = 50 - \dfrac{0.5g}{Sin\theta}$ (iii) and $T_1 = 50 + \dfrac{0.5g}{Sin\theta}$

Forces acting on Ring:

The forces acting on the ring are shown in Fig. 3:

Vertical forces: $T_2 Sin\theta = m_2g = 0.4g$ (iv)

Horizontal forces: $R = T_2 Cos\theta$ (v)

(d) Calculate the tensions in the two strings.

Using the expression for $Sin\theta$ from equation (iv): $Sin\theta = \dfrac{0.4g}{T_2}$ in equation (iii):

$$\Rightarrow T_2 = 50 - \dfrac{0.5g}{\dfrac{0.4g}{T_2}} \Rightarrow T_2 = 22.22 \quad \text{Newtons} \quad \Rightarrow T_1 = 77.78 \quad \text{Newtons}$$

11. Motion in a vertical circle

A mass of m kg at point A is attached by a string r metres long to a fixed point B and is given a horizontal velocity u m/s so that it begins to describe a vertical circle around B. (See Fig. 1). Find the:

(a)   Minimum value of u so that the mass just reaches D

(b)   Minimum value of u if the particle describes a circle without the string becoming slack.

Solution

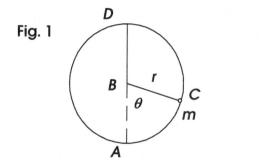

Fig. 1

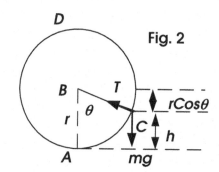

Fig. 2

(a) Minimum value of u so that the mass just reaches D
The mass is projected from A with initial velocity u m/s. Let v be the velocity of the mass
at a point C. When the mass is at C it has risen a vertical height, h where:
$h = r - r \cos \theta = r(1 - \cos \theta)$ (i)   (See Fig. 2)
Let: Kinetic Energy = K.E. and Potential Energy = P.E.
From the Principle of Conservation of Energy:
Total Energy at Point A = Total Energy at Point C
$\Rightarrow$ K.E. + P.E. (at Point A) = K.E. + P.E. (at Point C)
K.E. of the mass at point $A = \frac{1}{2} mu^2$ P.E. of the mass at point A = mg(0)   (say)

K.E. of the mass at point $C = \frac{1}{2} mv^2$ P.E. of the mass at point C = mgh

$\Rightarrow \frac{1}{2} mu^2 = \frac{1}{2} mv^2 + mgh$ (ii)
From equation (i): $u^2 = v^2 + 2gh = v^2 + 2gr(1 - \cos \theta) \Rightarrow v^2 = u^2 - 2gr(1 - \cos \theta)$ (iii)
If the mass is to have the minimum velocity to reach D assume that v = 0 at θ = 180°
From equation (iii): $0 = u^2 - 2gr(1 + 1) \Rightarrow u^2 = 4gr \Rightarrow u = 2\sqrt{gr}$

**(b) Minimum value of u if the particle describes a circle without the string becoming
slack.**
If the particle is to complete a full revolution the string must remain taut, this means that

the tension in the string must be $\geq$ 0. At point C: $\Rightarrow T - mg\cos\theta = \dfrac{mv^2}{r}$ (iv)

If the string is just on the point of becoming slack at its highest point D then

θ = 180°, $\cos \theta$ = -1 and, from equation (iv): $\Rightarrow g < \dfrac{v^2}{r} \Rightarrow v^2 > gr$ (v)

$\Rightarrow$ using equation (iii): $gr < u^2 - 2gr(1 - (-1)) < u^2 - 4gr \Rightarrow u^2 > 5gr$
$\Rightarrow u > \sqrt{5gr}$ to prevent slackness in the string

12. Motion in a vertical circle
A particle of mass m = 3 kg hangs freely from the end B of a light inextensible string of
length r = 1 metre attached to a fixed point O. The particle is then projected
horizontally with speed u = 2 m/s. At point B the particle has a speed v m/s and line OB
is inclined at an angle θ to the downward vertical.
(a)   Find v and the tension, T, in the string in terms of θ at point B.
(b)   Find the values of θ and the tension in the string at the point the particle comes to
an instantaneous rest.

Solution

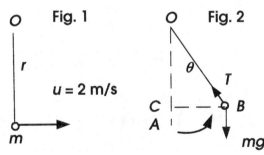

(a) Find v and the tension, T, in the string in terms of θ at point B.
When line OB is inclined at an angle θ to the downward vertical, the forces acting on
the particle are shown in Fig. 2:

Let $OB = r \Rightarrow T - mgCos\theta = \dfrac{mv^2}{r} \Rightarrow T = mgCos\theta + \dfrac{mv^2}{r}$ (i)

From the Principle of Conservation of Energy:
Total Energy of particle at A = Total Energy of particle at B
$\Rightarrow \frac{1}{2}mu^2 = \frac{1}{2}mv^2 + mg(r - rCos\theta) \Rightarrow u^2 = v^2 + 2gr(1 - Cos\theta)$
$\Rightarrow v^2 = u^2 - 2gr(1 - Cos\theta)$ (ii)

From equation (i): $T = mgCos\theta + \dfrac{mv^2}{r} = mgCos\theta + \dfrac{m(u^2 - 2gr(1 - Cos\theta))}{r}$ (iii)

Putting in values : $m = 3$ kg, $r = 1$ metre, $u = 2$ m/s   we get :

Equation (ii) becomes : $v^2 = 4 - 2g(1 - Cos\theta) \Rightarrow v = \sqrt{4 - 2g(1 - Cos\theta)}$
Equation (iii) becomes : $T = 3gCos\theta + 3(4 - 2g(1 - Cos\theta)) = 3gCos\theta + 12 - 6g(1 - Cos\theta)$
$\Rightarrow T = 9gCos\theta + 12 - 6g = 88.3Cos\theta - 46.9$ Newtons

(b) Find the values of θ and the tension in the string at the point the particle comes to an instantaneous rest

When the particle comes to instantaneous rest $\Rightarrow v = 0 \Rightarrow v^2 = 0$. From equation (ii):
$v = 0 = \sqrt{u^2 - 2gr(1 - Cos\theta)} \Rightarrow u^2 - 2gr + 2grCos\theta = 0$

$\Rightarrow Cos\theta = \dfrac{2gr - u^2}{2gr} = \dfrac{2g - 4}{2g} \Rightarrow \theta = 37.24°$

The particle comes to instantaneous rest when: $\theta = 37.24°$. At this point the tension in the string, T, is, from equation (i): $T = mgCos\theta + \dfrac{mv^2}{r} = 3 \times 9.81 \times 0.796 + 0 = 23.42$ Newtons

13. Motion of a pendulum
One end of a light inextensible string of length r ($r = 1$ metres) is fixed at point O. The other end is attached to a mass M_1 of 1 kg which is held at point A (where A is level with O) and allowed to fall (See Fig. 1):
(a)   Find the speed of M_1 when it has fallen through an angle θ and is at point B.
(b)   Find also the tension of the string at point B.
(c)   If M_1 collides and coalesces with a mass M_2 of 2 kg at rest at point C. The combined masses rise to point D. Find the vertical height of D above C (see Figs. 2 and 3).

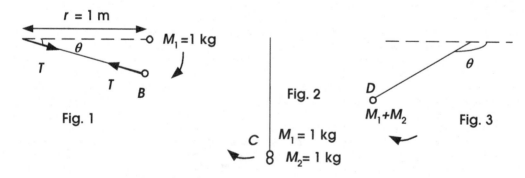

Fig. 1   Fig. 2   Fig. 3

Solution
(a) Find the speed of M_1 when it is at point B.
Let Kinetic Energy = K.E. and Potential Energy = P.E.   From the Principle of Conservation of Energy: Total Energy at Point A = Total Energy at Point B

$\Rightarrow$ K.E. + P.E. (at Point A) = K.E. + P.E. (at Point B)

(i) At point A:
K.E. of the mass at point $A = 0$
P.E. of the mass at point $A = M_1g(r)$ (say) $\Rightarrow$ Total Energy of M_1 at Point $A = M_1gr$
(ii) At point B:
After falling through angle θ, speed = v and energy is:
K.E. of the mass at point $B = \frac{1}{2} M_1v^2$
P.E. of the mass at point $B = M_1g(r - r\,Sin\,\theta) = M_1gr(1 - Sin\,\theta)$
$\Rightarrow$ Total Energy of M_1 at Point $B = \frac{1}{2} M_1v^2 + M_1gr(1 - Sin\,\theta)$

But: Total Energy at point A = Total Energy at point B
$\Rightarrow M_1gr = \frac{1}{2} M_1v^2 + M_1gr(1 - Sin\,\theta) \Rightarrow v = \sqrt{2gr\,Sin\,\theta}$ But: $r = 1$ m $\Rightarrow v = \sqrt{2g\,Sin\,\theta}$ m/s

(b) Find also the tension of the string at point B.
When M_1 has fallen through an angle θ the forces acting on the particle are:

$T - M_1g\,Sin\,\theta = \dfrac{M_1v^2}{r}$ where v = speed of particle (See Fig. 2)

Putting in values : $M_1 = 1$kg, $r = 1$ metre :

$T - (1)g\,Sin\,\theta = \dfrac{M_1v^2}{r} = \dfrac{1\left(\sqrt{2g\,Sin\,\theta}\right)^2}{1} = 2g\,Sin\,\theta \Rightarrow T = 3g\,Sin\,\theta$ Newtons

(c) Find the vertical height of D above C (see Figs. 2 and 3).
At the lowest point, C, $\theta = 90°$, M_1 has a speed of $v = \sqrt{2g\,Sin\,\theta} = \sqrt{2g}$ m/s. M_1 then collides with M_2, of 2 kg: the masses coalesce into a combined mass $(M_1 + M_2) = 3$ kg. Using the Principle of Conservation of Linear Momentum: $M_1v_1 + M_2v_2 = (M_1 + M_2)V$ where V = velocity of combined mass after collision at C.

Thus: if $M_1 = 1$ kg, $M_2 = 2$ kg, $v_1 = \sqrt{2g}$ m/s, $v_2 = 0$ m/s $\Rightarrow V = \dfrac{\sqrt{2g}}{3}$ m/s

The combined mass will rise until it reaches point D. Then:
Total Energy at point C = Potential Energy at highest point D
But, at point C the Potential Energy = $(M_1 + M_2)gr(1 - Sin\,\theta)$ and $\theta = 90°$
$\Rightarrow$ Potential Energy = 0 $\Rightarrow$ At point C the Total Energy = K.E. = $\frac{1}{2}(M_1 + M_2)V^2$
At point D, the velocity $v = 0$ ($\Rightarrow$ K.E. = 0) and P.E. = $(M_1 + M_2)gh$
where h = vertical distance between C and D

$\Rightarrow \frac{1}{2}(M_1 + M_2)V^2 = (M_1 + M_2)gh \Rightarrow \frac{1}{2}(3)\left(\dfrac{\sqrt{2g}}{3}\right)^2 = (3)gh \Rightarrow h = \dfrac{1}{9}$ metre

= maximum vertical height reached by the combined mass.

14. Particle sliding on a spherical surface
A particle of mass m kg slides from rest down the outside of a smooth sphere of radius r metres until it leaves the sphere.
(a) Find the value of angle θ where particle slides from rest at point B and leaves sphere at point D (See Figs. 1 and 2)
(b) Find the value of angle θ where particle slides from rest at point C, at a vertical height of $r/2$ metres above O and leaves the sphere at point E (See Figs. 3, 4).

In the case of (b) above, find:

(c)   Horizontal distance between E and the vertical axis, OB and vertical distance between E and O. (See Figs. 3 and 4)

(d)   The particle's distance from vertical axis OB when it strikes the ground at F.

Solution

(Assume potential energy at $X = 0$)

(a) Particle slides from rest at point B and leaves sphere at point D

The particle is initially at B (given). See Figs. 1 and 2

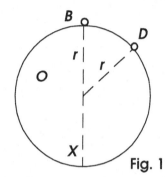

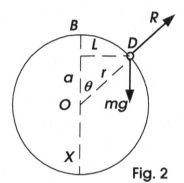

Fig. 1   Fig. 2

Total Energy of particle at any point = Kinetic Energy (K.E.) + Potential Energy (P.E.)

At point B the particle is at rest and its energy = Potential Energy (P.E.) only:
P.E. = $mg(2r)$ (where mg = weight of particle)   (i)

At point D, the particle is moving with velocity v:

⇒ The energy of the particle at $D = \frac{1}{2}mv^2 + mg(r + rCos\theta)$ (ii)

⇒ from equations (i) and (ii): $2mgr = \frac{1}{2}mv^2 + mg(r + rCos\theta)$

⇒ $4gr = v^2 + 2gr + 2grCos\theta$ ⇒ $v^2 = gr(2 - 2Cos\theta)$ (iii)

When the particle is at point D it is moving with velocity v and the forces acting on it are shown in Fig. 2:

R = Reaction between particle and sphere ⇒ $mgCos\theta - R = \dfrac{mv^2}{r}$ (iv)

Putting the value for v^2 derived in equation (iii) into equation (iv) gives:

$$mgCos\theta - R = \left(\frac{m}{r}\right)(gr(2 - 2Cos\theta)) \Rightarrow R = mgCos\theta - 2mg + 2mgCos\theta = 3mgCos\theta - 2mg$$

Particle leaves sphere when $R = 0 \Rightarrow 0 = 3mgCos\theta - 2mg \Rightarrow Cos\theta = \frac{2}{3} \Rightarrow \theta = Cos^{-1}\frac{2}{3}$

(b) Particle slides from rest at point C and leaves sphere at point E

The particle is initially at C where $AB = AO = r/2$ (given) (where A and C are in the same horizontal line). See Figs. 3 and 4.

The Total Energy of the particle at any point = Kinetic Energy (K.E.) + Potential Energy (P.E.)

At point C the particle is at rest and its energy = Potential Energy (P.E.) only:

P.E. = $mg(|OB| - |AB|) = mg(r + \frac{1}{2}r) = \frac{3}{2}mgr$ (where mg = weight of particle)   (v)

At point E, the particle is moving with velocity v:

⇒ The energy of the particle at $E = \frac{1}{2}mv^2 + mg(r + rCos\theta)$ (vi)

72

$\Rightarrow$ from equations (i) and (ii): $\frac{3}{2} mgr = \frac{1}{2} mv^2 + mg(r + rCos\theta) \Rightarrow gr = v^2 + 2grCos\theta$

$\Rightarrow v^2 = gr(1 - 2Cos\theta)$ (vii)

When the particle is at point E it is moving with velocity v and the forces acting on it are shown in Fig. 3 :

R = Reaction between particle and sphere $\Rightarrow mgCos\theta - R = \dfrac{mv^2}{r}$ (viii)

From equations (i) and (ii):
Putting the value for v^2 derived in equation (vii) into equation (viii) gives :

$$mgCos\theta - R = \left(\frac{m}{r}\right)(gr(1 - 2Cos\theta)) \Rightarrow R = mgCos\theta - mg + 2mgCos\theta = 3mgCos\theta - mg$$

Particle leaves sphere when $R = 0 \Rightarrow 0 = 3mgCos\theta - mg \Rightarrow Cos\theta = \frac{1}{3} \Rightarrow \theta = Cos^{-1}\frac{1}{3}$

(c) Horizontal and vertical distances between D and the vertical axis, OB and O respectively.

From above: $Cos\theta = \dfrac{a}{r} \Rightarrow a = \dfrac{r}{3} \Rightarrow L = \sqrt{r^2 - \left(\dfrac{r}{3}\right)^2} = \dfrac{2r\sqrt{2}}{3}$

$\Rightarrow$ the particle leaves the sphere when it is at a vertical height of $\dfrac{r}{3}$ metres

above the centre of the sphere and $\dfrac{2r\sqrt{2}}{3}$ metres from the vertical axis, OB.

(d) When the particle strikes the ground at E find its distance from axis OB.

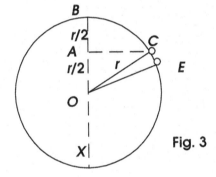

Fig. 3

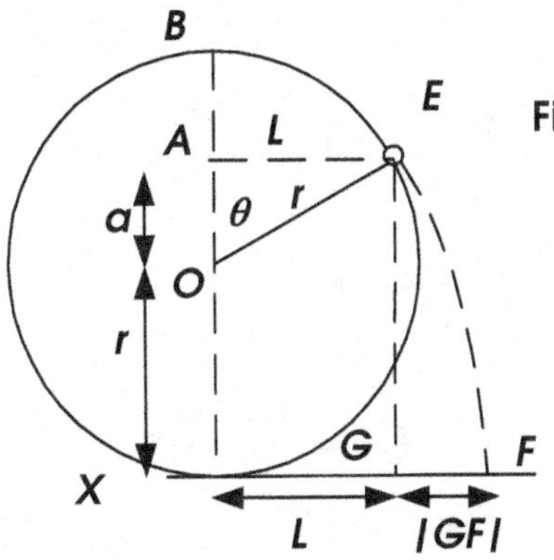

Fig. 4

When the particle leaves the sphere at E it travels as a projectile (See Fig. 4). Assume that the particle leaves the sphere at time $t = 0$ and strikes the ground at F in time t.

From (c) its initial distance from vertical diameter OB at $t = 0$ is $L = \dfrac{2r\sqrt{2}}{3}$ metres   (ix)

From equation (vii), the speed of the particle when it leaves the sphere is

$$v^2 = gr\left(1 - 2\left(\frac{1}{3}\right)\right) = \frac{gr}{3} \Rightarrow v = \sqrt{\frac{gr}{3}} \quad (x)$$

where v is perpendicular to OE. The components of v are:
Horizontal component = $v_x = v\,Cos\,\theta$
Vertical (i.e. downward) component = $v_y = v\,Sin\,\theta + gt$

The horizontal distance travelled in time t from $E = |GF| = v_x t$ (where G is vertically below E). In time t the vertical displacement of the particle = $\frac{4}{3} r$

But, the vertical component of displacement of the particle $= vt\,Sin\,\theta + \frac{1}{2}gt^2$

i.e. vertical component of displacement $= vt\,Sin\,\theta + \frac{1}{2}gt^2 = vt\dfrac{2\sqrt{2}}{3} + \frac{1}{2}gt^2 = \frac{4}{3}r$

$\Rightarrow$ Re-writing gives : $t^2 + \dfrac{4\sqrt{2}}{3\sqrt{3}}\left(\sqrt{\dfrac{r}{g}}\right)t - \dfrac{8r}{3g} = 0 \Rightarrow t^2 + 0.3476\left(\sqrt{r}\right)t - 0.2718r = 0$

$\Rightarrow t = \dfrac{-0.3476\sqrt{r} \pm \sqrt{(0.3476)^2 r + 4(0.2718)r}}{2} = 0.3757\sqrt{r}$ (for $t > 0$)   (xi)

Using equations (x) and (xi): Horizontal distance travelled in time, $t = |GF| = v_x t$

$\Rightarrow |GF| = \dfrac{1}{3}\sqrt{\dfrac{gr}{3}}\left(0.3757\sqrt{r}\right) = 0.226r$

$\Rightarrow$ Distance from vertical axis $= L + 0.226r = 0.9428r + 0.226r = 1.169r$

15. Particle moving in vertical circle

A particle of mass m kg is moving in a vertical circle on the end of a string of length 0.9 metres. See Fig. 1. If the maximum string tension permissible is $8mg$ Newtons find the maximum and minimum velocities of the particle.

Solution

Let string length $= L$ metres and maximum string tension permissible $= T_P$

The maximum tension occurs in the string at point A while the minimum tension occurs at point B: (See Figs 2 and 3). These tensions are:

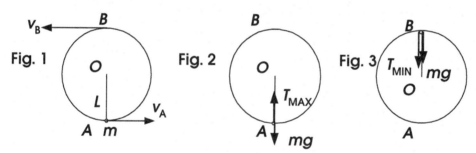

$T_{MAX} - mg = \dfrac{mv_A^2}{L} \Rightarrow T_{MAX} = \dfrac{mv_A^2}{L} + mg = 8mg \Rightarrow v_A = \sqrt{7gL} = \sqrt{6.3g} = 7.861\,m/s$

Energy at A = Energy at B

Energy at A = Potential Energy $(= 0, say)$ + Kinetic Energy $= \frac{1}{2}mv_A^2$

Energy at B = Potential Energy + Kinetic Energy $= 2mgL + \frac{1}{2}mv_B^2$

$\Rightarrow \frac{1}{2}mv_A^2 = 2mgL + \frac{1}{2}mv_B^2 \Rightarrow v_B^2 = v_A^2 - 4gL = 26.487 \Rightarrow v_B = 5.15$ m/s

16. Particles sliding over pulley

A string with two particles each of mass m kg at each end lies over a smooth fixed pulley. At equilibrium each particle is level with the centre of the pulley (see Fig. 1). Particle 1 is slightly displaced.

(a)   Find the reaction exerted by particle 2 on the pulley at point A

(b)   If the particle leaves the sphere at point C find the value of θ (See Fig. 3)

Solution

(a) Find the reaction exerted by the second particle on the pulley at point A

Total Energy of particle at any point = Kinetic Energy (K.E.) + Potential Energy

(P.E.). At equilibrium, (as shown in Fig. 1), the two particles are at rest. Therefore, the Total Energy of the system consists only of Potential Energy

$\Rightarrow$ Total Energy = Potential Energy =

$mg(3r + rCos45°)$ per particle $\times$ 2 particles = $mgr(6 + \sqrt{2})$ (i)

When the second particle is at its highest point, A, the Total Energy of the system is K.E.$_1$ + P.E.$_1$ + K.E.$_2$ + P.E.$_2$ (see Fig. 2):

$\Rightarrow$ Total Energy = $\frac{1}{2}mv_1^2 + mg(3r) + \frac{1}{2}mv_2^2 + mg(4r)$

But the particles are connected $\Rightarrow v_1 = v_2 \Rightarrow$ Total Energy = $mv^2 + 7mgr$ (ii)

But the Principle of Conservation of Energy holds that the Total Energy remains unchanged $\Rightarrow (6 + \sqrt{2})mgr = mv^2 + 7mgr$ which gives : $v^2 = gr(\sqrt{2} - 1)$ (iii)

R = Reaction between particle and pulley (at point A say)

$\Rightarrow mg - R = \dfrac{mv^2}{r} \Rightarrow R = mg - \dfrac{mv^2}{r}$

Using equation (iii) we have : $R = mg - \dfrac{m}{r}gr(\sqrt{2} - 1) = mg(2 - \sqrt{2})$

(b) If the particle leaves the sphere at point C find θ.

Assume the line CO makes an angle of θ to the vertical OA. At point C the particles are moving at v_1 m/s and the forces acting on the pulley are shown in Fig. 3. To find when the second particle leaves the pulley, consider it at angle θ from the vertical (see Fig.

3). The forces acting at point C are: $mgCos\theta - R = \dfrac{mv_1^2}{r}$ (iv)

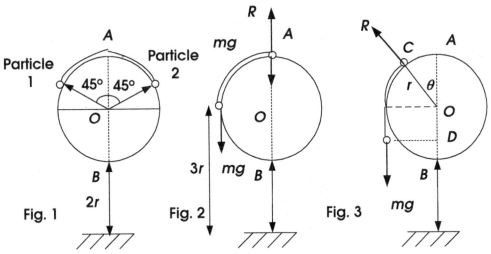

Fig. 1   **Fig. 2**   **Fig. 3**

Energy of this system at this point is (K.E.$_1$+ P.E.$_1$) (of m_1) + (K.E.$_2$ + P.E.$_2$) (of m_2)

$\frac{1}{2}m_1v_1^2 + PE_1 + \frac{1}{2}m_1v_1^2 + PE_2 = mv_1^2 + PE_1 + PE_2$ (as $m_1 = m_2 = m$)

Total string length = πr metres

In Fig. 1, the total string length in contact with the pulley = $\dfrac{\pi r}{2}$ metres

In Fig. 2, the total string length in contact with the pulley $= \dfrac{\pi r}{2}$ metres

In Fig. 3, the total string length in contact with the pulley $= \dfrac{\pi r}{2} - $ arc AC

$= \dfrac{\pi r}{2} - (\text{radius} \times \text{angle in radians}) = \dfrac{\pi r}{2} - \dfrac{\theta \pi r}{180}$

$\Rightarrow$ String length not in contact with pulley $= |OD| = \dfrac{\pi r}{2} - \left(\dfrac{\pi r}{2} - \dfrac{\theta \pi r}{180} \right) = \dfrac{\theta \pi r}{180}$

$\Rightarrow PE_1 = mg\left(3r - \left(\dfrac{\theta \pi r}{180} \right) \right)$ and $PE_2 = mg(3r + r Cos\theta)$

$\Rightarrow$ Total energy of system is : $mv_1^2 + mg\left(3r + r Cos\theta + 3r - \pi r\left(\dfrac{\theta}{180} \right) \right)$

and this is equal to the initial energy of the system, $mgr\left(6 + \sqrt{2} \right)$

$\Rightarrow mgr\left(6 + \sqrt{2} \right) = mv_1^2 + mgr\left(6 + Cos\theta - \left(\dfrac{\pi \theta}{180} \right) \right)$

$\Rightarrow mgr\sqrt{2} = mv_1^2 + mgr Cos\theta - mgr\left(\dfrac{\pi \theta}{180} \right) \Rightarrow v_1^2 = gr\sqrt{2} + \left(\dfrac{gr\pi \theta}{180} \right) - gr Cos\theta$

But the particle leaves the sphere when $R = 0 \Rightarrow$ from equation (iv)

$mg Cos\theta = \dfrac{mv_1^2}{r} \Rightarrow v_1^2 = gr Cos\theta \Rightarrow 2gr Cos\theta = gr\sqrt{2} + \left(\dfrac{gr\pi \theta}{180} \right) \Rightarrow 2 Cos\theta = \sqrt{2} + \left(\dfrac{\pi \theta}{180} \right)$

$\Rightarrow Cos\theta = \dfrac{1}{\sqrt{2}} + \dfrac{\pi \theta}{360}$ Solving for θ: θ = approximately 23.8°

Chapter 5 Collisions

1. Change in momentum of a mass
A mass of 3 kg is moving at 10 m/s. A force of 6 Newtons acts for 4 seconds in the same line as the direction of motion. Find the resulting velocity of the mass and the change in its momentum.

Solution
Resulting velocity: $m(v - u)/t = ma = F$: $\Rightarrow F = 3(v - 10)/4 = 6 \Rightarrow v = 18$ m/s
Change in momentum = $m(v - u) = 3(18 - 10) = 24$ kg.m/s

2. Change in momentum of a mass
A mass of 10 kg is moving at 3 m/s. A force F acts for 6 seconds opposite to its direction of motion. If the resulting velocity is – 2 m/s. Find the magnitude of F and the change in momentum.

Solution
Resulting velocity: $m(v - u)/t = ma = F$: $\Rightarrow F = 10(-2 - 3)/6 = -8.333$ Newtons (minus sign indicating that force is applied in opposite direction to original motion)
Change in momentum = $m(v - u) = 10(-2 - 3) = -50$ kg.m/s

3. Change in momentum of a mass
A mass of 6 kg is accelerating at 4 m/s². Find the force F which is acting and the change of momentum after 7 seconds.

Solution
Resulting velocity: $m(v - u)/t = ma = F$: $\Rightarrow F = 6(v - u)/t = (6)(4) = 24$ Newtons
Change of momentum = $m(v - u) = Ft = (24)(7) = 168$ kg.m/s

4. Change in momentum of a mass
A mass of 10 kg falls from a height of 4.905 metres. If acceleration due to gravity is $g = 9.81$ m/s² find the momentum of the body when it hits the ground.

Solution
Initial velocity = $u = 0$
Use general equation: $v^2 = u^2 + 2as$: $v^2 = 0 + 2(9.81)(4.905) \Rightarrow v = 9.81$ m/s
$\Rightarrow$ Momentum on hitting the ground = $mv = (10)(9.81) = 98.1$ kg.m/s

5. Impulse
A bullet of mass 0.015 kg is fired into a block of wood with a velocity of 650 m/s. It comes to rest in 0.01 seconds. If the resisting force of the wood is constant find the magnitude of this force and its impulse.

Solution
$Ft = m(v - u) \Rightarrow$ Impulse $= Ft = m(v - u) = 0.015(0 - 650) = -10.5$ kg./s
$\Rightarrow F(0.01) = -10.5 \Rightarrow F = 1,050$ Newtons

6. Impulse
A drop hammer of mass 50 kg strikes a rock with a velocity of 10 m/s and does not rebound. Find the impulse on the rock.

Solution
$Ft = m(v - u) \Rightarrow$ Impulse $= Ft = 50(0 - 10) = -500$ kg.m/s

7. Impulse
A ball of mass 0.06 kg strikes a surface with a velocity of 20 m/s and rebounds vertically with a velocity of 12 m/s. Find the impulse on the ball.
Solution
$Ft = m(v - u) \Rightarrow$ Impulse $= Ft = 0.06(-12 - 20) = -1.92$ kg.m/s

8. Impulse
A ball of mass 0.5 kg travelling at 20 m/s is struck and sent back at 32 m/s in the opposite direction. Find the impulse on the ball.
Solution
$Ft = m(v - u) \Rightarrow$ Impulse $= Ft = 0.5(-32 - 20) = -26$ kg.m/s

9. Impulse
A hammer of mass 7 kg travelling horizontally at 6 m/s hits a piece of timber and rebounds horizontally with a velocity of 2 m/s. If the impact lasted for 0.01 seconds find the average force exerted by the hammer on the timber block.

Solution

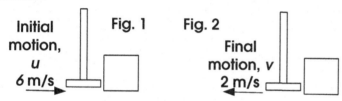

Note: u and v are in opposite directions. Assume u's direction is negative.
Using the general equation for impulse: $Ft = m(v - u)$, where:
$m = 7$ kg, $u = -6$ m/s, $v = 2$ m/s, $t = 0.01$ seconds
$\Rightarrow F(0.01) = 7(2 - (-6)) = 56$ N.s $\Rightarrow F = 5,600$ Newtons

10. Impulse
A ball of mass 1 kg falls from a height of 5 metres onto a flat solid surface and rebounds to a height of 4 metres. If the duration of contact between the ball and the surface was 0.1 seconds find the impulse of the force on the ball.

Solution

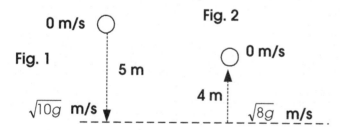

Speed of ball before impact (See Fig. 1):
Using the general equation of linear motion: $v^2 = u^2 + 2as$ where:
$u = 0$ m/s, $a = g = 9.81$ m/s^2, $s = 5$ metres $\Rightarrow v^2 = 10g \Rightarrow v = \sqrt{(10g)}$
Note: Let $v = u_{BEFORE}$ below

Speed of ball after impact (See Fig. 2):
Height reached = 4 metres. Final velocity = 0 m/s, $a = g = -9.81$ m/s^2
Using the general equation of motion: $v^2 = u^2 + 2as$
$\Rightarrow u^2 = 0 + 8g \Rightarrow u = \sqrt{(8g)}$ (Note: Let $u = v_{\text{AFTER}}$ below)

Assume downward motion is the negative direction.
Use the general equation for impulse: $Ft = m(v - u) = m(v_{\text{AFTER}} - u_{\text{BEFORE}})$ where:
$m = 1$ kg, $u = u_{\text{BEFORE}} = \sqrt{(10g)}$ m/s, $v = v_{\text{AFTER}} = \sqrt{(8g)}$ m/s, $t = 0.01$ seconds
$\Rightarrow$ Impulse $= 1(\sqrt{(8g)} - (-\sqrt{(10g)})) = 18.76$ N.s

11. Force acting on a body
A force of 100 Newtons acting on a body for 3 seconds increases its velocity from 8 m/s
to 20 m/s in the direction of the force. Calculate:
(a) The change in the momentum of the body
(b) The mass of the body
(c) Change in kinetic energy of the body
(d) The distance over which the force acts

Solution
Use the equations: $Ft = m(v - u)$ and $Fs = \frac{1}{2} mv^2 - \frac{1}{2} mu^2$
(a) The change in the momentum of the body
$Ft = m(v - u) \Rightarrow$ change in the momentum $= m(v - u) = (100)(3) = 300$ kg.m/s

(b) The mass of the body
$m(v - u) = 300 = m(20 - 8) \Rightarrow m = 25$ kg

(c) Change in kinetic energy of the body
Change in kinetic energy $= \frac{1}{2} mv^2 - \frac{1}{2} mu^2 = \frac{1}{2} (25)(20)^2 - \frac{1}{2} (25)(8)^2 = 4{,}200$ J

(d) The distance over which the force acts
$Fs = \frac{1}{2} mv^2 - \frac{1}{2} mu^2 = 4{,}200 \Rightarrow 100s = 4{,}200 \Rightarrow s = 42$ metres

12. Resisting force
An armour-piercing bullet of mass 0.03 kg strikes a steel plate which is 0.05 metres thick
an penetrates it. The striking velocity is 800 m/s and the exit velocity is 50 m/s. Calculate:
(a) The change in momentum of the bullet
(b) The loss in kinetic energy of the bullet
(c) The resisting force (assume constant)
(d) The time taken to penetrate the steel plate

Solution
Use the equations: $Ft = m(v - u)$ and $Fs = \frac{1}{2} mv^2 - \frac{1}{2} mu^2$
(a) The change in momentum of the bullet
Change in momentum $= m(v - u) = 0.03(50 - 800) = -22.5$ kg.m/s

(b) The loss in kinetic energy of the bullet
Loss in kinetic energy = Initial kinetic energy – Final kinetic energy $= \frac{1}{2} mu^2 - \frac{1}{2} mv^2$
$= \frac{1}{2} (0.03)(800)^2 - \frac{1}{2} (0.03)(50)^2 = -9{,}562..5$ J

(c) The resisting force (assume constant)
$Fs = \frac{1}{2} mv^2 - \frac{1}{2} mu^2 = = -9{,}562..5 \Rightarrow F(0.05) = -9{,}562..5 \Rightarrow F = -191{,}250$ Newtons

(d) The time taken to penetrate the steel plate
$Ft = m(v - u) = -22.5 \Rightarrow -191,250\, t = -22.5 \Rightarrow t = 0.0001176$ seconds

13. Bullet fired into block of wood
A bullet of mass 0.015 kg is fired into a block of wood of mass 5 kg. If the bullet remains embedded in the block and gives it a velocity of 3 m/s, find the initial velocity of the bullet.

Solution
From the Principle of Conservation of Linear Momentum:
$(m_1 + m_2)v = m_1u_1 + m_2u_2$
But : (given) $m_1 = 0.015$ kg, $m_2 = 5$ kg and $u_2 = 0$ m/s, $v = 3$ m/s
$\Rightarrow (0.015 + 5)3 = 0.015(u) + 5(0) \Rightarrow u = 1,003$ m/s

14. Railway wagons in yard
A wagon of mass 5,000 kg moving at 5 m/s catches up on and shunts into another wagon of mass 2,000 kg which is moving at 2 m/s (in the same direction). Find the common velocity of the joined trucks.
Solution
From the Principle of Conservation of Linear Momentum:
$(m_1 + m_2)v = m_1u_1 + m_2u_2$
But : (given) $m_1 = 5,000$ kg, $m_2 = 2,000$ kg and $u_1 = 5$ m/s, $u_2 = 2$ m/s
$\Rightarrow (5,000 + 2,000)v = 5,000(5) + 2,000(2) \Rightarrow 7,000v = 29,000 \Rightarrow v = 4.143$ m/s

15. Railway wagons in yard
A wagon of mass 5,000 kg moving at 5 m/s collides with another wagon of mass 2,000 kg which is moving at 2 m/s (in the opposite direction). If the two trucks remain joined after the collision find their common velocity.

Solution
From the Principle of Conservation of Linear Momentum:
$(m_1 + m_2)v = m_1u_1 + m_2u_2$
But : (given) $m_1 = 5,000$ kg, $m_2 = 2,000$ kg and $u_1 = 5$ m/s, $u_2 = -2$ m/s
$\Rightarrow (5,000 + 2,000)v = 5,000(5) - 2,000(2) \Rightarrow 7,000v = 21,000 \Rightarrow v = 3$ m/s

16. Railway wagons in yard
A wagon of mass 8,000 kg moving at 2 m/s catches up on and shunts into another wagon of mass 3,000 at rest. If the relative velocities of the wagons after impact is 0.9 m/s find the velocities of both wagons after impact and the loss of kinetic energy due to the impact.

Solution
From the Principle of Conservation of Linear Momentum:

$$m_1 u_1 + m_2 u_2 = m_1 v_1 + m_2 v_2$$

But : (given) $m_1 = 8,000$ kg, $m_2 = 3,000$ kg and $u_1 = 2$ m/s, $u_2 = 0$ m/s

$\Rightarrow 8,000(2) + 3,000(0) = 8,000 v_1 + 3,000 v_2 \Rightarrow 8 v_1 + 3 v_2 = 16$

$\Rightarrow$ Also : $v_2 - v_1 = 0.9$ m/s   (As $v_2 > v_1$)

Solving these simultaneous equations : $v_1 = 1.209$ m/s $v_2 = 2.109$ m/s

Kinetic energy before impact : $\frac{1}{2} m_1 u_1^2 + \frac{1}{2} m_2 u_2^2$

Kinetic energy before impact $= \frac{1}{2} m_1 v_1^2 + \frac{1}{2} m_2 v_2^2$

Loss in Kinetic Energy : $\frac{1}{2} m_1 u_1^2 + \frac{1}{2} m_2 u_2^2 - \frac{1}{2} m_1 v_1^2 - \frac{1}{2} m_2 v_2^2$

$= \frac{1}{2}(8,000)(2)^2 + \frac{1}{2}(3,000)(0)^2 - \frac{1}{2}(8,000)(1.209)^2 - \frac{1}{2}(3,000)(2.109)^2$

$= 3,481.5$ J

17. Two masses colliding

A mass of 1 kg travelling with a velocity of $2i - 5j$ collides with a mass of 2 kg travelling with a velocity $6i + 8j$. If both masses coalesce after the collision find the magnitude and direction of the combined masses' velocity and the loss in Kinetic Energy which occurs.

Solution

From the Principle of Conservation of Linear Momentum:

Horizontal (i.e. i) velocity components : $m_1 u_{H1} + m_2 u_{H2} = (m_1 + m_2) v_H$

Vertical (i.e. j) velocity components : $m_1 u_{V1} + m_2 u_{V2} = (m_1 + m_2) v_{V2}$

But : (given) $m_1 = 1$ kg, $m_2 = 2$ kg

$u_{H1} = 2$ $u_{H2} = 6$ $u_{V1} = -5$ $u_{V2} = 8$ m/s

Horizontally:

Horizontal (i.e. i) velocity components : $m_1 u_{H1} + m_2 u_{H2} = (m_1 + m_2) v_H$

$\Rightarrow (1)(2) + (2)(6) = (1 + 2) v_H \Rightarrow v_H = \frac{14}{3}$ m/s

Vertically:

Vertical (i.e. j) velocity components : $m_1 u_{V1} + m_2 u_{V2} = (m_1 + m_2) v_{V2}$

$\Rightarrow (1)(-5) + (2)(8) = (1 + 2) v_V \Rightarrow v_V = \frac{11}{3}$ m/s

Combined mass $= 1 + 2 = 3$ kg

Velocity of combined mass $= V = v_H i + v_V j = \frac{14}{3} i + \frac{11}{3} j$ m/s

Magnitude $= \sqrt{\frac{14}{3}^2 + \frac{11}{3}^2} = 5.935$ m/s

Direction : $Tan^{-1} \left(\frac{\frac{11}{3}}{\frac{14}{3}} \right) = 38.16°$ with the $i-$ direction

18. Impulse

Find the magnitude of the impulse which will cause a mass of m kg moving with speed u m/s to travel with speed u m/s at an angle of θ to its original direction. (See Fig. 1)

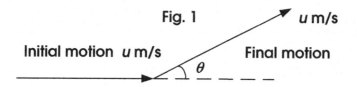

Fig. 1

u m/s
Initial motion u m/s
Final motion
θ

Solution

Initial velocity $= u\,i + 0\,j$, Final velocity $= u\,Cos\,\theta\ i + u\,Sin\,\theta\ j$

Impulse $= m$ (Final velocity) $- m$ (Initial velocity) $= m(u - u\,Cos\,\theta)\ i + m(0 - u\,Sin\,\theta)\ j$

Impulse $= \sqrt{m^2(u - uCos\theta)^2 + m^2(0 - uSin\theta)^2} = \sqrt{2m^2u^2(1 - Cos\theta)}$

$= mu\sqrt{2(1 - Cos\theta)}$ Newton - seconds

19. Bullet fired into block of wood

A bullet of mass 0.015 kg is fired into a block of wood of mass 5 kg. If it is embedded in the block and gives it a velocity of 1.5 m/s. find the initial velocity of the bullet.

Solution

From the Principle of Conservation of Linear Momentum: $(m_1 + m_2)v = m_1u_1 + m_2u_2$

But : (given) $m_1 = 0.015$ kg, $m_2 = 5$ kg and $u_2 = 0$ m/s, $v = 2$ m/s

$\Rightarrow (0.015 + 5)2 = 0.015(u) + 5(0) \Rightarrow u = 668.7$ m/s

20. Bullet and Timber block

A mass m_1 kg is travelling at speed u m/s strikes a mass m_2 kg at rest on a smooth table. If the bodies coalesce, find the speed of the combined masses, V, and the loss of kinetic energy due to impact.

Fig. 1 Fig. 2

m2
m1 + m2
m1
u m/s
0 m/s
V m/s

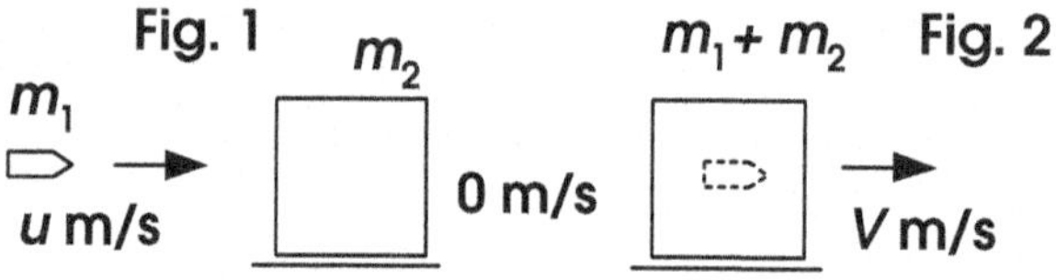

Solution

Mass m_1 has initial speed u m/s. Mass m_2 has initial speed 0 m/s.

From the Principle of Conservation of Linear Momentum:

$m_1u + m_2(0) = (m_1 + m_2)V$ (V = velocity of combined masses after impact)

$$\Rightarrow V = \frac{m_1}{m_1 + m_2}u$$

Initial kinetic energy $= \frac{1}{2}m_1u^2$

Final Kinetic Energy $= \frac{1}{2}(m_1 + m_2)(V)^2 = \frac{1}{2}(m_1 + m_2)\left(\frac{m_1 u}{m_1 + m_2}\right)^2 = \frac{1}{2}\frac{(m_1 u)^2}{m_1 + m_2}$

Loss of kinetic energy $= \frac{1}{2}m_1 u^2 - \frac{1}{2}\frac{(m_1 u)^2}{m_1 + m_2} = \frac{1}{2}m_1 u^2\left(1 - \frac{m_1}{m_1 + m_2}\right)$

$= \frac{1}{2}m_1 u^2\left(\frac{m_1 + m_2 - m_1}{m_1 + m_2}\right) = \frac{1}{2}\left(\frac{m_1 m_2}{m_1 + m_2}\right)u^2$

21. Bullet and Timber block

A bullet of mass m kg moving with speed u m/s strikes, and becomes embedded in, a stationary block of timber, of mass M kg. The combined mass then moves along a rough table where the coefficient of friction $= \mu$. (See Fig. 1). Find the distance travelled by the combined mass before it comes to rest.

Solution

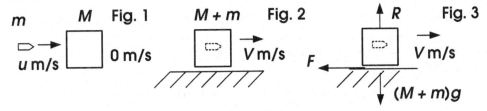

The situations before and after impact are shown in Figs. 1 and 2 respectively.
From the Principle of Conservation of Linear Momentum:

$$mu + M(0) = (M + m)V \quad \Rightarrow V = \left(\frac{m}{M + m}\right)u \quad \text{(i)}$$

The forces acting on the combined mass in Newtons are shown in Fig. 3:
$(M + m)g =$ Weight of combined mass, $R =$ Reaction on surface, $F =$ Frictional force
From Fig. 3: Vertical forces : $R = (M + m)g$, Horizontal forces : $-F$
(Note : the frictional force F opposes the motion). $F = \mu R = \mu(M + m)g$

Horizontally : $(M + m)(\text{Acceleration}) = -F = -\mu(M + m)g$

$\Rightarrow (\text{Acceleration}) = \frac{-\mu(M + m)g}{(M + m)} = -\mu g$

Using the standard equation :

$(\text{Final velocity})^2 = (\text{Initial velocity})^2 + 2(\text{Acceleration})(\text{Distance travelled})$

where : Final velocity $= 0$, Initial velocity $= V = \left(\frac{m}{M + m}\right)u$ m/s

we get : Distance travelled $= \frac{V^2}{2\mu g} = \frac{1}{2\mu g}\left(\frac{m}{M + m}\right)^2 u^2$ metres

22. Bullet and Timber block

A timber block, of mass M kg, is suspended from a fixed point by a light inextensible string of length L. A bullet of mass m kg moving with speed u m/s strikes, and becomes embedded in, the block and sets the combined mass in motion. Find the vertical height, h metres, reached before the combined mass comes to rest. (Assume $h < L$)

Solution

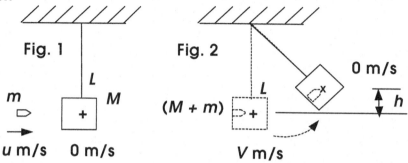

Fig. 1   Fig. 2

m M $(M + m)$ 0 m/s h

u m/s   0 m/s V m/s

The situations before and after impact are shown in Figs. 1 and 2 respectively.
From the Principle of Conservation of Linear Momentum:

$$mu + M(0) = (M+m)V \quad \Rightarrow V = \left(\frac{m}{M+m}\right)u \qquad (i)$$

Consider the energy of the combined mass just after impact and when it comes to rest:
(See Fig. 3)
Initial Energy = Potential Energy (P.E.) + Kinetic Energy (K.E.)
Assume Potential Energy in this position = 0

$$\Rightarrow \text{Initial Energy} = \text{K.E.} = \tfrac{1}{2}(M+m)V^2 = \tfrac{1}{2}(M+m)\left(\frac{m}{M+m}u\right)^2 = \tfrac{1}{2}\frac{m^2u^2}{M+m}$$

When combined mass comes to rest : K.E. = 0
Final Energy = P.E. + K.E. = P.E. = $(M+m)gh$ But : Initial Energy = Final Energy

$$\Rightarrow (M+m)gh = \tfrac{1}{2}\frac{m^2u^2}{M+m} \Rightarrow h = \frac{m^2u^2}{2g(M+m)^2} \quad \text{metres}$$

23. Direct impact: two spheres moving

Sphere 1, moving with velocity $2u$ m/s impinges directly on sphere 2, moving in the same direction with velocity u m/s. Both spheres have mass m kg. Find the resulting velocities of the spheres and show that if $u = 1$ and coefficient of restitution between the spheres, $e = 0.5$, the impact causes a loss in kinetic energy of $\frac{3}{16}m$ Joules

$v_1 \longrightarrow$ $v_2 \longrightarrow$ **Fig. 1**

Sphere 1 c_1 c_2 Sphere 2

$u_1 = 2u$ m m $u_2 = u$

Solution
Fig. 1 shows the spheres' velocities at moment of impact.
Principle of Conservation of Linear Momentum:

$\Rightarrow mv_1 + mv_2 = m(2u) + m(u) = 3mu \Rightarrow v_1 + v_2 = 3u$ (i)

From Newton's Experimental Law :$\Rightarrow v_1 - v_2 = -e(2u - u) = -eu$ (ii)

Add equations (i) and (ii) $\Rightarrow 2v_1 = u(3 - e) \Rightarrow v_1 = \dfrac{u}{2}(3 - e) \Rightarrow v_2 = \dfrac{u}{2}(3 + e)$

The loss of Kinetic Energy (K.E.) caused by the impact can be calculated as follows:

Initial K.E $= \frac{1}{2}m_1 u_1^{\,2} + \frac{1}{2}m_2 u_2^{\,2} = \frac{1}{2}m(2u)^2 + \frac{1}{2}mu^2 = \frac{5}{2}mu^2$

Final K.E. $= \frac{1}{2}m_1 v_1^{\,2} + \frac{1}{2}m_2 v_2^{\,2} = \frac{1}{2}m\left[\dfrac{u}{2}(3 - e)\right]^2 + \frac{1}{2}m\left[\dfrac{u}{2}(3 + e)\right]^2 = \frac{1}{2}m\dfrac{u^2}{4}\left(18 + 2e^2\right)$

Loss in K.E = Initial K.E. − Final K.E. $= \frac{5}{2}mu^2 - \frac{1}{2}m\dfrac{u^2}{4}\left(18 + 2e^2\right)$ Joules

$= \frac{1}{2}mu^2\left(5 - \left(\dfrac{18 + 2e^2}{4}\right)\right) = \frac{1}{2}mu^2\left(\dfrac{20 - 18 - 2e^2}{4}\right) = \frac{1}{2}mu^2\left(\dfrac{2 - 2e^2}{4}\right)$

$= \dfrac{mu^2\left(1 - e^2\right)}{4} = \dfrac{m(1)^2\left(1 - 0.5^2\right)}{4} = \dfrac{3}{16}m$ Joules

24. Direct impact: Two spheres moving in opposite directions

Two smooth spheres of masses m and $2m$ kg and centres C_1 and C_2 collide directly when moving in opposite directions with speeds u_1 and u_2 m/s respectively. (See Fig. 1). If the heavier sphere is brought to rest by the impact find the coefficient of restitution, e.

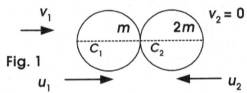

Fig. 1

Solution

Fig.1 shows the initial and final velocities in m/s along the line joining $C_1 C_2$.

From the Principle of Conservation of Linear Momentum:

$mv_1 + 2m(0) = m(u_1) - 2m(u_2) \Rightarrow v_1 = u_1 - 2u_2$ (i)

From Newton's Experimental Law: $v_1 - v_2 = -e(u_1 - (-u_2)) \Rightarrow v_1 = -e(u_1 + u_2)$ (ii)

As equations (i) = equation (ii) $\Rightarrow u_1 - 2u_2 = -e(u_1 + u_2) \Rightarrow e = -\left(\dfrac{u_1 - 2u_2}{u_1 + u_2}\right)$

25. Three balls colliding

Three balls each of mass m kg are shown in Fig 1. If Ball 1 is released from rest and strikes Ball 2 which then strikes Ball 3, find the velocity of each ball after the second collision has occurred. The coefficient of restitution is $e = 0.75$.

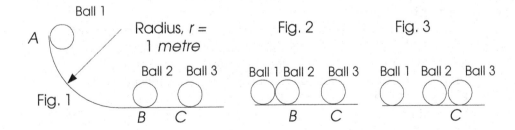

Solution
(a) For Ball 1: (See Fig. 1)
Total Energy at A = Kinetic Energy at A + Potential Energy at A
But the ball is at rest $\Rightarrow$ Total Energy at A = Potential Energy only = mgr

The velocity reached by Ball 1 at point B is v (say) $\Rightarrow$ Total Energy at B
= Kinetic Energy at B ($\frac{1}{2} mv^2$)+Potential Energy at B (0) = $\frac{1}{2} mv^2$

But : Total Energy at A = Total Energy at B
$\Rightarrow mgr = \frac{1}{2} mv^2 \Rightarrow v^2 = 2gr \Rightarrow v = \sqrt{2gr}$; But $r = 1$ (given) $\Rightarrow v = \sqrt{2g}$ (i)

(b) First collision (between Ball 1 and Ball 2 at B): (See Fig. 2)
Ball 1: initial velocity = $u_1 = \sqrt{2g}$ from equation (i).   Ball 2: initial velocity = 0
From the Principle of Conservation of Linear Momentum
$\Rightarrow mv_1 + mv_2 = mu_1 + mu_2 = m\sqrt{2g} + 0 = m\sqrt{2g} \Rightarrow v_1 + v_2 = \sqrt{2g}$ (ii)
From Newton's Experimental Law : $v_1 - v_2 = -e(v-0) = -0.75\sqrt{2g}$ (iii)
Adding equations (ii) and (iii) we get : $2v_1 = 0.25\sqrt{2g} \Rightarrow v_1 = 0.125\sqrt{2g}$
$\Rightarrow v_2 = \sqrt{2g} - (0.125)\sqrt{2g} = 0.875\sqrt{2g}$ from equation (ii)
Thus Ball 2 has a velocity of $0.875\sqrt{2g}$ when it collides with Ball 3

(c) Second collision (between Ball 2 and Ball 3 at C): (See Fig. 3)
From the Principle of Conservation of Linear Momentum:
V_3 = final velocity of Ball 2, v_4 = final velocity of Ball 3
$mv_3 + mv_4 = m$ (initial velocity, Ball 2) + m (Initial velocity, Ball 3)
$mv_3 + mv_4 = m\,0.875\sqrt{2g} + m(0)$ $\Rightarrow v_3 + v_4 = 0.875\sqrt{2g}$ (iv)
and from Newton's Experimental Law :
$v_3 - v_4 = -e(0.875\sqrt{2g} - 0) \Rightarrow v_3 - v_4 = -(0.75)(0.875)\sqrt{2g}$ (v)
Adding equations (iv) and (v) :
$\Rightarrow 2v_3 = (0.25)(0.875)\sqrt{2g} \Rightarrow v_3 = (0.125)(0.875)\sqrt{2g}$ and
$v_4 = (0.875)\sqrt{2g} - v_3 = (0.875)\sqrt{2g} - (0.125)(0.875)\sqrt{2g} = (0.875)^2\sqrt{2g}$
$\Rightarrow$ Resulting velocities are :
After first collision (Ball 1 and Ball 2)velocities are :
Ball 1: $0.125\sqrt{2g} = 0.554$ m/s, Ball 2 : $0.875\sqrt{2g} = 3.88$ m/s
After second collision (Ball 2 and Ball 3), velocities are :
Ball 2 : $(0.125)(0.875)\sqrt{2g} = 0.484$ m/s, (Ball 3): $(0.875)^2\sqrt{2g} = 3.39$m/s

26. Oblique impact: one sphere at rest
A sphere of mass M kg moving with speed u m/s collides obliquely with a second
smooth sphere, of mass m, at rest. The direction of motion of the moving sphere is
inclined at 45° to the line of centres at impact, and the coefficient of restitution is e.
See Fig. 1. After impact the direction of motion of the spheres are at right angles.
(a)   Express the velocities of the two spheres after impact in terms of u
(b)   Express m in terms of M
(c)   Find the kinetic energy lost as a result of the impact.

Solution

(a) Express the velocities of the two spheres after impact in terms of u

See Fig. 2. Let v_1, v_2 be the components of velocity along the line of centres, $C_1 C_2$, for Spheres 1, 2 respectively. From the Principle of Conservation of Linear Momentum:

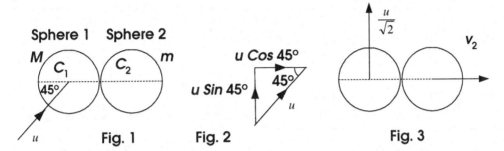

Sphere 1   Sphere 2

M   m   u Cos 45°

Fig. 1   Fig. 2   Fig. 3

Note: The components of velocity $\perp C_1 C_2$ are not altered by the impact.

$$Mv_1 + mv_2 = Mu\,Cos\theta + m(0) \quad \text{But}: \theta = 45°(\text{given}) \Rightarrow Mv_1 + mv_2 = \frac{Mu}{\sqrt{2}} \quad \text{(i)}$$

From Newton's Experimental Law we have :

$$v_1 - v_2 = -e(uCos\theta - 0) \text{ and since } \theta = 45° :\Rightarrow v_1 - v_2 = -\frac{eu}{\sqrt{2}} \quad \text{(ii)}$$

Also, multiplying equation (ii) by m gives : $mv_1 - mv_2 = -\dfrac{meu}{\sqrt{2}}$ (iii)

Add equations (i) and (iii) to eliminate $v_2 \Rightarrow (M+m)v_1 = \dfrac{u}{\sqrt{2}}(M - em)$

$$\Rightarrow v_1 = \frac{u}{\sqrt{2}}\left(\frac{M - em}{M+m}\right) \quad \text{(iv)}$$

Putting this value into equation (ii): $v_2 = v_1 + \dfrac{eu}{\sqrt{2}} = \dfrac{u}{\sqrt{2}}\left(\dfrac{M - em}{M+m}\right) + \dfrac{eu}{\sqrt{2}}$

$$\Rightarrow v_2 = \frac{u}{\sqrt{2}}\left(\frac{M - m}{M+m} + e\right) = \frac{u}{\sqrt{2}}\left(\frac{M + eM}{M+m}\right) \quad \text{(v)}$$

(b) Express m in terms of M

The velocities after impact are at 90° (given). Now the final velocity of Sphere 2 is along line of centres, $C_1 C_2$. (See Fig. 3) Thus, Sphere 1 must have a final velocity with a component $\perp$ line of centres $C_1 C_2$ only.

$\Rightarrow v_1 = $ component of velocity of Sphere 1 along $C_1 C_2 = 0$

$$\Rightarrow v_1 = \frac{u}{\sqrt{2}}\left(\frac{M - em}{M+m}\right) = 0 \Rightarrow m = \frac{1}{e}M \Rightarrow \text{Rewrite equation (v) as}: \ v_2 = \frac{eu}{\sqrt{2}}$$

(c) Find the Kinetic Energy lost as a result of the impact

Initial kinetic energy $= \frac{1}{2}Mu^2$ Final kinetic energy $= M\left(\dfrac{u}{\sqrt{2}}\right)^2 + \frac{1}{2}mv_2^2$

$$= \frac{1}{2}M\left(\frac{u}{\sqrt{2}}\right)^2 + \frac{1}{2}m\left(\frac{eu}{\sqrt{2}}\right)^2 = \frac{1}{4}Mu^2 + \frac{1}{4}\frac{M}{e}u^2e^2$$

$$\Rightarrow \text{Loss in Kinetic Energy} = \frac{1}{2}Mu^2 - \frac{1}{4}Mu^2 - \frac{1}{4}Mu^2e = \frac{1}{4}Mu^2(1-e)$$

27. Oblique impact with stationary sphere

Sphere A of mass m kg travelling at u m/s collides obliquely with stationary sphere B of equal mass. (See Fig. 1). The coefficient of restitution, $e = 1$.
(a)   Show that the paths of the spheres after collision are at right angles.
(b)   Prove that there is no loss in kinetic energy.

Solution

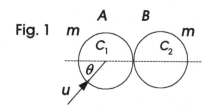
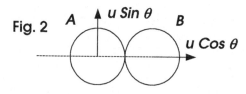

(a) Show that the paths of the spheres after the collision are at right angles

Let sphere A have a velocity u at angle θ to line joining the centres C_1C_2.
The components of u are: $u\cos\theta$ along C_1C_2 and $u\sin\theta \perp$ to C_1C_2
Let v_1 and v_2 be the components of velocity along the line of centres, C_1C_2, for Spheres A and B respectively.

Principle of Conservation of Linear Momentum:
$mv_1 + mv_2 = mu\cos\theta + m(0) \Rightarrow v_1 + v_2 = u\cos\theta$ (i)
From Newton's Experimental Law :
$v_1 - v_2 = -e(u\cos\theta - 0) = -(1)(u\cos\theta - 0) \Rightarrow v_1 - v_2 = -u\cos\theta$ (ii)

Adding equations (i) and (ii) gives: $2v_1 = 0 \Rightarrow v_1 = 0 \Rightarrow v_2 = u\cos\theta$
The velocity components after collision are:
A: along C_1C_2: 0; $\perp C_1C_2$: $u\sin\theta$ B: along C_1C_2: $u\cos\square$; $\perp C_1C_2$: $u\sin\theta$
$\Rightarrow$ after impact the final velocities of Spheres A and B make a right angle to each other.

(b) Prove that there is no loss in kinetic energy (K.E.)

Initial K.E. $= \frac{1}{2}mu^2$; Final K.E. $= \frac{1}{2}m(u\sin\theta)^2 + \frac{1}{2}m(u\cos\theta)^2 = \frac{1}{2}mu^2 \Rightarrow$ no loss in K.E.

28. Two spheres in oblique impact

A sphere of mass 1 kg moving at $\sqrt{34}$ m/s collides with a sphere of mass 2 kg moving at 4 m/s. The directions of motion of the spheres make angles of $\theta = \text{Tan}^{-1}(3/5)$ and $\alpha = \text{Tan}^{-1}(1)$ respectively, with the line of centres, both angles being measured in the same sense. The coefficient of restitution is 0.8.
(a)   Find the spheres' speeds and directions of motion after impact and

(b)   Calculate the kinetic energy lost in the collision.

Solution

The directions of the spheres before impact are shown in Fig. 1:

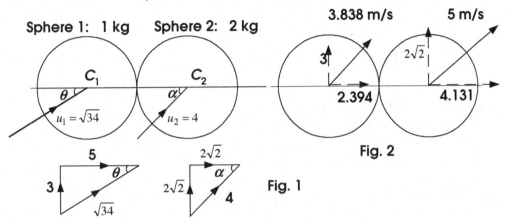

(a) Find the spheres' speeds and directions of motion after impact

From the Principle of Conservation of Linear Momentum:

$m_1v_1 + m_2v_2 = m_1u_1Cos\theta + m_2u_2Cos\alpha$

But : $u_1Cos\theta = \sqrt{34}\times\dfrac{5}{\sqrt{34}} = 5$ and $u_2Cos\alpha = 4\times\dfrac{1}{\sqrt{2}} = 2\sqrt{2}$

$\Rightarrow v_1 + 2v_2 = 1\times5 + 2\times2\sqrt{2} = 5 + 4\sqrt{2} \Rightarrow v_1 + 2v_2 = 5 + 4\sqrt{2}$ (i)

From Newton's Experimental Law :

$v_1 - v_2 = -e(u_1Cos\theta - u_2Cos\alpha) = -0.8(5 - 2\sqrt{2}) \Rightarrow v_1 - v_2 = -4 + 1.6\sqrt{2}$ (ii)

Add equations (i) and $2\times$ (ii) to get $:\Rightarrow 3v_1 = -3 + 7.2\sqrt{2} \Rightarrow v_1 = -1 + 2.4\sqrt{2} = 2.394$

$\Rightarrow v_2 = v_1 + 4 - 1.6\sqrt{2} \Rightarrow v_2 = 4.131$

The resultant velocities and directions of motion of the spheres after impact are shown in Fig. 2 (note: components of velocities $\perp C_1C_2$ are unaltered by the impact).

(b) Calculate the kinetic energy lost in the collision.

Initial Kinetic Energy (K.E) $= \frac{1}{2}(1)(\sqrt{34})^2 + \frac{1}{2}(2)(4)^2 = 33$ Joules

Final K.E $= \frac{1}{2}(1)(3.838)^2 + \frac{1}{2}(2)(5.007)^2 = 32.43$ Joules $\Rightarrow$ Loss in K.E $= 0.6$ Joules approx.

29. Oblique impact: two moving spheres

A smooth sphere A collides obliquely with another smooth sphere B of equal mass which is at rest. The direction of motion of A makes angles with the line of centres at impact of α and β before and after impact respectively. (see Fig. 1). If the coefficient of restitution is e, prove that: $Tan\,\beta = \dfrac{2Tan\alpha}{(1-e)}$

Solution

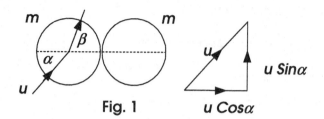

Fig. 1

Let initial velocity of $A = u$. Let initial velocity of $B = 0$.
From Principle of Conservation of Linear Momentum:
$m_1 v_1 + m_2 v_2 = m_1 u_1 + m_2 u_2$ But : $m_1 = m_2 = m$ (say), $u_1 = uCos\alpha$, $u_2 = 0$
$\Rightarrow mv_1 + mv_2 = muCos\alpha + m(0) \Rightarrow v_1 + v_2 = uCos\alpha$ (i)
From Newton's Experimental Law :
$v_1 - v_2 = -e(uCos\alpha - 0) = -euCos\alpha \Rightarrow v_1 - v_2 = -euCos\alpha$ (ii)
Add equations (i) and (ii) : $\Rightarrow 2v_1 = uCos\alpha(1 - e) \Rightarrow v_1 = \dfrac{uCos\alpha(1 - e)}{2}$

But: $Tan\beta = \dfrac{uSin\alpha}{v_1} = \dfrac{uSin\alpha}{\dfrac{uCos\alpha(1 - e)}{2}} \Rightarrow Tan\beta = \dfrac{2 Tan\alpha}{(1 - e)}$

30. Two spheres, oblique impact

Two smooth spheres P and Q of mass $2m$ kg and m kg respectively, collide obliquely
and the coefficient of restitution for the collision is 0.25. The velocity of P before impact
is $3v\,i + 5v\,j$ and the velocity of Q before impact is $-4v\,i + 3v\,j$ where i points along the
line of the centres at impact, $C_1 C_2$. Find:
(a)   The velocities of the spheres after impact
(b)   The loss in kinetic energy.

Solution
(a)   **The velocities of the spheres after impact**
The directions of velocities are shown in Fig. 1:
Initially:
For P: $v_p = 3v\,i + 5v\,j \Rightarrow$ Velocity along $C_1 C_2$: $3v$, Velocity $\perp C_1 C_2$: $5v$
For Q: $v_q = -4v\,i + 3v\,j \Rightarrow$ Velocity along $C_1 C_2$: $-4v$, Velocity $\perp C_1 C_2$: $3v$

After impact:
Let: $v_p = v_{hp} + v_{vp}$ (i) where:
v_{hp} = Resultant velocity component of P along $C_1 C_2$
$v_{vp} = 5v$, i.e.: resultant velocity component of P perpendicular to $C_1 C_2$ is unchanged.

Solution

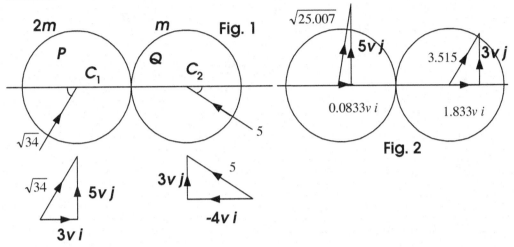

Let: $v_q = v_{hq} + v_{vq}$ (ii) where:
v_{hq} = Resultant velocity component of Q along C_1C_2 (i.e. "horizontal" component).
$v_{vq} = 3v$ = resultant velocity component of Q perpendicular to C_1C_2 is unchanged.

From the Principle of Conservation of Linear Momentum:
(Mass P)v_{hp} + (Mass Q)v_{hq} = (Mass P)$3v$ + (Mass Q)($-4v$)
$\Rightarrow 2m(v_{hp}) + m(v_{hq}) = 2m(3v) + m(-4v) = 0 \Rightarrow 2v_{hp} + v_{hq} = 2v$ (iii)

From Newton's Experimental Law: $v_{hp} - v_{hq} = -e(3v - (-4v)) \Rightarrow v_{hp} - v_{hq} = -0.25 (7v)$ (iv)
Adding equations (iii) and (iv) $\Rightarrow 3v_{hp} = -0.25 v \Rightarrow v_{hp} = 0.0833v \Rightarrow v_{hq} = 1.833v$

The directions of the final velocities are shown in Fig. 2.

(b) The loss in kinetic energy (K.E.)

Initial K.E. $= \frac{1}{2}m_p v_p^2 + \frac{1}{2}m_q v_q^2 = \frac{1}{2}(2m)\left((3v)^2 + (5v)^2\right) + \frac{1}{2}(m)\left((3v)^2 + (-4v)^2\right)$

$= mv^2(34) + mv^2\left(\frac{25}{2}\right) = mv^2\left(\frac{93}{2}\right) = \frac{1}{2}mv^2(93) = 46.5mv^2$ Joules

Final K.E. $= \frac{1}{2}m_p v_p^2 + \frac{1}{2}m_q v_q^2 = \frac{1}{2}(2m)\left((0.0833v)^2 + (5v)^2\right) + \frac{1}{2}(m)\left((1.833v)^2 + (3v)^2\right)$

$= \frac{1}{2}mv^2(62.37) = 31.187$ Joules

$\Rightarrow$ Loss in K.E. = Initial K.E - Final K.E. $= mv^2(46.5 - 31.187) = 15.31mv^2$ Joules

31. Oblique impact: two moving spheres

Spheres A and B of equal size and masses 3 kg and 1 kg and velocities $4i + 3j$ m/s and $2i$ m/s respectively collide on a smooth horizontal surface. If $e = 1$ find the spheres' velocities immediately after the collision.

Solution

Given: $m_1 = 3$ kg, $m_2 = 1$ kg, $u_1 = 5$ m/s, $u_2 = 2$ m/s
The directions of motion of the spheres before impact are given in Fig. 1:
$u_1 = 5$ m/s, $u_1 \cos\theta = 4i$, $u_1 \sin\theta = 3j$ $u_2 = 2$ m/s, $u_2 \cos\theta = 2i$, $u_2 \sin\theta = 0j$
From the Principle of Conservation of Linear Momentum:

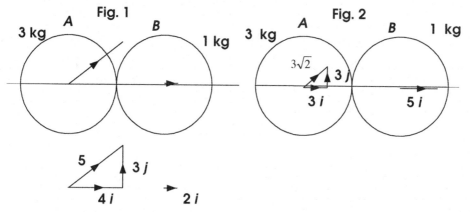

$m_1 v_1 + m_2 v_2 = m_1 u_1 Cos\theta + m_2 u_2 Cos\alpha$ (Note : $\alpha = 0$)

$\Rightarrow 3(v_1) + 1(v_2) = (3)(4) + (1)(2) = 14 \Rightarrow 3v_1 + v_2 = 14$ (i)

From Newton's Experimental Law : $v_1 - v_2 = -e(u_1 Cos\theta + u_2 Cos\alpha) \Rightarrow v_1 - v_2 = -1(4-2) = -2$ (ii)

Add equations (i) and (ii) $\Rightarrow 4v_1 = 12 \Rightarrow v_1 = 3m/s \Rightarrow v_2 = 5m/s$

Velocities of A, B after impact (see Fig.2): $v_A = \sqrt{3^2 + (3)^2} = 4.243$ m/s, $v_B = 5m/s$

32. Two smooth spheres, oblique collision

A smooth sphere of mass $2m$ and velocity u collides with another sphere of mass m kg which is at rest as shown in Fig. 1. If the coefficient of restitution is 0.5 find the spheres' velocities after the collision. Find the angle, α, made by Sphere 1's final velocity with the line of centres, $C_1 C_2$.

Solution

The components of u before impact are shown in Fig. 1:

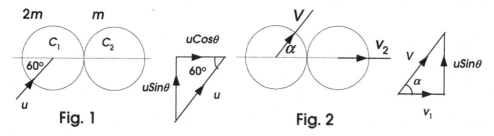

Along line joining $C_1 C_2 : uCos\theta = \dfrac{u}{2}$

Perpendicular to line $C_1 C_2 : uSin\theta = \dfrac{u\sqrt{3}}{2}$ This is unaltered by the impact.

From the Principle of Conservation of Linear Momentum:
If v_1 and v_2 are the components of velocities along $C_1 C_2$ after impact then:

$2mv_1 + mv_2 = m\dfrac{u}{2} - m(0) \Rightarrow 2v_1 + v_2 = u$ (i)

From Newton's Experimental Law : $v_1 - v_2 = -e(u_1 Cos\theta - (u_2))$

$$v_1 - v_2 = -\frac{1}{2}\left(\frac{u}{2} + 0\right) = -\frac{u}{4} \Rightarrow v_1 - v_2 = -\frac{u}{4} \quad \text{(ii)}$$

Adding equations (i), (ii) to eliminate $v_2 \Rightarrow 3v_1 = \frac{3u}{4} \Rightarrow v_1 = \frac{u}{4}$ (iii) (See Fig. 2)

Also, from equation (ii) : $v_2 = v_1 + \frac{u}{4} = \frac{u}{2}$

Thus, the velocities after impact are (See Fig. 2) :

Sphere 1 : Velocity $v_1 = \frac{u}{4}$ along C_1C_2; Velocity $\frac{u\sqrt{3}}{2} \perp C_1C_2$.

Sphere 2 : Velocity v_2 along C_1C_2 (See Fig. 2). Thus : $\alpha = Tan^{-1}\left(\frac{u\sqrt{3}}{2} \div \frac{u}{4}\right) = Tan^{-1}(2\sqrt{3})$

33. Ball falls onto inclined plane

A ball falls a height of 4 metres from point A to strike a plane which is inclined at 45° to the horizontal at point B (See Fig. 1). The coefficient of restitution is: $e = 0.8$. Find the distance from the bottom of the inclined plane at D to E the point at which the ball rebounds.

Solution

The speed of the ball at B, v, is, from equation : $v^2 = u^2 + 2as$ where :
u = initial velocity = 0 s = distance travelled = 4   metres
a = acceleration, $g \Rightarrow v = \sqrt{8g} = 2\sqrt{2g}$ m/s

Consider a ball striking a horizontal plane at 45°. See Fig. 2. The components of the velocity can be split into:

A component perpendicular to plane (heading into the plane) = $v_{PER.}$ = $- v Cos\,45° = -$

$2\frac{\sqrt{2g}}{\sqrt{2}} = -2\sqrt{g}$ (Note: " - " sign indicates direction of component)

A component parallel to plane = v_{PLANE} = $v Sin\,45° = \frac{2\sqrt{2g}}{\sqrt{2}} = 2\sqrt{g}$

The ball will rebound with speed v_f whose components are (See Fig. 3):

A component perpendicular to plane (heading out of the plane = $e(v_{PER.}) = 1.6\sqrt{g}$

A component parallel to plane = v_{PLANE} = $v Sin\,45° = \frac{2\sqrt{2g}}{\sqrt{2}} = 2\sqrt{g}$

Therefore : $v_f = \sqrt{(v_{PER.})^2 + (v_{PLANE})^2} \Rightarrow v_f = \sqrt{(1.6\sqrt{g})^2 + (2\sqrt{g})^2} = 8.02$ m/s

and : $Tan\theta = \frac{v_{PER}}{v_{PLANE}} = \frac{1.6\sqrt{g}}{2\sqrt{g}} \Rightarrow \theta = 38.66° \Rightarrow \alpha = -6.34°$

Using this information, it can be seen that the ball will rebound from the inclined plane as a projectile with components of velocity as shown in Fig. 3.
But: (given) C is 1.5 metres vertically below B, so that the ball will reach E in the time it takes to travel vertically downwards the distance BC:

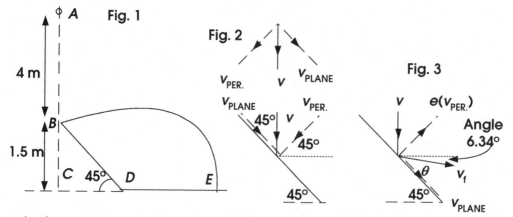

If : $|BC|$ = 1.5 metres and initial velocity in downward direction = u,

where : $u = 8.02 \, Sin(6.34°) = 0.886$ m/s $\Rightarrow v^2 = u^2 + 2gs = (0.887)^2 + (2)(9.81)(1.5) = 30.2$
$\Rightarrow v = 5.5$ m/s = velocity in downward direction after falling 1.5 metres

But : $v = u + gt \Rightarrow t = \dfrac{v - u}{g} = \dfrac{5.5 - 0.886}{9.81} = 0.47$ (between striking B and E)

Horizontal distance travelled in 0.47 seconds : CE = horizontal velocity × time
Horizontal velocity = $8.02 \, Cos(6.34°) = 7.97$ m/s $\Rightarrow |CE| = 7.97 × 0.471 = 3.75$ metres
But : CD = 1.5 metres $\Rightarrow DE = 3.75 - 1.5 = 2.25$ metres

34. Ball within cylinder

An open-ended steel cylinder is placed on a horizontal table with its cylindrical axis in a vertical plane. Its horizontal diameter = line AC. An elastic ball is projected from point A on the inside of a steel cylinder of radius r with initial velocity u m/s at an angle θ to AC. It strikes the cylinder wall at point B and rebounds to strike point C (See Figs 1, 2, looking vertically down). Find the magnitude of coefficient of restitution, e.

Solution

(Draw OB where O = centre of cylinder and $|OB|$ = the radius.
Angle $\angle CAB = \theta$ (given) $\Rightarrow \angle ABO = \theta$ Also, angle $\angle CBO = \angle BCO = \alpha$
But from Fig. 2: $2\theta + 2\alpha = 180°$ $\Rightarrow \theta + \alpha = 90°$ Draw: $BD \perp OB \Rightarrow \angle CBD = \theta$

Before impact at point B the ball travels towards B and its velocity has components (See Fig. 3):
(i) Perpendicular to plane BD: $u_{PER} = u \, Cos\theta$ along OB (towards B)
(ii) Parallel to plane BD: $u_{PLANE} = u \, Sin\theta$ along BD heading towards D

After impact the ball travels towards C with velocity components (See Fig. 4):
(i) Perpendicular to plane BD: $e \, u_{PER} = eu \, Cos\theta$ (towards O along OB). Note: direction is reversed.
(ii) Parallel to plane BD: $u_{PLANE} = u \, Sin\theta$ along BD heading towards D
But the resulting direction of motion must pass through C

$$\Rightarrow Tan \angle CBD = Tan\theta = \frac{euCos\theta}{uSin\theta} = \frac{e}{Tan\theta} \Rightarrow e = Tan^2\theta$$

94

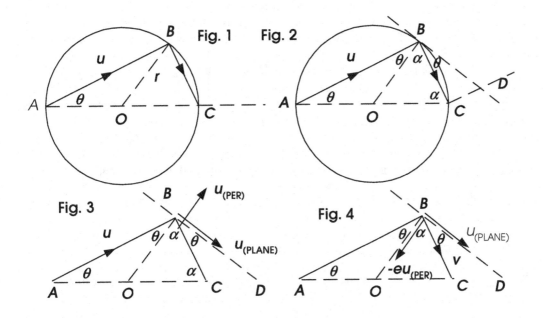

Fig. 1   **Fig. 2**

Fig. 3   **Fig. 4**

35. Ball striking a wall

A ball is thrown from point A with a speed of 19.62 m/s at an angle of 45°. It strikes a vertical wall 9.81 metres away at point B, rebounds and strikes the ground at point C. If the coefficient of restitution between the ball and the wall, $e = \frac{1}{3}$, find the distance between A and C.

Solution

The ball will behave as a projectile launched at $\theta = 45°$ to the horizontal $\Rightarrow$ after travelling a horizontal distance of 9.81 metres the following will apply:

(i) Distance travelled horizontally:

$$x = 9.81 = ut\,Cos\,\theta \Rightarrow t = \frac{9.81}{(19.62)\left(\dfrac{1}{\sqrt{2}}\right)} = \frac{1}{\sqrt{2}} \quad \text{seconds}$$

(ii) Vertical height reached in this time, t:

$$= y = utSin\theta - \tfrac{1}{2}gt^2 = (19.62)\left(\frac{1}{\sqrt{2}}\right)\left(\frac{1}{\sqrt{2}}\right) - (9.81)\left(\frac{1}{2}\right)\left(\frac{1}{\sqrt{2}}\right)^2 = (9.81)\left(\frac{3}{4}\right)$$

At time of impact:

(iii) Horizontal velocity: $v_x = uCos\theta = 19.62\left(\dfrac{1}{\sqrt{2}}\right) = 9.81\sqrt{2} \quad$ m/s

(iv) Vertical velocity: $v_y = uSin\theta - gt = 19.62\left(\dfrac{1}{\sqrt{2}}\right) - (9.81)\left(\dfrac{1}{\sqrt{2}}\right) = \dfrac{9.81}{\sqrt{2}} \quad$ m/s

The direction of the ball after impact are shown in Fig. 1. The components of the velocity at impact are shown in Fig. 2.

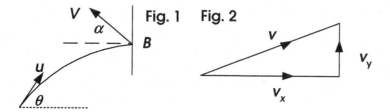

Fig. 1 Fig. 2

Let the ball rebound from the wall with velocity V at angle α to the horizontal:

$$V\cos\alpha = e\,v_x = e9.81\sqrt{2}; \quad V\sin\alpha = v_y = \frac{9.81}{\sqrt{2}}; \quad Tan\alpha = \frac{V\sin\alpha}{V\cos\alpha} = \frac{1}{2e} = \frac{3}{2}$$

$$\Rightarrow V = \sqrt{\left(e9.81\sqrt{2}\right)^2 + \left(\frac{9.81}{\sqrt{2}}\right)^2} = 9.81\sqrt{\frac{2}{9}+\frac{1}{2}} = 9.81\sqrt{\frac{13}{18}}$$

Now, consider a particle projected with velocity V at angle α to the horizontal from point B, where say, $(x, y) = (0,0)$, and hitting the ground at a point C with co-ordinates $\left(x, -\left(\frac{3}{4}\right)(9.81)\right)$. Use the following equation (See Chapter 3):

$$y = x Tan\alpha - \frac{gx^2}{2V^2 \cos^2\alpha} = x Tan\alpha - \frac{gx^2}{2V^2}\left(\sec^2\alpha\right) = x Tan\alpha - \frac{gx^2}{2V^2}\left(1 + Tan^2\alpha\right)$$

$$\Rightarrow -\left(\frac{3}{4}\right)(9.81) = \left(x\left(\frac{3}{2}\right)\right) - \frac{(9.81)x^2\left(1+\left(\frac{3}{2}\right)^2\right)}{2(9.81)^2\left(\frac{13}{18}\right)}$$

$$\Rightarrow \frac{(9.81)x^2\left(\frac{13}{4}\right)}{2(9.81)^2\left(\frac{13}{18}\right)} - x\left(\frac{3}{2}\right) - \left(\frac{3}{4}\right)(9.81) = 0 \Rightarrow \frac{9x^2}{4(9.81)} - x\left(\frac{3}{2}\right) - \left(\frac{3}{4}\right)(9.81) = 0$$

$$\Rightarrow 9x^2 - 6(9.81)x - 3(9.81)^2 = 0 \Rightarrow 3x^2 - 2(9.81)x - (9.81)^2 = 0$$

Solving this equation : $x = 9.81$ metres (For $x > 0$)

$\Rightarrow$ Point C is located at Point $A \Rightarrow$ distance $|AC| = 0$

36. Particle falls onto inclined plane

A particle is dropped from a height of h metres on to a smooth plane inclined at an angle α (= 30°) to the horizontal. The coefficient of restitution, $e = 1$. If, after the impact, the particle has a velocity with a positive vertical component, find how far down the plane its next point of impact (see Fig. 1)

Solution

After falling a distance h, the particle speed is: $v^2 = 0 + 2gh \Rightarrow v = \sqrt{2gh}$ (i)

See Fig. 2. The components of the velocity at A can be split into:

A component perpendicular to plane = $v_{PER.}$ = $-v\cos\alpha = -\sqrt{\dfrac{3gh}{2}}$

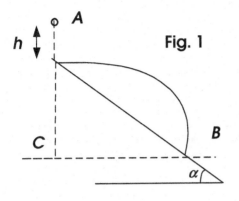

Fig. 1 **Fig. 2**

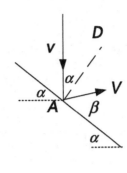

(Note: " - " sign indicates direction – in this case, into the plane)

A component parallel to plane = $v_{PLANE} = v\sin\alpha = \sqrt{\dfrac{gh}{2}}$

Consider the velocity of the particle after impact. See Fig. 2.
Let AD be $\perp$ to plane AB. Let the particle rebound with velocity V at an angle β to the plane; The components of V are:

Perpendicular to plane (out of the plane) = $-e(v_{PER.})$ = $-e\left(-\sqrt{\dfrac{3gh}{2}}\right) = \sqrt{\dfrac{3gh}{2}}$

Parallel to plane = $v_{PLANE} = \sqrt{\dfrac{gh}{2}}$

$\Rightarrow Tan\beta = \dfrac{\sqrt{\dfrac{3gh}{2}}}{\sqrt{\dfrac{gh}{2}}} = \sqrt{3}$ and $Tan\alpha = \dfrac{1}{\sqrt{3}} \Rightarrow \beta = 60°, \alpha = 30°$

Now, consider a particle projected with velocity V at angle $(60° - 30°) = 30°$ to the horizontal from point $(0,0)$, and hitting a point B with, say, co-ordinates $(x, y) = (S\sqrt{3}, -S)$ on the inclined plane AB. Note: $|AC| = S$. From Fig. 1:

$Sin30° = \dfrac{|AC|}{|AB|} = 0.5$ But $|AB|$ = Range $\Rightarrow$ Range = $2|AC|$ = $2S$, (say)

Use the following equation:

$y = xTan\alpha - \dfrac{gx^2}{2V^2Cos^2\alpha} \Rightarrow -S = S\sqrt{3}\dfrac{1}{\sqrt{3}} - \dfrac{g3S^2 4}{2(2gh)3} \Rightarrow -S = S - \dfrac{S^2}{h} \Rightarrow \dfrac{S^2}{h} = 2S \Rightarrow S = 2h$

$\Rightarrow$ Range = $2S = 4h$ = distance down plane to next point of impact

Chapter 6 The Laws of Motion

1. Resistance to a boat's motion

After its engine is switched off, a boat of mass m kg slows at a constant rate from a speed of u m/s to rest in a distance of s metres. Find the resistance to the boat's motion.

Solution

Using the standard equation: $F = ma$ (Force = mass × acceleration), gives:
$F = -R$, where R = Resistance; The "-" sign indicates that the boat is slowing.

$$\Rightarrow ma = -R \Rightarrow a = -\frac{R}{m}$$

Also, use the standard equation for linear motion: $v^2 = u^2 + 2as$ where $v = 0$

$$\Rightarrow 0 = u^2 + 2\left(-\frac{R}{m}\right)s \Rightarrow R = \frac{mu^2}{2s} \quad \text{Newtons}$$

2. Force acting on a mass

Find the magnitude of a force F Newtons which acts on a 1 kg mass, accelerating it from 2 m/s to 5 m/s in 4 seconds.

Solution

$v = u + at \Rightarrow 5 = 2 + a(4) \Rightarrow a = 0.75$ m/s²; $F = ma \Rightarrow F = 1(0.75) = 0.75$ Newtons

3. Multiple forces acting on a mass

Four separate forces act on a body of mass 10 kg (Fig. 1). If the body is accelerated vertically downwards at 2 m/s² as a result find forces F and P.

Fig. 1

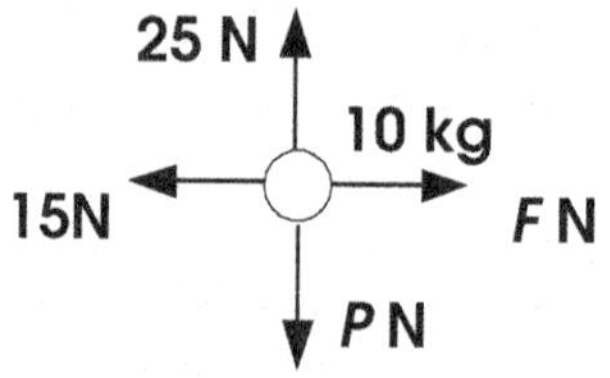

Solution

Given: no horizontal motion $\Rightarrow F - 15 = 0 \Rightarrow F = 15$ Newtons
Given: Net vertical force downwards $= P - 25 =$ mass x acceleration
$\Rightarrow P - 25 = (10)(2) = 20 \Rightarrow P = 45$ Newtons

4. Force exerted by car engine

Find the minimum force necessary to accelerate a car of mass 600 kg uniformly from 0 – 20 m/s while travelling over a distance of 40 metres.

Solution

Using general equation: $v^2 = u^2 + 2as \Rightarrow 20^2 = 0^2 + 2(a)(40) \Rightarrow a = 5$ m/s²
Using equation: $F = ma \Rightarrow F = ma = (600)(5) = 3,000$ Newtons

5. Braking force to stop a train

A braking force of 50 kN slows a 70 Tonnes train from 30 m/s to 0 m/s. How far will the train travel before stopping.

Solution

Using equation: $F = ma \Rightarrow -50,000 = 70,000(a) \Rightarrow a = -5/7$ m/s²

Using equation: $v^2 = u^2 + 2as \Rightarrow 0^2 = 30^2 + 2(-5/7)(s) \Rightarrow s = 630$ metres

6. Lifting force
Find the force required to lift a 150 kg steel beam up a lift shaft with an acceleration of 0.25 m/s².

Solution
The sketch shows the forces acting on the steel beam:

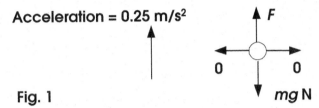

Acceleration = 0.25 m/s²

F

0   **0**

mg N

Fig. 1

The forces in Newtons acting on the mass are:
Horizontally: $\sum F_{H\text{-}RES} = 0$ Vertically: $\sum F_{V\text{-}RES} = F - mg$
$\Rightarrow F - mg = ma \Rightarrow F = mg + ma = 150(9.81) + 150(0.25) = 1,509$ Newtons

7. Acceleration of a mass
A mass of 1000 kg is suspended on the end of a steel cable. If the mass can be accelerated form 0 m/s to 3 m/s over a distance of 8 metres find the force exerted by the cable (= tension T) when the mass is (a) ascending (b) descending.

Solution

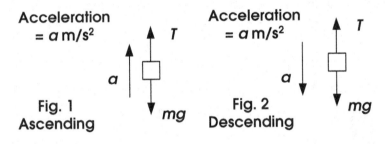

Acceleration = a m/s² T a **mg**
Fig. 1 Ascending

Acceleration = a m/s² T a **mg**
Fig. 2 Descending

Ascending
Use equation: $v^2 = u^2 + 2as \Rightarrow 3^2 = 0^2 + 2(a)(8) \Rightarrow a = 0.5625$ m/s²
Use equation: $F = ma$ Vertically: $\sum F_{V\text{-}RES} = T - mg$ Net force, $F = \sum F_{V\text{-}RES}$
$\Rightarrow T - mg = ma \Rightarrow T = mg + ma = 1000(9.81) + 1000(0.5625) = 10,373$ Newtons

Descending
Use equation: $v^2 = u^2 + 2as \Rightarrow 3^2 = 0^2 + 2(a)(8) \Rightarrow a = 0.5625$ m/s²
Use equation: $F = ma$ Vertically: $\sum F_{V\text{-}RES} = mg - T$ Net force, $F = \sum F_{V\text{-}RES}$
$\Rightarrow mg - T = ma \Rightarrow T = mg - ma = 1000(9.81) - 1000(0.5625) = 9,248$ Newtons

8. Escalator moving vertically up and down
A man of mass 90 kg enters an escalator. Find the reaction exerted by the floor on the man when the escalator is: (a) Stationary (b) Accelerating down at 2 m/s²

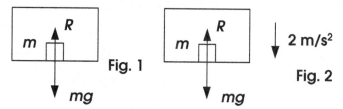

Fig. 1   Fig. 2

(a) Stationary (See Fig. 1)
Vertically: $\sum F_{\text{V-RES}} = mg - R = 0 \Rightarrow R = mg = (90)(9.81) = 882.9\text{Newtons}$
(b) Accelerating downwards at 2 m/s^2
Vertically: $\sum F_{\text{V-RES}} = mg - R = ma = (90)(2) = 180$ Newtons
$\Rightarrow R = mg - 180 = (90)(9.81) - 180 = 702.9$ Newtons

9. Person in lift

A person of mass 80 kg is standing in a lift. Ignoring the mass of the lift itself, find the reaction on the floor of the lift when the lift is:
(a) Ascending with acceleration $a = 5$ m/s^2
(b) Descending with acceleration $a = 5$ m/s^2

Solution
See Fig. 1 for both cases. The forces acting in Newtons are:
Ma = Force accelerating lift, R = Reaction on floor of lift, Mg = Weight of man
(a) Ascending with acceleration $a = 5$ m/s^2
Equation of motion: $Ma = R - Mg \Rightarrow 80a = R - 80g$
$\Rightarrow 80(5) = 400 = R - 80g \Rightarrow R = 400 + 80g = 1,184.8$ Newtons
(b) Descending with acceleration $a = 5$ m/s^2
Equation of motion: $Ma = Mg - R \Rightarrow 80a = 80g - R = 400$ Newtons
$\Rightarrow R = 80g - 400 = 384.8$ Newtons

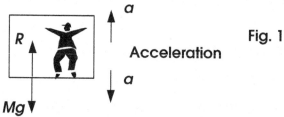

Acceleration   Fig. 1

10. Mass on weighing scales in lift

A mass of 1 kg is placed on a weighing scales which sits in a lift which is being accelerated upwards See Fig. 1. Ignoring the mass of the lift itself, find the acceleration if the reaction, R, between the mass and the weighing scales is 10 Newtons.

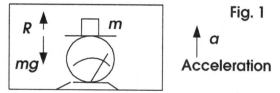

Fig. 1

Acceleration

Solution
Fig. 1 shows the forces in action. The forces acting in Newtons are:
ma = Force accelerating lift upwards
R = Reaction between mass and weighing scales

mg = Weight of mass on weighing scales

The equation of motion is: $ma = R - mg \Rightarrow a = \dfrac{R - mg}{m} = \dfrac{10 - g}{1} = 0.19$ m/s²

11. Mass on spring balance in lift

A mass of m kg with a true weight W_1 is placed on a spring balance and appears to weigh W_2 where $W_2 = 0.9\ W_1$ in a moving lift. Ignoring the mass of the lift itself, find the acceleration of the lift.

Solution

As $W_2 < W_1$, the lift must be accelerating downwards
Fig. 1 shows the forces acting in Newtons:
ma = Force accelerating lift
T = Tension on spring of the spring balance
W = Weight of mass carried on spring balance

The equation of motion is: $mg - T = ma$.   But, $W_1 = mg$ and $T = W_2$

$\Rightarrow W_1 - W_2 = ma \Rightarrow a = \dfrac{W_1 - W_2}{m}$

But : $W_1 = mg \Rightarrow m = \dfrac{W_1}{g} \Rightarrow a = \dfrac{W_1 - W_2}{W_1}g = \dfrac{W_1 - 0.9W_1}{W_1}g = 0.1g$ m/s²

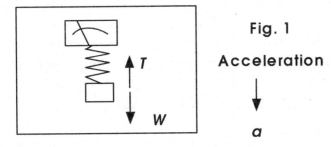

Fig. 1

Acceleration

a

12. Mass on spring balance in lift

A spring balance carrying a mass m kg is placed in a lift. When the lift is accelerated upwards at a m/s² the reading indicates a mass of 10 kg. When the lift is accelerated downwards at a rate $2a$ m/s², the reading indicates a mass of 7 kg. Ignoring the mass of the lift itself, find m and a.

Solution

Fig. 1 shows the forces in action. The forces in Newtons are:
ma = Force accelerating lift, T = Tension on spring of the spring balance
mg = Weight of mass on weighing machine

Travelling up: The equation of motion is: $T - mg = ma \Rightarrow 10g - mg = ma$ (i)
Travelling down: The equation of motion is: $mg - T = m\,(2a)$, giving:

$$mg - 7g = 2ma \quad \text{(ii)}; \quad \text{Adding equations (i)} + \text{(ii) gives}: 3g = 3ma \Rightarrow a = \frac{g}{m}$$

Put into equation (i): $10g - mg = g \Rightarrow mg = 9g \Rightarrow m = 9\,\text{kg}$ and $a = \frac{g}{9}\,\text{m/s}^2$

Fig. 1

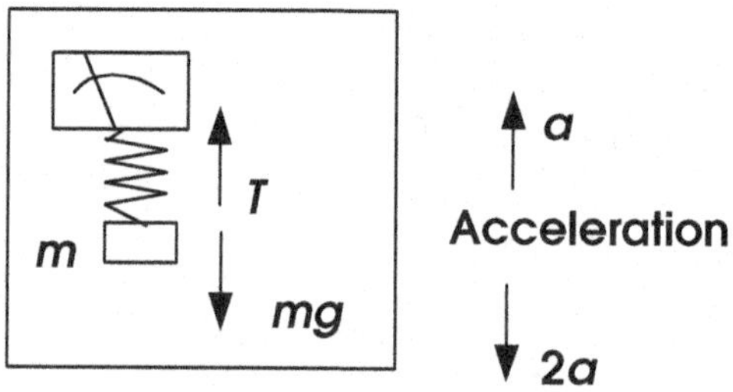

Acceleration

13. Vehicle travelling against resistive force

The engine in a vehicle of mass m = 2,000 kg provides a tractive force T = 3,000 Newtons. There is a constant resistive force R of 2,000 Newtons.

(a)   If it starts from rest on a horizontal surface find its velocity after 100 metres
(b)   If the engine is then switched off how far will the vehicle travel before stopping?

Solution

(a) If it starts from rest find its velocity after 100 metres

Let : a_1 = acceleration in m/s^2, u = initial velocity = 0, v = velocity after 100 metres

Equation of Motion : $ma_1 = T - R \Rightarrow a_1 = \dfrac{T - R}{m} = \dfrac{3,000 - 2,000}{2,000} = 0.5$

$\Rightarrow$ Using : $v^2 = u^2 + 2a_1s \Rightarrow v^2 = 0 + (2)(0.5)(100) \Rightarrow v = 10\quad\text{m/s}$

(b) If the engine is then switched off how far will the vehicle travel before stopping?

Let : a_2 = acceleration in m/s^2, u = initial velocity = 10, v = final velocity = 0, Tractive force = 0 Newtons, Resistive force = R Newtons

Equation of Motion : $ma_2 = T - R = -R \Rightarrow a_2 = \dfrac{-R}{m} = \dfrac{-2,000}{2,000} = -1\quad\text{m/s}^2$

i.e. it is slowing down. Use the equation: $v^2 = u^2 + 2a_2s$

$\Rightarrow 0^2 = 10^2 + 2(-1)s \Rightarrow s = \dfrac{100}{2} = 50\,\text{metres}$

14: Truck travelling up and down inclines

A truck of mass 10 Tonnes travels at constant velocity up an incline rising at 1 in L. The constant tractive and (opposing) resistance forces are 5,000 and 3,000 Newtons respectively. It reaches the top of the incline and begins to travel down a similar slope (See Fig. 1). Find its acceleration down the incline.

Solution

The forces acting (in Newtons) are: N = Force delivered by truck's engine, Q = Resistance force, R = Reaction between truck and horizontal surface
Mg = Weight of truck

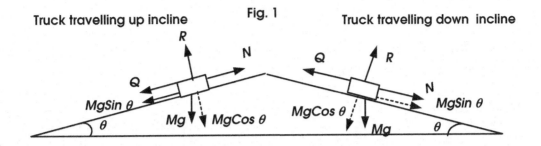

Truck travelling up incline   Fig. 1   **Truck travelling down incline**

Truck travelling up incline
The forces in Newtons acting perpendicular to and parallel to the incline respectively
are: Σ_{PERP} : $R - Mg \cos \theta = 0$ Σ_{PAR} : $N - Q - Mg \sin \theta$
Given: truck travels at constant velocity $\Rightarrow N = Q + Mg \sin \theta$
$\Rightarrow 5,000 = 3,000 + 10,000 g\left(\frac{1}{L}\right) \Rightarrow L = 49.05$

Truck travelling down incline
The forces in Newtons acting perpendicular to and parallel to the incline respectively
are: Σ_{PERP} : $R - Mg \cos \theta = 0$ Σ_{PAR} : $N + Mg \sin \theta - Q$
$= (Q + Mg \sin \theta) + Mg \sin \theta - Q = 2Mg \sin \theta \Rightarrow Ma = 2Mg \sin \theta \Rightarrow a = 2g \sin \theta$
$\Rightarrow a = 2g\left(\frac{1}{L}\right) = 2g\left(\frac{1}{49.05}\right) = 0.4 \quad m/s^2$

15. Runaway train
A train of mass $M = 50$ Tonnes rolls from rest down an incline of slope of 1 in 100 for a
distance of 1,500 metres and then continues rolling along a horizontal track. If it faces a
constant resistance force of $Q = 2,500$ Newtons throughout the motion, how far will the
train travel on the horizontal track before stopping.

Solution
The forces acting on the train on the incline are (Fig. 1):
Perpendicular to incline: Σ_{PERP} : $R - Mg \cos \theta = 0$
Parallel to incline: Σ_{PAR} : $Mg \sin \theta - Q$
Let a_1 = acceleration down the incline $\Rightarrow Ma_1 = Mg \sin \theta - Q$

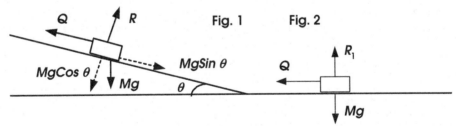

Fig. 1   Fig. 2

$$\Rightarrow a_1 = \frac{Mg\sin\theta - Q}{M} = \frac{50,000g\left(\frac{1}{100}\right) - 2,500}{50,000} = 0.0481 \quad m/s^2$$

Using *general* equation : $v^2 = u^2 + 2as \Rightarrow v^2 = 0^2 + 2(0.0481)L \Rightarrow v = 12 \ m/s$
The forces acting on the train on the horizontal track are (Fig. 2):
Vertical: Σ_{VER} : $R_1 - Mg \cos \theta = 0$ Horizontal: Σ_{HOR} : $- Q \Rightarrow Ma_2 = - Q$
Let a_2 = acceleration on the horizontal track

$$a_2 = -\left(\frac{Q}{M}\right) = -\frac{2,500}{50,000} = -0.05 \quad m/s^2$$

Using general equation : $v^2 = u^2 + 2as \Rightarrow 0^2 = u^2 - 2(0.05)\,s$

where : $u^2 = 144 \Rightarrow 0^2 = 144 - 2(0.05)s \Rightarrow s = 1,440$ metres

16. Work done by force

A force F Newtons is applied to an object of mass m kg and is opposed by a resistive force of 7.5 Newtons. If the force moves the object through a distance of 10 metres find the work done by the force.

Solution

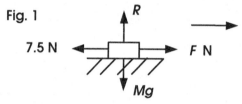

Fig. 1

The forces acting are:

Vertical: $\Sigma_{V\text{-RES}}$: $R - Mg = 0$

Horizontal: $\Sigma_{H\text{-RES}}$: $F - 7.5 \Rightarrow$ applied force must be ≥ 7.5 N to move the object

$\Rightarrow$ Work done $= (7.5)(10) = 75$ J

17. Energy gained by a mass

A force F Newtons acts on a mass m kg changing its velocity from u to v m/s with constant acceleration as it travels a distance s metres. Find the energy gained by m.

Solution

$$F = ma \Rightarrow a = \frac{F}{m} \quad \text{and} : v^2 = u^2 + 2as \Rightarrow v^2 = u^2 + 2\left(\frac{F}{m}\right)s$$

$$\Rightarrow mv^2 = mu^2 + 2Fs \Rightarrow Fs = \frac{1}{2}\left(mv^2 - mu^2\right)$$

= Energy required to accelerate the mass = Energy gained by the mass

18. Lifting a mass

Find the work required to lift a mass of m kg is lifted up a distance of x metres.

Solution

Fig. 1

Force = mg Newtons, Distance = x metres $\Rightarrow$ Work done = Force $\times$ Distance = mgx Joules

19. Cycling a horizontal distance

A cyclist can exert a force of P Newtons. Find the minimum work done to cycle a bicycle of mass 100 kg for 1,000 metres against a resistance $Q = 15$ Newtons. (See Fig. 1)

Solution

The forces acting on the bicycle in Newtons are (Fig. 2): mg = Weight of mass m (Newtons), R = Normal Reaction, Q = Resistance force, P = Force exerted by cyclist.

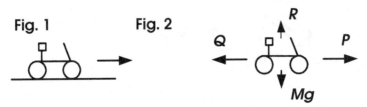

Fig. 1 **Fig. 2**

The forces acting are: Vertically: Σ_{VER}: $R - Mg = 0$ Horizontally: Σ_{HOR}: $P - Q$
For motion to occur the force exerted by the cyclist overcome the resistance $\Rightarrow P \geq Q$
$\Rightarrow P \geq 15$ Newtons. Minimum work input achieved when $P = 15$ Newtons $\Rightarrow$ Work Done = $Px = (15)(1000) = 15,000$ Joules

20. Mass sliding down a smooth slope
A mass of 10 kg at rest at the top of a smooth slope inclined at $Tan^{-1}(0.75)$ to the horizontal begins to slide. Find how far it has slid down the slope when it has a velocity of 8 m/s.

Solution
The energy is conserved: Initial P.E. + K.E. = Final P.E. + K.E.
Initial P.E. = $(10)gh$ Initial K.E. = $\frac{1}{2}(4)v^2 = \frac{1}{2}(4)(0)^2 = 0$
Initial P.E. + K.E. = $10gh$ J

Final P.E. = 0   Final K.E. = $\frac{1}{2}(4)v^2 = \frac{1}{2}(4)(8)^2 = 144$ J
Final P.E. + K.E. = 144 J $\Rightarrow 10gh = 144 \Rightarrow h = 144/10g = 1.47$ metres

But: $Tan\ \theta = \frac{3}{4} \Rightarrow Sin\ \theta = 3/5$
But: $Sin\ \theta = h/L \Rightarrow L = h/Sin\ \theta = 5\ h/3 = 2.45$ metres

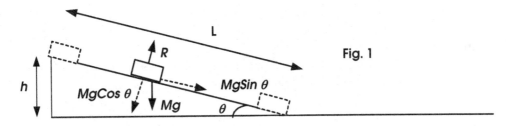

L

R

Fig. 1

MgSin θ

h

MgCos θ Mg

θ

21. Driving up an incline
Find the minimum energy consumed by the engine of a car of mass $m = 2,000$ kg, which is driven $x = 2,000$ metres up a surface inclined at $\square = 30°$ to the horizontal against a resistance $Q = 300$ Newtons.

Solution
The forces (in Newtons) are shown in Fig. 2: mg = Weight of mass m, R = Normal Reaction, Q = Resistance force, P = Applied force exerted by engine

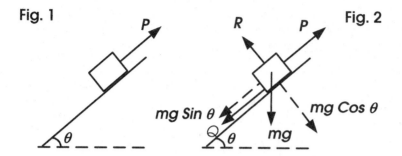

Fig. 1   R   P   Fig. 2

mg Sin θ   mg Cos θ

θ   θ   mg

(a) Derive an expression for the minimum energy consumed

Forces parallel to the surface: $\Sigma_{PAR} = P - Q - mg\,Sin\,\theta$ Equation (i)
Forces perpendicular to surface: $\Sigma_{PERP} = R - mg\,Cos\,\theta = 0$ Equation (ii)
The minimum force required is when: $P - Q - mg\,Sin\,\theta = 0 \Rightarrow P = Q + mg\,Sin\,\theta \Rightarrow$ Minimum
energy consumed (= work done) $= Px = (Q + mg\,Sin\,\theta)\,x$
$= (300 + 2,000g(0.5))\,(2,000) = 2,022,000$ Joules

22: Force in direction of motion

Find the power exerted by a force of 6 Newtons acting over a distance of 10 metres for
a period of 15 seconds.

Solution

Work done by the force = (6)(10) = 60 J
Power exerted = rate of doing work = 60/15 = 4 J/s = 4 Watts

23. Force acting at an angle to the motion

A force of 40 Newtons acting a 60° to the line of motion of an object for a distance of 8
metres over a period of 20 seconds. Find the power exerted by the force.

Solution

Force acting along line of motion = 40 Cos 60° = 20 Newtons
Work done by the force = (20)(8) = 160 J
Power exerted = rate of doing work = 160/20 = 8 J/s = 8 Watts

24. Power of an engine

A car is travelling at a constant speed u = 35 m/s. If the resistance to motion is Q = 700
Newtons find the power being developed by the engine.

Solution

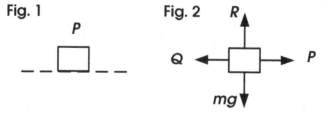

Fig. 1   Fig. 2   R

P   Q   P

mg

The forces are: Vertical: Σ_{VER}: $R - Mg = 0$ Horizontal: Σ_{HOR}: $P - Q$
For motion to occur: $P \geq Q$ Newtons $\Rightarrow P \geq 700$ Newtons. The minimum power
consumption occurs when the force is a minimum i.e. when P = 700 Newtons
$\Rightarrow$ Engine Power = Force × Velocity = Pu Watts = (700)(35) = 24,500 Watts

25: Fire hose

Firemen drop the suction end of a hose into a pond and then pump water to a height of $h = 10$ metres above the pond level at a rate of $k = 10$ litres/s. If the water leaves the nozzle at a velocity of $v = 20$ m/s find the pumping power being used.

Solution

NOTE: water is being lifted from the pond as well as being accelerated so an energy approach will be used:

Energy of water prior to being pumped:

Potential Energy = 0, Kinetic Energy = 0 $\Rightarrow$ Total Energy = 0

Volume of water being pumped = k litres/s = 10 l/s. As 1 litre of water has a mass of 1 kg the mass of water being pumped per second = k kg/s = 10 kg/s.

Energy of water after being pumped:

Potential Energy (applies to k kg/s per second) = kgH Joules/s = kgH Watts = 300g Watts.

Kinetic Energy (applies to k kg/s per second) = ½ kv^2 Joules/s = ½ kv^2 Watts = 2,000 Watts $\Rightarrow$ Total Energy = 4,943 Watts

26. Driving up an incline

A locomotive can pull a train of total mass $M = 100$ Tonnes along a horizontal track at a constant speed of $u = 10$ m/s against a constant resistance $Q = 100,000$ Newtons.

(a) Derive an expression for the power of the locomotive.

(b) If the locomotive, producing the same power, then starts up an incline (at an angle of $\theta = 5°$ to the horizontal), find the maximum speed achieved.

Solution

(a) Derive an expression for the power of the locomotive.

The arrangement is shown in Fig. 1 and the forces are shown in Fig. 2:

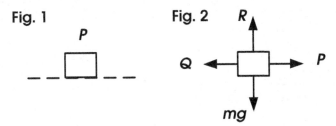

The forces are: Vertical: Σ_{VER} : $R - Mg = 0$ Horizontal: Σ_{HOR} : $P - Q$

For motion to take place the force exerted by the engine must be able to overcome the resistance $\Rightarrow P \geq Q$ Newtons . If the speed is constant then the minimum value of P = Q Newtons = 100,000 Newtons.

Power = Force × Velocity = Qu = (100,000)(10) = 1,000,000 Watts = 1 MW

(b) If the locomotive, producing the same power, then starts up an incline (at an angle of $\theta = 5°$ to the horizontal), find the maximum speed achieved.

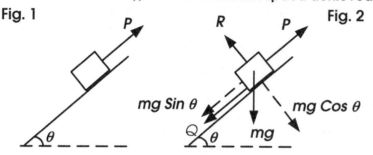

Fig. 1 **Fig. 2**

$mg\,Sin\,\theta$ $mg\,Cos\,\theta$ mg

The forces (in Newtons) are shown in Fig. 2: mg = Weight of mass m, R = Normal Reaction, Q = Resistance force, P = Applied force exerted by engine

Forces parallel to the surface: $\Sigma_{PAR} = P - Q - mg\,Sin\,\theta$ Equation (i)
Forces perpendicular to surface: $\Sigma_{PERP} = R - mg\,Cos\,\theta = 0$ Equation (ii)

When the train is travelling at maximum speed (v m/s, say) $\Rightarrow$ acceleration = 0
$\Rightarrow$ Upward force = Downward force $\Rightarrow P - Q - mg\,Sin\,\theta = 0 \Rightarrow P = Q + mg\,Sin\,\theta$
$\Rightarrow$ Power = Force $\times$ Velocity = $Pv = (Q + mg\,Sin\,\theta)v$
But, from above: Power = Force $\times$ Velocity = Qu Watts. The power output of the locomotive is constant $\Rightarrow (Q + mg\,Sin\,\theta)v = Qu$

$$\Rightarrow v = \frac{Qu}{Q + MgSin\theta} = \frac{(100,000)(10)}{100,000 + (100,000)g(0.0872)} = 5.39 \quad m/s$$

27. Maximum speed down an incline
A truck travels through three stages of a journey – on a horizontal surface, up an incline (at an angle $\theta = 5°$ to the horizontal) and down an incline (at an angle $\theta = 5°$ to the horizontal). See Fig. 1. The power output of its engine and the resistance to motion are constant throughout the journey. If the maximum speeds during the first two stages are $u = 8$ m/s and $v = 5$ m/s respectively derive an expression for the maximum speed w m/s down the incline.

Solution
The forces (in Newtons) are shown in Fig. 2: Mg = Weight of mass m, R = Normal Reaction, Q = Resistance force, N = Applied force exerted by engine

Stage 1: Truck travelling along horizontal surface
The forces acting are: Vertical: Σ_{VER} : $R - Mg = 0$ Horizontal: Σ_{HOR} : $P - Q$
If maximum velocity is u m/s then this velocity occurs when the horizontal acceleration
$= 0 \Rightarrow P - Q = 0 \Rightarrow P = Q$ Newtons
$\Rightarrow$ Power = Force $\times$ Velocity = Pu Watts = Qu Watts Equation (i)
Stage 2: Truck travelling up incline
The forces in Newtons acting perpendicular to, and parallel to, the incline respectively
are: Σ_{PERP} : $R - Mg\,Cos\,\theta$ Σ_{PAR} : $N - Q - Mg\,Sin\,\theta$
Given: If maximum velocity is v m/s then this velocity occurs when the acceleration
parallel to the plane $= 0 \Rightarrow N = Q + Mg\,Sin\,\theta$
Power = Force $\times$ Velocity = Nv Watts = $(Q + Mg\,Sin\,\theta)\,v$ Watts Equation (ii)

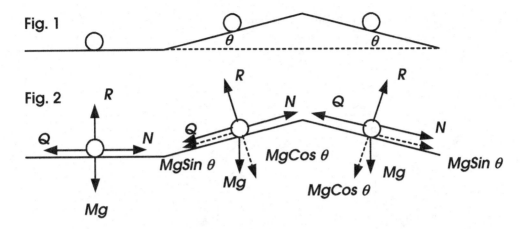

Fig. 1

Fig. 2

MgSin θ   Mg   MgCos θ   Mg   MgCos θ   MgSin θ

Stage 3: Truck travelling down incline

The forces in Newtons acting perpendicular to, and parallel to, the incline respectively
are: Σ_{PERP} : $R - Mg\,Cos\,\theta = 0$ Σ_{PAR} : $N + Mg\,Sin\,\theta - Q$
Given: If maximum velocity is w m/s then this velocity occurs when the acceleration
parallel to the plane $= 0 \Rightarrow N = Q - Mg\,Sin\,\theta$
Power = Force × Velocity = Nv Watts = $(Q - Mg\,Sin\,\theta)\,w$ Watts Equation (iii)

From equation (i): $Q = \dfrac{Power}{u}$

From equation (ii): $mgSin\theta = \dfrac{Power}{v} - Q = \dfrac{Power}{v} - \dfrac{Power}{u}$

From equation (iii):
$$w = mgSin\theta = \dfrac{Power}{Q - mgSin\theta} = \dfrac{Power}{\dfrac{Power}{u} - \left(\dfrac{Power}{v} - \dfrac{Power}{u}\right)} = \dfrac{uv}{2v - u} = \dfrac{40}{2} = 20\ m/s$$

28. Motion on an inclined plane

A mass m kg is projected at initial speed u m/s from point A up a plane inclined at an
angle θ to the horizontal.
(a) If the plane is smooth find the maximum vertical height, h metres, reached before
     the mass comes to rest on the plane at point B
(b) If the plane is rough and has a coefficient of friction μ, find the maximum vertical
     height reached when the mass comes to rest on the plane and the loss in energy
     of the mass at this point.

Solution

Figs. 1,2 show the forces acting on the mass in cases (a),(b) respectively:

(a) The plane is smooth

Let acceleration up the plane = a m/s² ($a < 0$ as mass is decelerating) $\Rightarrow a = -gSin\,\theta$
But: $v^2 = u^2 + 2aS$ (where S = distance travelled along surface of plane)
$\Rightarrow v^2 = u^2 + 2aS = u^2 + 2(-g\,Sin\theta)S$

But final velocity, $v = 0 \Rightarrow 0 = u^2 - 2SgSin\,\theta \Rightarrow S = \dfrac{u^2}{2gSin\theta}$

But: $Sin\,\theta = \dfrac{Maximum\ vertical\ height\ reached}{S} = \dfrac{h}{S} \Rightarrow S = \dfrac{u^2S}{2gh} \Rightarrow h = \dfrac{u^2}{2g}$

(b) The plane is rough

The forces acting on the mass are shown in Fig. 2.

The frictional force $= F = \mu R \Rightarrow$ Equation of motion :

$ma = -F - mgSin\theta = -\mu R - mgSin\theta$ and $R = mgCos\theta$

$\Rightarrow ma = -\mu mgCos\theta - mgSin\theta \Rightarrow a = -\mu gCos\theta - gSin\theta$

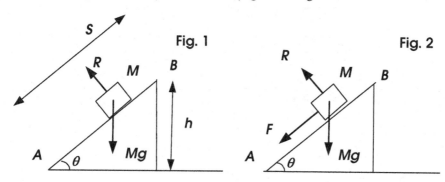

Fig. 1   Fig. 2

But $v^2 = u^2 + 2aS \Rightarrow v^2 = u^2 + 2(-\mu gCos\theta - gSin\theta)S$

When the mass reaches its maximum vertical height, h, then $v = 0$

$$\Rightarrow S = \frac{u^2}{2(\mu gCos\theta + gSin\theta)} \qquad \text{But}: \quad h = S(Sin\theta) \Rightarrow h = \frac{u^2 Sin\theta}{2(\mu gCos\theta + gSin\theta)}$$

Energy at A = Potential Energy at A + Kinetic Energy at A
Assume Potential Energy at A = 0 $\Rightarrow$ Energy at A = Kinetic Energy at A

Energy at B = Potential Energy at B + Kinetic Energy at B
Kinetic Energy at B = 0 (as $v = 0$) $\Rightarrow$ Energy at B = Potential Energy at B
Energy at $A = \frac{1}{2}mu^2$ Joules;   Energy at $B = mgh$ Joules   (See Fig. 2)
Loss of energy $= \frac{1}{2}mu^2 - mgh$ which gives, (substituting for h):

$$\text{Loss of energy} = \frac{1}{2}mu^2\left(\frac{\mu Cos\theta}{\mu Cos\theta + Sin\theta}\right) \quad \text{Joules}$$

29. Motion of two connected masses

Two masses, A and B, each of mass m kg, are connected by a light string of length 8 metres. A is initially held at rest in a position midway between two fixed pegs X and Y and is at the same level as X and Y. B hangs vertically below particle A (See Fig. 1). If A is released find:

(a)   The vertical distance below XY at which the masses collide
(b)   The relative velocity between the masses at the point of collision.

Solution

The dimensions of the system when at rest are shown in Fig. 2. The dimensions at the point of collision are shown in Fig. 3.

(a) Where the masses collide

When the masses collide the length of the strings between the masses and the pegs will be equal $\Rightarrow$ 4 × length = 8 metres $\Rightarrow$ length = 2 metres. (See Fig. 3) It can be seen that the collision will occur at a distance $\sqrt{3}$ metres vertically below A's initial position.

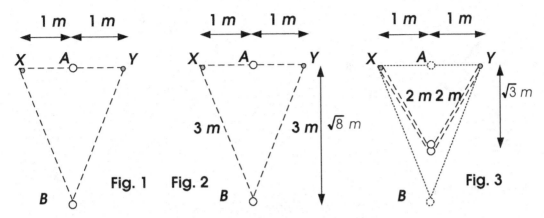

Fig. 1   **Fig. 2**   **Fig. 3**

(b) The relative velocity between the masses at the point of collision.

The total energy of the system is unchanged, so: Initial Energy = Final Energy.

Initial Energy

As the system is at rest there will be no kinetic energy component only a potential energy component. Assume the potential energy of particle B at its initial point is 0. The potential energy of particle A is: $mgh = mg\sqrt{8}$ Joules

Final Energy = Kinetic Energy + Potential Energy

Kinetic Energy

Note: particles are connected so their speeds at the point of collision will be the same. Assume each particle has a velocity of v metres per second at the point of collision. Then the relative velocity of the particles will be $2v$. But the kinetic energy of each particle will be $\frac{1}{2}mv^2 \Rightarrow$ the kinetic energy of the system = $2 \times (\frac{1}{2}mv^2) = mv^2$.

Potential Energy

The potential energy of each particle at the point of collision is: $mg(\sqrt{8} - \sqrt{3})$
$\Rightarrow$ Total potential energy = $2mg(\sqrt{8} - \sqrt{3}) \Rightarrow$ Final Energy = $mv^2 + 2mg(\sqrt{8} - \sqrt{3})$

But: Initial Energy = Final Energy
$\Rightarrow mg\sqrt{8} = mv^2 + 2mg(\sqrt{8} - \sqrt{3}) \Rightarrow g\sqrt{8} = v^2 + 2g(\sqrt{8} - \sqrt{3})$
$\Rightarrow v^2 = 2g\sqrt{3} - g\sqrt{8} = 2g\sqrt{3} - 2g\sqrt{2} \Rightarrow v = \sqrt{2g\sqrt{3} - 2g\sqrt{2}}$
$\Rightarrow$ Relative velocity = $2v = 2\sqrt{2g\sqrt{3} - 2g\sqrt{2}}$ m/s

30. Travelling along a slope

A mass of m kg is projected up a rough slope (inclined at an angle θ to the horizontal) which has a coefficient of friction μ.
(a)   Find its acceleration.
(b)   If the same mass is sliding down the slope, find its acceleration.

Solution

The forces acting on the mass in Newtons are:
Mg = Weight of mass, F = Frictional force, R = Reaction on plane
Fig. 1 shows the forces acting when the mass is travelling up the plane.
Fig. 2 shows the forces acting when the mass is travelling down the plane.

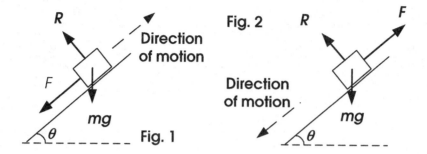

Fig. 2

Direction of motion

Fig. 1

Direction of motion

(a) Find its acceleration

Upward motion: (See Fig. 1) The forces acting are:

Parallel to plane : $ma = -mgSin\theta - F$ Perpendicular to plane : $R = mgCos\theta$

But, $F = \mu R \Rightarrow F = \mu mgCos\theta \Rightarrow ma = -mgSin\theta - \mu mgCos\theta$

$\Rightarrow a = -gSin\theta - \mu gCos\theta$

(b) If the same mass is sliding down the slope, find its acceleration.

Parallel to plane : $ma = mgSin\theta - F$ Perpendicular to plane : $R = mgCos\theta$

But, $F = \mu R = \mu mgCos\theta \Rightarrow ma = mgSin\theta - \mu mgCos\theta \Rightarrow a = gSin\theta - \mu gCos\theta$

31. Travelling along a slope

When a cart is travelling down a rough slope (inclined at an angle θ to the horizontal) it experiences a resistance to its motion of P. If the cart travels up the slope with initial speed u m/s find how far it travels before coming to rest , where: (a) $P = mgSin\theta$, (b) $P = \frac{1}{2}\ mgSin\theta$.

Solution

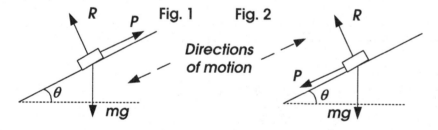

Fig. 1   Fig. 2

Directions of motion

The forces in Newtons are:

R = Reaction between slope and cart, P = Resistive force, mg = Weight of cart

(a) $P = mgSin\theta$

Downward motion: (See Fig. 1)

Parallel to the plane : $ma = mgSin\theta - P = mgSin\theta - mgSin\theta \Rightarrow a = 0$

Upward motion: (See Fig. 2)

Parallel to the plane : $ma = -mgSin\theta - P = -mgSin\theta - mgSin\theta$

$\Rightarrow ma = -2mgSin\theta \Rightarrow a = -2gSin\theta$

Initial speed $= u$, Final speed $= v = 0$ (given)

$\Rightarrow$ using $v^2 = u^2 + 2as$ where : s = distance travelled by cart

before coming to rest $\Rightarrow u^2 = 2(2gSin\theta)s = 4gsSin\theta \Rightarrow s = \dfrac{u^2}{4gSin\theta}$

(b) $P = \frac{1}{2}\ mgSin\theta$

Downward motion:

Parallel to the plane : $ma = mgSin\theta - P = mgSin\theta - \frac{1}{2}mgSin\theta = \frac{1}{2}mgSin\theta$

Upward motion:

Parallel to the plane : $ma = -mgSin\theta - P = -mgSin\theta - \frac{1}{2}mgSin\theta = -\frac{3}{2}mgSin\theta$

$\Rightarrow a = -\frac{3}{2}gSin\theta$

Initial speed $= u$,　Final speed $= v = 0 \Rightarrow$ using : $v^2 = u^2 + 2as$

$\Rightarrow u^2 = 2(\frac{3}{2}gSin\theta)s = 3gsSin\theta \Rightarrow s = \dfrac{u^2}{3gSin\theta}$

32. Single Pulley

Two masses, m_1 and m_2 kg (where $m_1 > m_2$) are connected by a light inextensible string and suspended from a pulley as shown in Fig. 1. Write down the equations of motion of the masses and find the acceleration, a m/s², of the masses and the tension in the string, T Newtons. (Assume no friction).

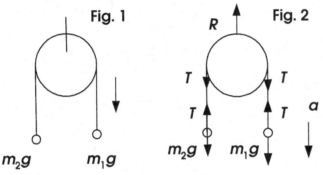

Solution

The forces in Newtons acting on the system are shown in Fig. 2 where:

T = Tension in the string

$R = 2T$ = Force supporting the pulley system

m_1g, m_2g = Weights of masses

a = Acceleration of m_1 and m_2 in m/s²

The equations of motion are:

For m_1 : $m_1g - T = m_1a$　Equation (i).　For m_2 : $T - m_2g = m_2a$　　Equation (ii)

Adding equations (i) and (ii) gives: $m_1g - m_2g = m_1a + m_2a$

Thus: $(m_1 - m_2) \times g = (m_1 + m_2) \times a \Rightarrow a = \dfrac{(m_1 - m_2)}{m_1 + m_2} \times g$　m/s²

But, equation (ii) gives: $T = m_2a + m_2g$

$T = \dfrac{m_2(m_1 - m_2)g}{m_1 + m_2} + m_2g = \dfrac{m_2(m_1 - m_2)g}{m_1 + m_2} + \dfrac{(m_1 + m_2)m_2g}{m_1 + m_2} = \dfrac{2m_1m_2g}{m_1 + m_2}$　Newtons

33. Multiple Pulleys

In the pulley system shown in Fig. 1, pulley A has a mass of M_A kg. Particle C has a mass of M_C kg. Pulley B is fixed in position but free to revolve. Assume there is no friction involved. Write down the equations of motion and find the accelerations of M_A and M_C (if $M_C > M_A$) and the tension, T Newtons, of the cord.

Solution

The forces in Newtons acting on the system are shown in Fig. 2 where:
T = Tension in the string
$R = 2T = T'$ = Force supporting the pulley system
$M_A g$, $M_C g$ = Weights of pulley and particle respectively.

a_A = Acceleration of Pulley A a_C = Acceleration of Particle C
The equations of motion are:
For A: $2T - M_A g = M_A a_A$ (i)   For C: $M_C g - T = M_C a_C$ (ii)
Adding (i) and 2 x (ii): $2M_C g - M_A g = M_A a_A + 2M_C a_C$ (iii)

But the distance moved by pulley A is half that moved by pulley C

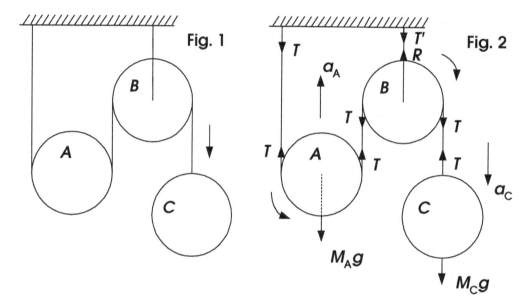

Fig. 1   Fig. 2

$\Rightarrow 2a_A = a_C$ Equation (iii) can be re-written: $(2M_C - M_A) \times g = M_A a_A + 4M_C a_A$

$\Rightarrow a_A = \dfrac{(2M_C - M_A)}{M_A + 4M_C} \times g$ and $a_C = 2\dfrac{(2M_C - M_A)}{M_A + 4M_C} g$ m/s^2

From equation (i): $T = \dfrac{M_A}{2} \times (a_A + g)$

$\Rightarrow T = \dfrac{M_A}{2}\left[\left(\dfrac{(2M_C - M_A)}{M_A + 4M_C}\right)g\right) + g\right] = \dfrac{M_A}{2}\left[\dfrac{(2M_C - M_A)g + (M_A + 4M_C)g}{M_A + 4M_C}\right] = \dfrac{3M_A M_C g}{M_A + 4M_C}$ Newtons

34. Pulley and Mass on smooth table

A pulley system consists of a movable pulley of mass m_2 kg around which passes a cord with one end fixed at point B and the other connected to mass m_1. If the table on which m_1 rests is smooth, find the accelerations of the mass and pulley, A and a m/s^2, respectively, and the tension of the cord, T.

Solution

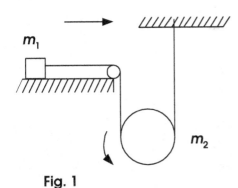

Fig. 1

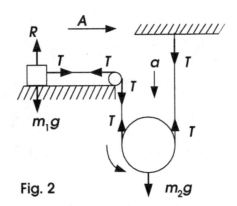

Fig. 2

The forces in Newtons involved are shown Fig. 2:
R = Reaction between mass m_1 and the table
T = Tension in the cord
m_1g, m_2g = Weights of masses
Also: A = Acceleration of mass m_1 m/s² $\quad a$ = Acceleration of pulley m_2 m/s²

Distance travelled by pulley = ½ distance travelled by mass in same time $\Rightarrow A = 2a$ m/s²
The equations of motion are:
For m_1 : $T = m_1A$ (i)   For m_2 : $m_2g - 2T = m_2a$ (ii)
Adding 2 ×equation (i) to equation (ii) gives:
$m_2g = 2m_1A + m_2a = 4m_1a + m_2a \Rightarrow m_2g = a(4m_1 + m_2)$ (iii)

$$\Rightarrow a = \frac{m_2g}{4m_1 + m_2} \text{ m/s}^2, \quad A = \frac{2m_2g}{4m_1 + m_2} \text{ m/s}^2 \Rightarrow T = m_1A = \frac{2m_1m_2g}{4m_1 + m_2} \text{ Newtons}$$

35. Two masses
In the pulley system shown in Fig. 1, mass m_2 kg hangs over a pulley and is connected
by a string to a mass m_1 kg which moves on a table. Find the acceleration, a m/s², and
tension, T Newtons, in the system when the table is:
(a) Smooth (i.e. when there is no friction)
(b) Rough (assume that the frictional force which is acting as m_1 is pulled towards the

edge = F Newtons where the coefficient of friction, $\mu = \dfrac{m_2}{m_1}$)

Solution
The forces in Newtons involved are shown in Fig. 2:
T = Tension in the string, R = Reaction between m_1 and the table, F = Frictional force
m_1g, m_2g = Weights of masses
Also: a = Acceleration of m_1 and m_1 in m/s²

The equations of motion are: For m_1 : $T - F = m_1a$ (i)   For m_2 : $m_2g - T = m_2a$ (ii)

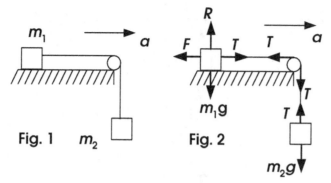

Fig. 1 m_2 Fig. 2

Adding equations (i) and (ii) to get equation (iii):

$$m_2g - F = m_1a + m_2a \quad \text{(iii)} \quad \Rightarrow m_2g - F = a(m_1 + m_2) \Rightarrow a = \frac{m_2g - F}{m_1 + m_2}$$

But $F = \mu R$ where μ = coefficient of friction

$$\Rightarrow F = \mu m_1g \Rightarrow \text{Acceleration in the system, } a = \frac{m_2g - \mu m_1g}{m_1 + m_2} = g\frac{(m_2 - \mu m_1)}{m_1 + m_2} \quad \text{m/s}^2 \quad \text{(iv)}$$

From equation (i): Tension in the system, T, is:

$$T = m_1a + F = m_1g\frac{(m_2 - \mu m_1)}{m_1 + m_2} + \mu m_1g = \frac{m_1m_2g}{m_1 + m_2}(1 + \mu) \quad \text{Newtons} \quad \text{(v)}$$

For the two cases specified in the question:

(a) When there is no friction:
Using equations (iv) and (v):

$$F = 0 \text{ then } \mu = 0 \Rightarrow a = \frac{m_2g}{m_1 + m_2} \quad \text{m/s}^2 \text{ and } T = \frac{m_1m_2g}{m_1 + m_2} \quad \text{Newtons}$$

(b) When the coefficient of friction, $\mu = \dfrac{m_2}{m_1}$

$$\text{Acceleration, } a = \frac{m_2g - \mu m_1g}{m_1 + m_2} \quad \text{If } \mu = \frac{m_2}{m_1} \Rightarrow a = g\frac{(m_1m_2 - m_2m_1)}{m_1 + m_2} = 0 \quad \text{m/s}^2$$

$$\text{Also, using equation (v): } T = \frac{m_1m_2g}{(m_1 + m_2)}\left(1 + \frac{m_2}{m_1}\right) = m_2g \quad \text{Newtons}$$

36. Three masses and rough table
A particle of mass $2M$ kg resting on a horizontal table is attached by two inelastic strings to particles of masses $3M$ and M kg hanging over smooth light pulleys at opposite edges of the table (See Fig. 1). When the system is released from rest, find the system's acceleration, a m/s^2, tensions in the strings and the distance, s, fallen in one second, where the table is:
(a) Rough and the coefficient of friction between particle and table is $\mu = \frac{1}{4}$
(b) Smooth i.e. there is no friction

Solution
The forces in Newtons acting in the system are: R = Reaction between $2M$ particle and the table, F = Frictional force, T_1, T_2 = Tensions in the strings, Mg, etc = Weights of particles

Also: a = Acceleration of the particles in m/s^2

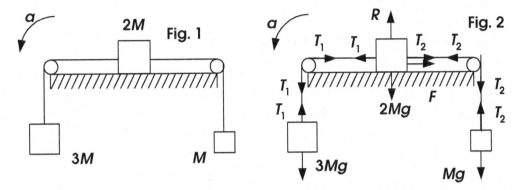

Fig. 1

Fig. 2

(a) Coefficient of friction between particle and table is $\mu = \frac{1}{4}$

Fig. 2 shows the forces acting on the particles. The equations of motion are:

For $3M$: $3Mg - T_1 = 3Ma$ (i)   For M : $T_2 - Mg = Ma$ (ii)

For $2M$: $T_1 - T_2 - F = 2Ma$ (iii)

But $F = \mu R$ and $R = 2Mg \Rightarrow F = \frac{1}{4}(2Mg) = \frac{1}{2}Mg \Rightarrow T_1 - T_2 - \frac{1}{2}Mg = 2Ma$ (iv)

Adding equations (i), (ii) and (iv) gives:

$$3Mg - Mg - \frac{1}{2}Mg = 6Ma \Rightarrow \frac{3}{2}Mg = 6Ma \Rightarrow a = \frac{g}{4} \ \text{m/s}^2$$

$\Rightarrow$ Equation (i) gives : $T_1 = 3Mg - 3Ma = \frac{9}{4}Mg$ Newtons

Equation (ii) gives : $T_2 = Ma + Mg = \frac{5}{4}Mg$ Newtons

For $3M$ particle:

Initial velocity = $u = 0$ m/s   Distance fallen in time t seconds

$= s = ut + \frac{1}{2}at^2 = \frac{1}{2}at^2 \Rightarrow s = \frac{1}{2}\left(\frac{g}{4}\right)(1)^2 = \frac{g}{8}$ metres

(b) The table is smooth i.e. there is no friction

When $F = 0$ the equations of motion become:

(v) For $3M$: $3Mg - T_1 = 3Ma$ (vi) For M : $T_2 - Mg = Ma$

(vii) For $2M$: $T_1 - T_2 - 0 = 2Ma$

Adding : equations (v) + (vi) + (vii) : $2Mg = 6Ma \Rightarrow a = \frac{g}{3}$ m/s^2

$\Rightarrow T_1 = 2Mg$ Newtons, $T_2 = \frac{4}{3}Mg$ Newtons and $s = \frac{1}{2}\left(\frac{g}{3}\right)(1)^2 = \frac{g}{6}$ metres

37. Two masses on rough double inclined plane

Find the acceleration and tension in the pulley system shown in Fig.1. where two masses ($M_1 > M_2$, both in kg) rest on the rough surfaces of a double inclined plane.

Solution

The forces acting on the system in Newtons are shown in Fig 2 (note: increased scale of diagram): R_1, R_2 = Reactions between M_1, M_2 and inclined planes respectively, F_1, F_2 = Frictional forces, T = Tension in the string, M_1g (etc.) = Weights of masses

$M_1g\,Sin\,\alpha$ and $M_1g\,Cos\,\alpha$ = components of M_1g acting parallel and perpendicular to the plane respectively. Similar reasoning applies for M_2g.
Also: a = Acceleration of the masses in m/s²
But: $R_1 = M_1gCos\alpha$ and $F_1 = \mu R_1$ and $R_2 = M_2gCos\beta$ and $F_2 = \mu R_2$
$\Rightarrow F_1 = \mu M_1gCos\alpha$ and $F_2 = \mu M_2gCos\beta$

The equations of motion are:
For M_1 : $M_1gSin\alpha - T - \mu M_1gCos\alpha = M_1a$ (i)
For M_2 : $T - M_2gSin\beta - \mu M_2gCos\beta = M_2a$ (ii)
Adding equations (i) and (ii):
$M_1gSin\alpha - \mu M_1gCos\alpha - M_2gSin\beta - \mu M_2gCos\beta = M_1a + M_2a$
$\Rightarrow a = \dfrac{M_1g(Sin\alpha - \mu Cos\alpha) - M_2g(Sin\beta + \mu Cos\beta)}{M_1 + M_2}$ m/s²

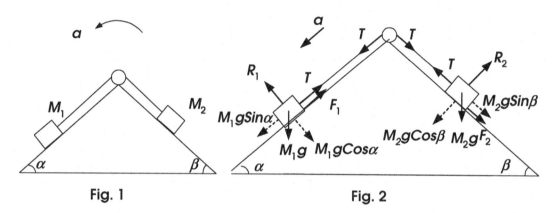

Fig. 1   **Fig. 2**

Putting this value of a in equation (i) it can be shown that:
$$T = \frac{M_1M_2g}{M_1 + M_2}\left[(Sin\alpha + Sin\beta) - \mu(Cos\alpha - Cos\beta)\right]\ \text{Newtons}$$

38. Two masses on a wedge
The pulley system shown in Fig. 1 consists of a mass, M_2 kg, connected by a string to a mass M_1 . The string passes over a pulley at the vertex of a fixed wedge, and M_1 lies on the rough inclined plane of the wedge where μ = coefficient of friction. Find the acceleration, a m/s², of M_1 and the Tension, T Newtons, in the string.

Solution
The forces in Newtons are shown in Fig 2:
R = Reaction between M_1 and the inclined plane
F = Frictional force
T = Tension in the string
Also: μ = Coefficient of friction, M_1g (etc.) = Weights of masses

$R = M_1g\,Cos\theta$ and $F = \mu R$ so that $F = \mu M_1gCos\theta$
The equations of motion are:
For M_1 : $M_1gSin\theta - T - \mu M_1gCos\theta = M_1a$ (i)   For M_2 : $T - M_2g = M_2a$ (ii)

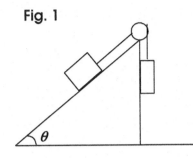

Fig. 1

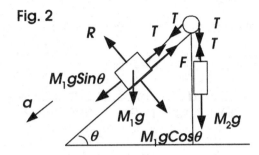

Fig. 2

Adding equations (i) and (ii):

$M_1gSin\theta - M_2g - \mu M_1gCos\theta = M_1a + M_2a = (M_1 + M_2)a$

$$\Rightarrow a = \frac{M_1g(Sin\theta - \mu Cos\theta) - M_2g}{M_1 + M_2} \quad m/s^2$$

From equation (ii): $T = M_2a + M_2g \Rightarrow T = \left(\frac{M_1M_2}{M_1 + M_2}\right)g(1 + Sin\theta - \mu Cos\theta)$ · Newtons

39. Multiple pulleys

A light inelastic string passes over a fixed pulley B, connecting a particle A of mass $5M$ kg to a light movable pulley C (Assume nil mass). A second light inelastic string passes over pulley C connecting particles D and E of masses $2M$ and M kg respectively. Find:

(a)   The acceleration, a m/s², of particle A
(b) T and S, the tensions in the strings connecting A, C and D, E respectively.
(c)   The acceleration, d m/s² of particle D relative to pulley C.

Solution

The forces and accelerations are shown in Fig.2
a = Acceleration of particle A and pulley C in m/s²
d = Acceleration of particles D and E relative to pulley C in m/s²
T = Tension in string connecting particle A and pulley C in Newtons
S = Tension in string connecting D and E in Newtons

The equations of motion are:
(i) For particle A: $5Mg - T = 5Ma$
(ii) For pulley C (Note: this pulley has no mass): $T - 2S = (Mass)(a) = (0)(a) = 0$
(iii) For particle D: $S - 2Mg = 2M(a - d)$
(iv) For particle E: $S - Mg = M(a + d)$

From equation (ii): $T = 2S$ Adding equation (iii) to $2 \times$ equation (iv) gives:
$3S - 4Mg = 4Ma$ i.e. $6S - 8Mg = 8Ma$ (v)

(a) The acceleration, a m/s², of particle A

Re-write equation (i) using relationship in equation (ii): $5Mg - 2S = 5Ma$ (vi)

Multiplying equation (vi) × 3 gives: $15Mg - 6S = 15Ma$ (vii)

Adding : equations (v) + (vii) : $7Mg = 23Ma \Rightarrow a = \frac{7}{23}g$ m/s²

(b) T and S, the tensions in the strings connecting A, C and D, E respectively.

To find S re-write equation (v) to get $30S - 40Mg = 40Ma$ (viii) and re-write equation (vi) to get $40Mg - 16S = 40Ma$ (ix)

Adding equations (viii) and (ix) gives: $S = \dfrac{80}{14}Ma = \dfrac{40}{7}Ma$

$\Rightarrow S = \dfrac{40}{7}M\left(\dfrac{7}{23}\right)g = \dfrac{40Mg}{23}$ Newtons   From equation (ii): $T = 2S = \dfrac{80Mg}{23}$ Newtons

(c) The acceleration, d m/s², of particle D relative to pulley C.

From equation (iii): for particle D: $S - 2Mg = 2M(a - d)$

it can be deduced that: $d = \dfrac{10}{23}g$ m/s²

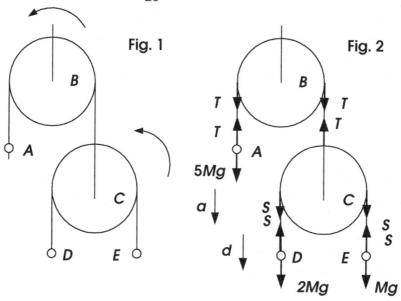

Fig. 1   Fig. 2

40. Pulley and Mass

A mass of $4M$ kg lies on a smooth table and is connected by a light string to a second pulley of mass $3M$ kg over which hangs another light string which carries masses of M and $2M$ kg. Find the common acceleration A m/s² of the $4M$ and $3M$ masses and the common acceleration a m/s² of the M and $2M$ masses relative to the $3M$ mass.

Solution

The forces and accelerations are shown in Fig 2:

S = Tension in the string joining the $4M$ and $3M$ masses in Newtons

T = Tension in the string joining the M and $2M$ masses in Newtons.

Mg (etc.) = Weights of masses

A = Acceleration of $4M$ and $3M$ masses in m/s²

a = Acceleration of M and $2M$ masses relative to the $3M$ mass in m/s².

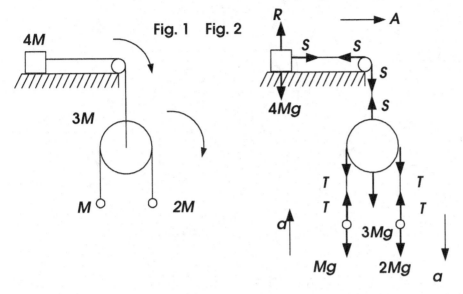

The equations of motion are:

(i) For $4M$ mass: $S = 4MA$ (ii) For $3M$ pulley: $3Mg + 2T - S = 3MA$
(iii) For M mass: $T - Mg = M(a - A)$ (iv) For $2M$ mass: $2Mg - T = 2M(a + A)$
Add equations (i) and (ii) to get: $3Mg + 2T = 7MA$ (v)
But, adding $2 \times$ equation (iii) to equation (iv) gives: $T = 4Ma$
Thus equation (v) becomes $3Mg + 8Ma = 7MA$ (vi)
Also equation (iv) becomes $2Mg - 4Ma = 2M(a + A)$ i.e: $2Mg - 2MA = 6Ma$ (vii)

Using equations (vi) and (vii) it can be shown that: $a = \dfrac{4}{29}g$ m/s^2, $A = \dfrac{17}{29}g$ m/s^2

41. Multiple Pulleys

In the pulley system shown in Fig. 1, pulleys A and C are fixed. Pulley B has a mass of M kg. Particle C has a mass of M kg. Assume there is no friction involved. Two masses, m and $3m$ kg are connected by a light inextensible string which also passes over pulleys A and C and under pulley B. If the system is released from rest:
(i) Show the forces acting on the system
(ii) Express the tension in the string, T, in terms of m, M and g.
(iii) Show that if $M = 3m$ then pulley B remains at rest while the two masses are in motion.

Solution

121

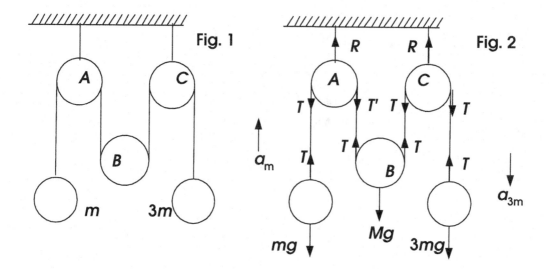

Fig. 1   Fig. 2

The forces in Newtons acting on the system are shown in Fig. 2 where:
T = Tension in the string, $R = 2T$ = Forces supporting pulleys A and C,
mg, $3mg$, Mg = Weights of and particles and pulley respectively.

a_m = Acceleration of particle of mass m a_{3m} = Acceleration of particle of mass $3m$
a_P = Acceleration of pulley of mass M
The equations of motion are:
For m: $T - mg = ma_m$ Equation (i)   For $3m$: $3mg - T = 3ma_{3m}$ Equation (ii)
For Pulley: $Mg - 2T = M(½(a_m - a_{3m})$ Equation (iii)
From equations (i) and (ii): $a_m = \dfrac{T - mg}{m}$, $a_{3m} = \dfrac{3mg - T}{3m}$

Re-write equation (iii):
$$Mg - 2T = \frac{M}{2}\left(\left(\frac{T - mg}{m}\right) - \left(\frac{3mg - T}{3m}\right)\right) = \frac{M}{2}\left(\frac{4T - 6mg}{3m}\right) \Rightarrow T = \left(\frac{3mM}{M + 3m}\right)g \quad \text{Newtons}$$

Re-write equation (iii):
$$\text{Acceleration of Pulley} = \tfrac{1}{2}\left(a_m - a_{3m}\right) = \frac{1}{2}\left(\frac{2Mg - 3mg - 3mg}{M + 3m}\right) = \left(\frac{M - 3m}{M + 3m}\right)g$$

$\Rightarrow$ If $M = 3m$ then acceleration $= 0 \Rightarrow$ Pulley will not move

42. Pulley system

In the pulley system shown in Fig. 1, the smooth pulley has a mass of $2m$ kg. Particles A and B each sit on a rough surface and are connected by a light inextensible string. They each have a mass of m kg. The coefficients of friction between particles A and B and their support surfaces are μ and ¼ respectively.
(i) Show the forces acting on the system
(ii) Express the tension in the string, T, in terms of m and g.
(iii) Find the minimum value of μ to prevent any movement of particle A.

Solution

(i) The forces in Newtons acting on the system are shown in Fig. 2 where:
T = Tension in the string
R = Reaction Force on mass = mg
mg, $2mg$ = Weights of particles and pulley respectively.
F_A, F_B = Friction forces acting on particles A and B respectively

122

Note: $F_A = \mu R = mg$ $F_B = \frac{1}{4} R = \frac{1}{4} mg$

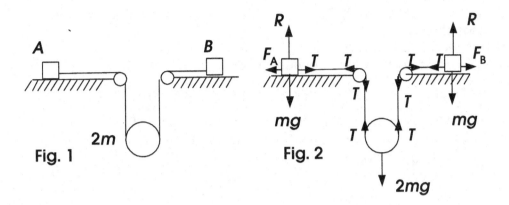

Fig. 1

Fig. 2

(ii) Express the tension in the string, T, in terms of m and g.
Accelerations are as follows: a_A (mass A), a_B (mass B), a_P (Pulley)
Assume that in a given time, t: Particle A moves a distance x (to the right) and
Particle B movesa distance y to the left.
$\Rightarrow$ the pulley moves down by a distance $\frac{1}{2} (x + y)$ $\Rightarrow$ $a_P = \frac{1}{2}(a_A + a_B)$

The equations of motion are:
Horizontal motion:
For A: $T - F_A = ma_A \Rightarrow T - \mu mg = ma_A$ Equation (i)
For B: $T - F_B = ma_B \Rightarrow T - \frac{1}{4} mg = ma_B$ Equation (ii)
Vertical motion for pulley: $2mg - 2T = 2ma_P \Rightarrow 2mg - 2T = ma_A + ma_B$
Adding equations (i) and (ii): $2T - \mu mg - \frac{1}{4} mg = ma_A + ma_B$ (iii)

$$\Rightarrow 2mg - 2T = 2T - \mu mg - \frac{1}{4} mg \Rightarrow T = \frac{1}{4}\left(\frac{9}{4} mg + \mu mg\right) = \left(\frac{9 + 4\mu}{16}\right) mg \quad Newtons$$

(iii) Find the minimum value of μ to prevent any movement of particle A.
For no movement of particle A: $a_A = 0 \Rightarrow T - \mu mg = 0 \Rightarrow T = \mu mg$
$$\Rightarrow \left(\frac{9 + 4\mu}{16}\right) mg = \mu mg \Rightarrow 9 + 4\mu = 16\mu \Rightarrow \mu = \frac{9}{12} = \frac{3}{4}$$

43. Mass on rough horizontal plane
A mass M kg rests on a rough horizontal plane which has a coefficient of friction μ and
angle of friction λ and it is acted upon by a force X inclined at an angle θ to the
horizontal as in Fig. 1. Find the value of X in terms of θ and λ if the mass just moves.

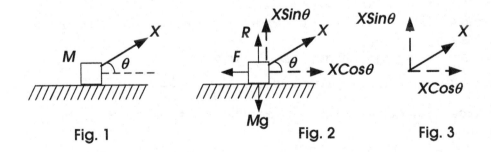

Fig. 1

Fig. 2

Fig. 3

Solution

The forces in Newtons acting are:

X = Force acting on M in Newtons ($X Cos\,\theta$ and $X Sin\,\theta$ are its horizontal and vertical components respectively)

R = Reaction between M and plane in Newtons

F = Frictional force opposing motion in Newtons = μR

μ = Coefficient of friction, λ = Angle of friction where $\mu = Tan\lambda = \dfrac{Sin\lambda}{Cos\lambda}$

When the mass is on the point of moving the forces are (See Figs. 2, 3):

Vertical forces: $Mg = R + XSin\theta \Rightarrow R = Mg - XSin\theta$ Horizontal forces: $F = XCos\,\theta$

But : $F = \mu R \Rightarrow \mu R = XCos\theta \Rightarrow \mu(Mg - XSin\theta) = XCos\theta$

$$\mu Mg = X(Cos\theta + \mu Sin\theta) \Rightarrow X = \frac{\mu Mg}{Cos\theta + \mu Sin\theta} = \frac{\left(\dfrac{Sin\lambda}{Cos\lambda}\right)}{Cos\theta + \dfrac{Sin\lambda}{Cos\lambda}Sin\theta} Mg$$

$$\Rightarrow X = \frac{MgSin\lambda}{Cos\theta\,Cos\lambda + Sin\theta\,Sin\lambda} = Mg\frac{Sin\lambda}{Cos(\lambda - \theta)}$$

44. Mass sliding on wedge

A body of mass M kg rests on a rough plane inclined at θ degrees to the horizontal (See Fig. 1). Find:

(a) The value of the coefficient of friction, μ, if the minimum force acting downwards along the plane, which will not move the mass, is D Newtons.

(b) The minimum force, U Newtons, which acting along the slope of the plane, will just cause the mass to move upwards.

(c) The values of μ and U if M = 1 kg, D = 0.1g Newtons and θ = 30°

Solution

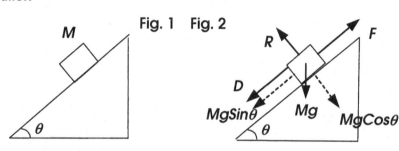

Fig. 1   Fig. 2

(a) The value of the coefficient of friction, μ

The forces in Newtons acting on mass M are shown in Fig. 2: D = Downward force, R = Reaction between M and the plane, Mg = Weight of M

$MgCos\,\theta$ and $MgSin\,\theta$ are the components of Mg acting perpendicular and parallel to the plane respectively.

The forces parallel to the inclined plane are: $D + MgSin\,\theta = F$

But: $F = \mu R = \mu Mg\,Cos\theta \Rightarrow D + MgSin\theta = \mu MgCos\theta \Rightarrow \mu = \dfrac{D + MgSin\theta}{MgCos\theta}$ (i)

(b) The minimum force which acting along the slope of the plane, will just cause the mass to move upwards.

Let the least force acting upwards along the inclined plane be U. Then the forces acting on mass M are shown in Fig. 3. Thus:

$$U = F + MgSin\theta \Rightarrow U = \mu MgCos\theta + MgSin\theta \Rightarrow U = Mg(Sin\theta + \mu Cos\theta)$$

(c)   The values of μ and U if $M = 1$ kg , $D = 0.1g$ Newtons and $\theta = 30°$

Fig. 3

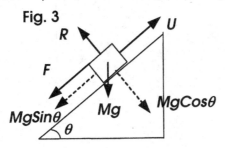

$$\mu = \frac{0.1g + \dfrac{g}{2}}{\dfrac{\sqrt{3}}{2}g} = \frac{1.2}{\sqrt{3}} = 0.693, \quad U = (1)(g)(0.5 + (0.693)(0.866)) = 10.79 \text{ Newtons}$$

45. Mass on rough inclined plane

A particle of mass M kg rests on a rough plane making angle θ with the horizontal as shown in Fig. 1. A force U Newtons directed at an angle of α to the plane acts on the particle. Find the minimum value of U which brings the particle just to the point of moving up the plane. If λ = angle of friction, find:
(a) U, in terms of θ, α and λ
(b)   The force acting (up) parallel to plane that would bring the particle to this point.

Solution

The forces in Newtons acting are:
R = Reaction between particle and plane F = Frictional force
Mg = Weight of particle, U = Force applied to mass, λ = Angle of friction
μ = Coefficient of friction = $Tan\lambda = \dfrac{F}{R}$

(a) Find U in terms of θ, α and λ

Resolving forces perpendicular to plane: $R + USin\alpha = MgCos\theta \Rightarrow R = MgCos\theta - USin\alpha$
Resolving forces parallel to the plane:
$UCos\alpha = Mg \, Sin\,\theta + F = Mg \, Sin\,\theta + \mu R = MgSin\theta + \mu(MgCos\theta - USin\alpha)$
$\Rightarrow U(Cos\alpha + \mu Sin\alpha) = Mg(Sin\theta + \mu Cos\theta)$
$\Rightarrow U\left(Cos\alpha + \dfrac{Sin\lambda}{Cos\lambda}Sin\alpha\right) = Mg\left(Sin\theta + \dfrac{Sin\lambda}{Cos\lambda}Cos\theta\right)$
$\Rightarrow U\left(\dfrac{Cos\alpha Cos\lambda + Sin\lambda Sin\alpha}{Cos\lambda}\right) = Mg\left(\dfrac{Sin\theta Cos\lambda + Sin\lambda Cos\theta}{Cos\lambda}\right)$
$\Rightarrow U Cos(\lambda - \alpha) = MgSin(\lambda + \theta) \Rightarrow U = Mg\dfrac{Sin(\lambda + \theta)}{Cos(\lambda - \alpha)}$ Newtons

(b) The force acting (up) parallel to the plane that would bring the particle to this point.

But, if $\alpha = 0$ then U becomes Up, say, i.e. a force acting upwards along the plane, and
$$Up = \frac{MgSin(\lambda + \theta)}{Cos\lambda} \text{ Newtons}$$

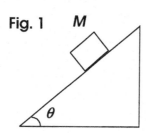

Fig. 1   M

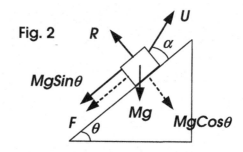

Fig. 2   R   U   α

MgSinθ

F   Mg   MgCosθ

46. Mass and wedge

A mass m kg slides down the surface of a smooth wedge of mass M kg which can slide on a smooth horizontal table (See Fig. 1). Find:

(a)   The acceleration of M, A m/s²
(b)   The acceleration of m relative to M, a m/s²
(c)   The relative values of a, A

Solution

The forces and accelerations are shown in Figs. 2 and 3:

a, A = Accelerations of m relative to M and M along table respectively in m/s²
R, S = Reactions between m, M and M and table respectively in Newtons

The equations of motion are:

(i) For m (vertical forces): $mg - R\cos\theta = ma\sin\theta$

(ii) For m (horizontal forces): $R\sin\theta = ma\cos\theta - mA$

(iii) For M (vertical forces): $S = Mg + R\cos\theta$

(iv) For M (horizontal forces): $R\sin\theta = MA$

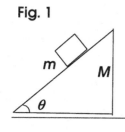

Fig. 1 m M θ

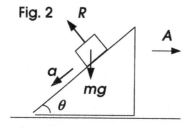

Fig. 2   R a mg A θ

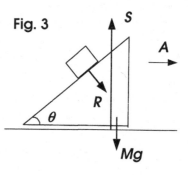

Fig. 3   S   A   R θ Mg

(a) The acceleration of M, A m/s²

Using equations (ii) and (iv) it can be seen that:

$$MA = ma\cos\theta - mA \Rightarrow (m + M)A = ma\cos\theta \Rightarrow a = \left(\frac{m + M}{m\cos\theta}\right)A \quad (v)$$

Also: (iv) gives $R = \dfrac{MA}{\sin\theta}$ thus (i) becomes $mg - \dfrac{MA\cos\theta}{\sin\theta} = ma\sin\theta$

$$\Rightarrow mg - \frac{MA\cos\theta}{\sin\theta} = m\sin\theta\left(\frac{m + M}{m\cos\theta}\right)(A) \quad \text{This gives:}$$

$$mg\sin\theta\cos\theta = A(m + M)\sin^2\theta + M\cos^2\theta) \Rightarrow A = \frac{mg\sin\theta\,\cos\theta}{m\sin^2\theta + M(\sin^2\theta + \cos^2\theta)} = \frac{mg\sin\theta\,\cos\theta}{m\sin^2\theta + M}$$

(b) The acceleration of m relative to M, a m/s²

Using equation (v): $a = \left(\dfrac{m + M}{m\cos\theta}\right) \times \dfrac{mg\,Sin\theta\,Cos\theta}{m\,Sin^2\theta + M} = \dfrac{(m + M)g\,Sin\theta}{m\,Sin^2\theta + M}$

(c) The relative values of a and A

From equation (v) above: $\dfrac{a}{A} = \dfrac{(m + M)}{m\,Cos\theta}$

47. Two connected masses on rough inclined plane

Two masses, M_1 and M_2 kg respectively, are connected by a light inextensible string and placed on a rough plane inclined at an angle θ to the horizontal. Find the value of θ if both masses are on the point of slipping. See Fig. 1.

Solution

The forces (in Newtons) acting on the system are shown in Fig 2 (note: increased scale of diagram to facilitate text display).

$R_1, R_2 =$ Reactions between inclined plane and M_1 and M_2 respectively

$F_1, F_2 =$ Frictional forces, $T =$ Tension in the string

$M_1g\,Sin\,\alpha$ and $M_1g\,Cos\,\alpha =$ components of weight M_1g acting parallel and perpendicular to the plane respectively. Similar reasoning applies for M_2g.

But, it is known that for M_1: $F_1 = \mu_1 R_1$ $R_1 = M_1g\,Cos\theta$ $F_1 = \mu_1 M_1g\,Cos\alpha$

$M_1g\,Sin\theta = F_1 + T = \mu_1 R_1 + T = \mu_1 M_1g\,Cos\theta + T$ (i)

Also, for M_2: $F_2 = \mu_2 R_2$ $R_2 = M_2g\,Cos\theta$ $F_2 = \mu_2 M_2g\,Cos\beta$

$M_2g\,Sin\theta + T = F_1 = \mu_2 R_2 = \mu_2 M_2g\,Cos\theta$ (ii)

From equations (i) and (ii): $T = M_1g\,Sin\theta - \mu_1 M_1g\,Cos\theta = \mu_2 M_2g\,Cos\theta - M_2g\,Sin\theta$

$\Rightarrow (M_1g + M_2g)Sin\theta = (\mu_2 M_2g + \mu_1 M_1g)Cos\theta$

$\Rightarrow Tan\theta = \dfrac{(\mu_1 M_1 + \mu_2 M_2)}{(M_1 + M_2)} \Rightarrow \theta = Tan^{-1}\left(\dfrac{\mu_1 M_1 + \mu_2 M_2}{M_1 + M_2}\right)$

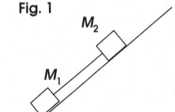

Fig. 1

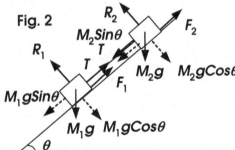

Fig. 2

Chapter 7   Jointed Rods

1. Object supported by two strings
An object of weight 100 Newtons is supported by two strings which are inclined to the vertical at 30° and 60° respectively (See Fig. 1). Calculate the tension in each string.

Solution
Horizontal forces: $T_1 \, Sin \, 30° = T_2 \, Sin \, 60° \Rightarrow T_1 = T_2 \, \sqrt{3}$
Vertical forces: $T_1 \, Cos \, 30° + T_2 \, Cos \, 60° = 100 \Rightarrow 2T_2 = 100$
$\Rightarrow T_2 = 50$ Newtons $\Rightarrow T_1 = 50\sqrt{3}$ Newtons

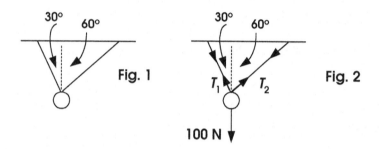

Fig. 1   **Fig. 2**

30° 60°   30° 60°

T_1 T_2

100 N

2. Forces acting on rod
A rod AB of length L metres and weight 6 Newtons is suspended from a smooth hinge at A. A horizontal force of 3 Newtons is applied to end B. See Fig. 1. Find the rod's inclination to the vertical and the reaction at the hinge.

Solution
The forces acting on the rod (in Newtons) on the rod are: Weight = 6, Horizontal Force at B = 3, R = Reaction at A (X, Y = Horizontal and vertical components of R)
The force components are:
Horizontal: $X - 3 = 0 \Rightarrow X = 3$ Vertical: $Y - 6 = 0 \Rightarrow Y = 6$
Moments about A (See Figs. 3, 4): $(6Sin \, \theta)L/2 = (3 \, Cos \, \theta)L \Rightarrow Tan \, \theta = 1 \Rightarrow \theta = 45°$

Magnitude of Reaction, $R = \sqrt{X^2 + Y^2} = \sqrt{9 + 36} = 3\sqrt{5}$ Newtons

Angle of R to vertical $= \alpha = Tan^{-1}\dfrac{X}{Y} = Tan^{-1}\dfrac{1}{2} \Rightarrow \alpha = 26.57°$ to the vertical

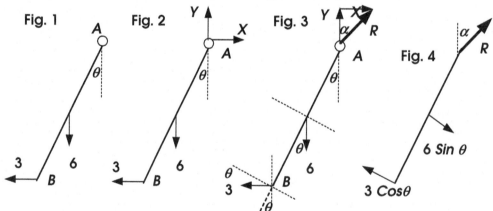

Fig. 1 A **Fig. 2** Y X **Fig. 3** Y X R α R

A θ A θ A θ **Fig. 4**

3   6   3   6 θ 6   6 $Sin \, \theta$

B B 3 B 3 $Cos\theta$

θ

3. Forces acting on a rod

A rod AB of length L metres and weight 6 Newtons is suspended from a smooth hinge at A. A horizontal force P Newtons is applied to B. See Fig. 1. If the inclination of the rod to the vertical is 30°, find and the reaction at the hinge.

Solution

The forces acting (in Newtons) on the rod are:
Horizontal force at $B = P$, Weight $= 6$
R = Reaction at A (X, Y = Horizontal and vertical components of R)

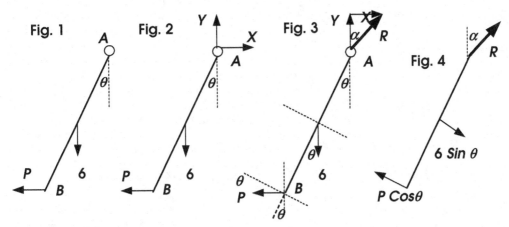

The forces acting are:
Horizontal: $X - P = 0 \Rightarrow X = P$ Vertical: $Y - 6 = 0 \Rightarrow Y = 6$
Moments about A (See Figs. 3,4):$(6Sin\ \theta)L/2 = (P\ Cos\ \theta)L$ But (given): $\theta = 30°$
$\Rightarrow 3L/2 = P\sqrt{3}L/2 \Rightarrow P = \sqrt{3}$ Newtons $(= X)$
Magnitude of Reaction $= \sqrt{X^2 + Y^2} = \sqrt{3 + 36} = \sqrt{39}$ Newtons

Angle to vertical $= \alpha = Tan^{-1}\dfrac{\sqrt{3}}{6} = Tan^{-1}\dfrac{1}{2\sqrt{3}} \Rightarrow \alpha = 16.1°$ to the vertical

4. Ladder

A ladder AB (4 metres) and mass M kg rests on a rough horizontal surface (at A) and makes contact with a smooth wall. If $\theta = 45°$ find the magnitude and direction of the resultant force at A. Find the value of μ.

Solution

The forces (in Newtons) acting are:
X = Reaction between the ladder and wall surface
R = Reaction between the ladder base and rough ground surface
F = Friction
Mg = Weight of ladder

The ladder is shown in Fig. 1. The forces are shown in Fig. 2.
For AB: Vertically: $R = Mg$ (i)   Horizontally: $F = X$ (ii)
(Also, regarding frictional forces: $F = \mu R$)
Taking moments about B for AB:
$RCos\ 45°\ (4) = MgCos\ 45°(2) + FSin45°(4) \Rightarrow 2RCos\ 45° = MgCos\ 45° + 2FSin45°$
$\Rightarrow 2RCos\ 45° = MgCos\ 45° + 2FSin45°$

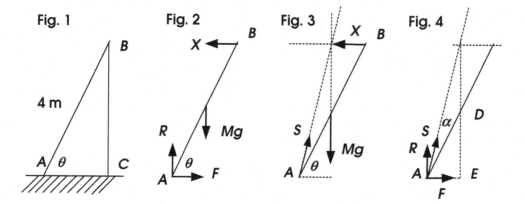

Fig. 1   Fig. 2   Fig. 3   Fig. 4

$$\Rightarrow 2R = Mg + 2F \Rightarrow 2Mg = Mg + 2F \Rightarrow F = \frac{1}{2}Mg$$

$$\Rightarrow \text{Resultant Force} = S = \sqrt{R^2 + F^2} = Mg\sqrt{\frac{5}{4}} = \frac{Mg\sqrt{5}}{2} \quad \text{See Fig. 3}$$

Direction of S (See Figs. 3, 4, 5): (Angle to horizontal plane)

$$F = \mu R \Rightarrow \mu = \frac{F}{R} = \frac{1}{2} \Rightarrow Tan(\theta + \alpha) = Tan(45° + \alpha) = \frac{R}{F} = 2 \Rightarrow 45° + \alpha = 63.43°$$

to the horizontal

5. Step Ladder

A step ladder consisting of two equal parts, *AB* and *BC*, each of mass *M* kg, is hinged at *B*. The mid-points of *AB* and *BC* are joined by a light rope. The legs rest on a smooth horizontal surface (assume no friction). A person of mass 5*M* kg stands on the ladder half way up side *AB* (See Fig. 1). When the rope is taut the angle between the two parts is 2α. If the tension in the taut rope is $\frac{3}{2}Mg$ Newtons find:

(a)   The values of the forces at *A*, *B* and *C*. (b) The value of the angle α.

Solution

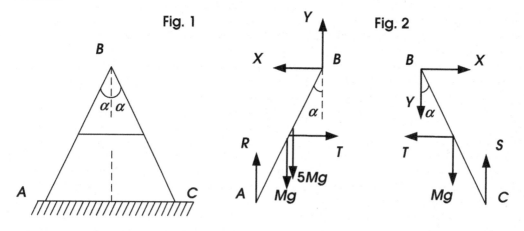

Fig. 1   Fig. 2

The forces (in Newtons) acting are:
X and *Y* = Components of the force acting at hinge *B*
R, S = Reactions between the ladder base and surface
T = Tension in the string

Mg = Weight of each half of the step ladder
$5Mg$ = Weight of person

(a) The values of the forces at A, B and C
The forces are shown in Fig. 2
For AB: Vertically: $R + Y = 5Mg + Mg = 6Mg$ (i)   Horizontally: $T = \frac{3}{2}Mg = X$ (ii)

For BC: Vertically: $S = Y + Mg$ (iii)   Horizontally: $T = \frac{3}{2}Mg = X$

Taking moments about B for AB:

$(\text{let } |AB| = |BC| = \ell) \Rightarrow R\,Sin\alpha\,\ell = 5Mg\frac{\ell}{2}Sin\alpha + Mg\frac{\ell}{2}Sin\alpha + T\,Cos\alpha\frac{\ell}{2}$

$\Rightarrow R\,Sin\alpha = 3Mg\,Sin\alpha + \frac{1}{2}T\,Cos\alpha$

But : $T = \frac{3}{2}Mg \Rightarrow R\,Sin\alpha = 3Mg\,Sin\alpha + \frac{3}{4}Mg\,Cos\alpha$ (iv)

Taking moments about B for BC:

$S\ell\,Sin\alpha = Mg\left(\frac{\ell}{2}\right)Sin\alpha + T\frac{\ell}{2}Cos\alpha$ But : $T = \frac{3}{2}Mg$

$\Rightarrow S\,Sin\alpha = \frac{1}{2}Mg\,Sin\alpha + \frac{1}{2}T\,Cos\alpha = \frac{1}{2}Mg\,Sin\alpha + \frac{3}{4}MgCos\alpha$ (v)

Re - writing the above equations :

(i) $R + Y = 6Mg$ (ii) $T = X = \frac{3}{2}Mg$ (iii) $S = Y + Mg$

(iv) $R\,Sin\alpha = 3Mg\,Sin\alpha + \frac{3}{4}Mg\,Cos\alpha$ (v) $S\,Sin\alpha = \frac{1}{2}Mg\,Sin\alpha + \frac{3}{4}MgCos\alpha$

Add equations (i) and (iii) to get : $R + S + Y = Y + 7Mg \Rightarrow R + S = 7Mg$ (vi)

Subtract equation (v) from equation (iv) we get : $(R - S)Sin\alpha = \frac{5}{2}MgSin\alpha$

$\Rightarrow R - S = \frac{5}{2}Mg$ (vii)

Adding equation (vi) and (vii) : $R = \frac{19}{4}Mg$ giving : $S = \frac{9}{4}Mg$, $Y = \frac{5}{4}Mg$

(b) The value of the angle α
From equation (iv) : $\frac{19}{4}MgSin\alpha = 3MgSin\alpha + \frac{3}{4}Mg\,Cos\alpha \Rightarrow \frac{7}{4}MgSin\alpha$

$= \frac{3}{4}Mg\,Cos\alpha \Rightarrow Tan\alpha = \frac{3}{7} \Rightarrow \alpha = Tan^{-1}\frac{3}{7}$

6. Beam and stay
A uniform beam AB of mass M kg and length L metres is hinged to a smooth wall at A. It is supported by a light stay CD, which joins at point C on the beam to a point D on the wall (see Fig. 1). If the beam carries a mass of $4M$ kg at B find:
(a)   The thrust in CD
(b)   The magnitude and direction of the reaction at A

Solution

(a) The thrust in CD
The forces (in Newtons) acting on the members are:

Let X and Y be the components of the reaction at hinge A and T be the thrust on CD. Take moments about A for AB:

$$Mg\frac{L}{2} + 4MgL = T\,Sin60°\,\frac{L}{4} \Rightarrow Mg + 8Mg = 2T\frac{\sqrt{3}}{2} \times \frac{1}{4}$$

$$\Rightarrow 9Mg = \frac{T\sqrt{3}}{4} \Rightarrow T = 12\sqrt{3}Mg \quad \text{Newtons}$$

(b) The magnitude and direction of the reaction at A
But the forces on AB are :

Vertically : $T\,Sin60° = Y + Mg + 4Mg = Y + 5Mg = \dfrac{T\sqrt{3}}{2}$

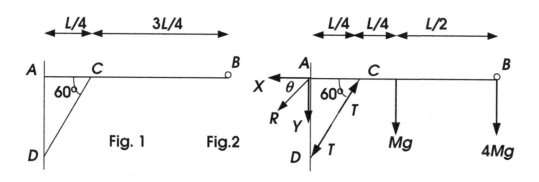

Fig. 1   Fig.2

$$\Rightarrow Y + 5Mg = \frac{12\sqrt{3}Mg\sqrt{3}}{2} = 18Mg \Rightarrow Y = 13Mg \quad \text{Newtons}$$

Horizontally : $X = T\,Cos60° \Rightarrow X = \dfrac{T}{2} = 6\sqrt{3}Mg \quad$ Newtons

Reaction at A : We have : Magnitude $= R = \sqrt{X^2 + Y^2}$

$$= \sqrt{\left(6\sqrt{3}Mg\right)^2 + \left(13Mg\right)^2} = Mg\sqrt{277}$$

$\Rightarrow R = 16.64\,Mg$ at an angle θ to the horizontal where :

$$Tan\theta = \frac{13Mg}{6\sqrt{3}Mg} = \frac{13}{6\sqrt{3}} \Rightarrow \theta = 51.36°$$

7. Two hinged rods

A rod AB of mass m kg is hinged to a fixed point at A and to rod BC of mass M kg at B (See Fig. 1). A horizontal force H Newtons is applied to C such that the inclinations of rods AB and BC to the vertical are θ and α respectively. If both rods are of length ℓ metres, show that: $\dfrac{Tan\alpha}{Tan\theta} = \dfrac{m + 2M}{M}$

Solution

Let the forces in Newtons acting on the rods be:
X_1, Y_1 = Horizontal and Vertical Reactions at A
X, Y = Horizontal and Vertical Reactions at B
mg, Mg = Weights of masses, H = Horizontal force

These forces are shown in Fig. 2

Let $|AB| = |BC| = \ell$

Take moments about A for AB: $mg\, Sin\theta\left(\dfrac{\ell}{2}\right) + Y\, Sin\theta(\ell) = X\ell\, Cos\theta$

$\Rightarrow mg\, Sin\theta + 2Y\, Sin\theta = 2X\, Cos\theta$ (i)

Take moments about C for BC:

$Mg\left(\dfrac{\ell}{2}\right)Sin\alpha + X\ell\, Cos\alpha = Y\ell Sin\alpha \Rightarrow Mg\, Sin\alpha + 2X\, Cos\alpha = 2Y\, Sin\alpha$ (ii)

Also: Horizontal forces: $H = X$; Vertical forces: $Mg = Y$

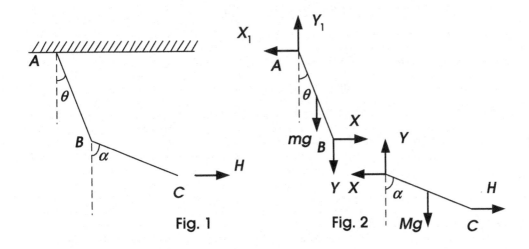

Fig. 1   **Fig. 2**

But, from BC we can see that :

$H = X$ and $Y = Mg \Rightarrow$ equation (ii) becomes :

$Mg\, Sin\alpha + 2H\, Cos\alpha = 2Mg\, Sin\alpha \Rightarrow 2H\, Cos\alpha = Mg\, Sin\alpha \Rightarrow Tan\alpha = \dfrac{2H}{Mg}$

and equation (i) becomes :

$mg\, Sin\theta + 2Mg\, Sin\theta = 2H Cos\theta \Rightarrow Sin\theta(mg + 2Mg) = 2H\, Cos\theta$

$\Rightarrow Tan\theta = \dfrac{2H}{mg + 2Mg} \Rightarrow \dfrac{Tan\alpha}{Tan\theta} = \dfrac{m + 2M}{M}$

8. Two hinged rods

Two rods AB, AC each of mass M kg and length ℓ metres are hinged together at A as shown in Fig. 1. AB is hinged at B and C (at same level as B) is supported by a light string which is kept vertical by the application of a horizontal force H Newtons. If the system is in equilibrium and the angle between the rods is 2θ, find the value of θ.

Solution

The forces in Newtons on the rods are shown in Fig. 2:

X_1, Y_1 = Components of the reaction at hinge B.

T = Tension in the string.

X, Y = Components of the reaction at A.

H = Horizontal force, Mg = Weights of masses

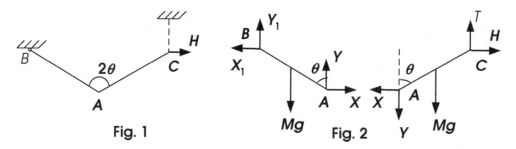

Fig. 1   Fig. 2

Taking moment about B for AB:

$$Mg\frac{\ell}{2}\,Sin\theta = Y\ell\,Sin\theta + X\ell\,Cos\theta \Rightarrow Mg\,Sin\theta = 2Y\,Sin\theta + 2X\,Cos\theta \quad (i)$$

Taking moments about C for AC

$$Mg\frac{\ell}{2}\,Sin\theta + Y\ell\,Sin\theta = X\ell\,Cos\theta \Rightarrow Mg\,Sin\theta + 2Y\,Sin\theta = 2X\,Cos\theta$$

$$\Rightarrow Mg\,Sin\theta = -2Y\,Sin\theta + 2X\,Cos\theta \quad (ii)$$

Adding the equations (i) and (ii) $\Rightarrow 2Mg\,Sin\theta = 4X\,Cos\theta \Rightarrow Tan\theta = \dfrac{2X}{Mg}$

But considering the forces acting on AC we see that $H = X$

$$\Rightarrow Tan\theta = \frac{2H}{Mg} \Rightarrow \theta = Tan^{-1}\left(\frac{2H}{Mg}\right)$$

9. Ladder and wall

One end of a uniform ladder of mass M kg rests against a smooth vertical wall and the other on rough horizontal ground so that it makes an angle:
$\theta = Tan^{-1}3/5$ with the horizontal (See Fig. 1).

(a)   Show that the ladder will start to slip if the coefficient of friction, $\mu < 5/6$
(b)   When $\mu = \frac{1}{2}$ the ladder is just prevented from slipping by a vertical string attached to a point a quarter of the ladder's length from point B. Find the tension in the string in terms of Mg.

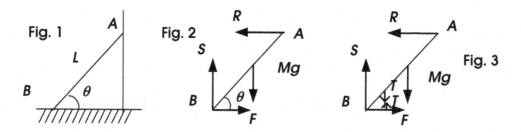

Fig. 1   Fig. 2   Fig. 3

Solution
(a) Show that the ladder will start to slip if the coefficient of friction, $\mu < 5/6$
The forces acting on the ladder are shown in Fig. 2. Let $|AB| = L$.
Horizontal forces: $F = R$ Also: $F = \mu R$
Vertical forces: $S = Mg$
Taking moments about A:

134

$$Mg\left(\frac{L}{2}\right)Cos\theta + FL\,Sin\theta = SL\,Cos\theta \Rightarrow MgCos\theta + 2F\,Sin\theta = 2S\,Cos\theta$$

$$\Rightarrow Mg + 2F\,Tan\theta = 2S \Rightarrow 2F\,Tan\theta = 2S - Mg \Rightarrow F = \frac{2S-Mg}{2\,Tan\theta} = \frac{S}{2\,Tan\theta} \quad (i)$$

From equation (i): $F = \dfrac{S}{2Tan\theta} = \mu\,S \Rightarrow \mu = \dfrac{1}{2Tan\theta} = \dfrac{1}{2\left(\dfrac{3}{5}\right)} = \dfrac{5}{6}$

$\Rightarrow$ The ladder is on the point of slipping when $\mu = \dfrac{5}{6}$

(b) Find the tension in the string in terms of *Mg*

If $\mu = 1/2$ and a vertical string is attached to a point a quarter of the ladder's length from point B, the forces acting are shown in Fig. 3:

Taking moments about A for AB:

$$TCos\theta\left(\frac{3L}{4}\right) + Mg\frac{L}{2}Cos\theta + FLSin\theta = SLCos\theta \Rightarrow 3T + 2Mg + 4FTan\theta = 4S \quad (iii)$$

But $\quad T + Mg = S \quad$ (iv) $\quad$ and $F = \mu S = \frac{1}{2}S \quad$ (v)

From equation (v) and the fact that $Tan\theta = \frac{3}{5}$, we can rewrite equation (iii):

$\Rightarrow 3T + 2Mg + 2S\frac{3}{5} = 4S \Rightarrow 3T + 2Mg = \frac{14}{5}S \quad$ (vi)

Subtracting : (vi) $- \frac{14}{5}\times$(iv) we get $\frac{1}{5}T - \frac{4}{5}Mg = 0 \Rightarrow T = 4Mg \quad$ Newtons

10. Two hinged rods

Two uniform rods AB, BC of length D and L metres and mass m and M kg respectively are smoothly hinged together at B. They stand in equilibrium with the end A resting on rough horizontal ground and the end C resting against a smooth vertical wall. AB, BC make angles of θ and 30° respectively with the horizontal (See Fig. 1). Find:

(a) The value of θ
(b) The coefficient of friction, μ, at A

Solution

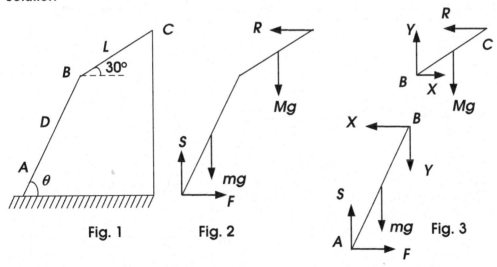

Fig. 1   Fig. 2   Fig. 3

(a) The value of θ

The forces in Newtons acting on the rods are:

R = Reaction between rod ABC and the smooth wall
X, Y = Components of reaction at hinge B
S = Vertical component of reaction between AB and the rough ground
F = Frictional force
mg, Mg = Weights of masses m, M

Take moments about B for AB:

$$mg\frac{D}{2} Cos\theta + FD\,Sin\theta = SD\,Cos\theta$$

$mg\,Cos\theta + 2F\,Sin\theta = 2S\,Cos\theta \Rightarrow mg + 2F\,Tan\theta = 2S$ (i)

Take moments about B for BC

$$RL\,Sin30° = Mg\frac{L}{2} Cos30° \Rightarrow R = \frac{\sqrt{3}}{2} Mg \quad \text{(ii)}$$

Consider the forces acting on ABC : $R = F, \quad S = (m + M)g$

From equation (ii) : $R = \dfrac{\sqrt{3}}{2} Mg = F$

Putting this information into equation (i) :

$$\Rightarrow mg + 2\left(\frac{\sqrt{3}}{2} Mg\right)Tan\theta = 2mg + 2Mg \Rightarrow \left(\sqrt{3}Mg\right)Tan\theta = mg + 2Mg$$

$$\Rightarrow Tan\theta = \frac{1}{\sqrt{3}}\left(\frac{m + 2M}{M}\right) \Rightarrow \theta = Tan^{-1}\frac{1}{\sqrt{3}}\left(\frac{m + 2M}{M}\right)$$

(b) The coefficient of friction, μ, at A

Also, $F = \mu S = \mu(m + M)g = \dfrac{\sqrt{3}}{2} Mg \Rightarrow \mu = \dfrac{\sqrt{3}}{2}\dfrac{M}{(m + M)}$

11. Three jointed rods

Rod AB of length L metres and mass M kg and rods BC, AC of mass M kg are freely jointed (See Fig. 1). AB is suspended by a string attached to its midpoint D. For both cases $\angle ACB = \theta = 60°$ and $90°$ find the horizontal and vertical components of the reaction at C.

Solution
(a) Case 1: $\theta = 60°$
The forces in Newtons acting are:
X, Y = Horizontal and Vertical components of Reactions at A, B, C
Mg = Weight of mass M

In this configuration, each of the rods has the same length, L.
In Fig. 1 each angle = 60°. The forces acting on the rods are shown in Fig. 2.
For AB: $3Mg = Mg + 2Y_1 \Rightarrow 2Mg = 2Y_1 \Rightarrow Mg = Y_1$ (i)
For AC: $Y_1 = Mg + Y_2$ But from equation (i) : $Y_1 = Mg \Rightarrow Y_2 = 0$

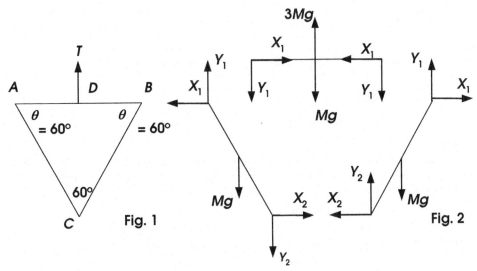

Fig. 1

Fig. 2

Take moments about A for AC :

$$\Rightarrow Mg\frac{L}{2}Cos60° = X_2LCos30° \quad \text{(Note : from above : } Y_2 = 0)$$

$$\Rightarrow Mg\frac{1}{2} = 2X_2\frac{\sqrt{3}}{2} = 0 \Rightarrow X_2 = \frac{Mg}{2\sqrt{3}}$$

Therefore, at C: Horizontal component $= \dfrac{Mg}{2\sqrt{3}}$ Vertical component $= Y_2 = 0$

(b) Case 2: : $\theta = 90°$

In this configuration, AB has the same length, L. BC, and AC have lengths of $\dfrac{L}{\sqrt{2}}$

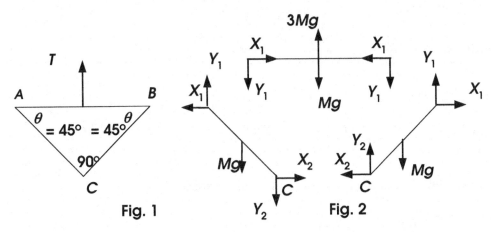

Fig. 1

Fig. 2

For AB: $3Mg = Mg + 2Y_1 \Rightarrow 2Mg = 2Y_1 \Rightarrow Mg = Y_1$ (i)
For AC: $Y_1 = Mg + Y_2$ But $Y_1 = Mg \Rightarrow Y_2 = 0$

Take moments about A for AC (See Fig. 2):

$$\Rightarrow Mg\left(\frac{1}{2}\right)\left(\frac{L}{\sqrt{2}}\right)Cos\,45° = X_2LCos45° \Rightarrow Mg\left(\frac{1}{2}\right)\left(\frac{1}{\sqrt{2}}\right) = X_2 \Rightarrow X_2 = \frac{Mg}{2\sqrt{2}}$$

Therefore, at C: Horizontal component $= \dfrac{Mg}{2\sqrt{2}}$ Vertical component $= Y_2 = 0$

12. Two hinged rods

Two rods AB, BC of equal length, L metres, but of mass M and $2M$ kg respectively, are freely hinged together at B. They stand in equilibrium in a vertical plane with the ends A and C on a rough horizontal plane as shown in Fig. 1. Find:

(a)   The values of the reactions at A and C.
(b)   Which of the rods will slip first.
(c)   The coefficient of friction, μ, if one rod is on the point of slipping.

Solution

The forces in Newtons acting are:
$X, Y, W, Z =$ Reactions at hinge B
R, S = Reactions between rods and the ground at A and C.
F_1, F_2 = Frictional forces at A and C.
$Mg, 2Mg$ = Weights of masses

$$|AB| = |BC| = L \qquad |AC| = L\sqrt{2}$$

(a) The values of the reactions at A and C

Considering the system ABC in Fig. 2: $R + S = 3\,Mg$

$$\Rightarrow \text{moments about } C: R(L\sqrt{2}) = Mg\left(\frac{3}{4}L\sqrt{2}\right) + 2Mg\left(\frac{1}{4}L\sqrt{2}\right)$$

$$\Rightarrow R = Mg\left(\frac{3}{4}\right) + 2Mg\left(\frac{1}{4}\right) = \frac{5Mg}{4} \Rightarrow S = \frac{7Mg}{4} \quad \text{Newtons}$$

(b) Which of the rods will slip first

But, $F_1 = \mu R, \qquad F_2 = \mu S$
From (a) we know that $S > R \Rightarrow F_2 > F_1 \Rightarrow$ rod AB will slip first

(c) The coefficient of friction, μ, if one rod is on the point of slipping

For AB:
Take moments about B for AB:

$$\Rightarrow (Mg)\frac{L}{2}Cos\,45° + F_1 LCos\,45° = RLCos\,45°$$

$$\Rightarrow Mg + 2F_1 = 2R = \frac{10}{4}Mg \Rightarrow F_1 = \frac{3}{4}Mg \quad \text{(i)}$$

But $F_1 = \mu R = \mu\frac{5}{4}Mg \Rightarrow \frac{3}{4}Mg = \mu\frac{5}{4}Mg \Rightarrow \mu = \frac{3}{5}$

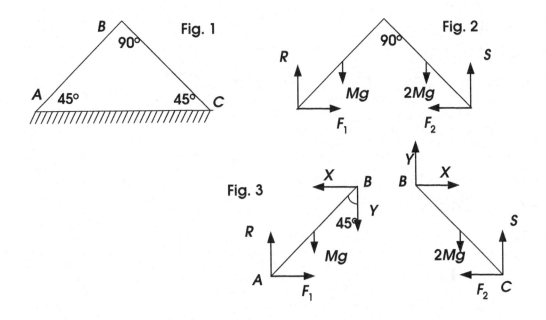

Fig. 1

Fig. 2

Fig. 3

13. Two pinned rods

Two rods AB and BC each of mass M kg and length L metres are pinned together at point B (See Fig. 1). The rods are held by a hinge at A. If the coefficient of friction at C is $\mu = 1/3$, find the angle 2θ for equilibrium.

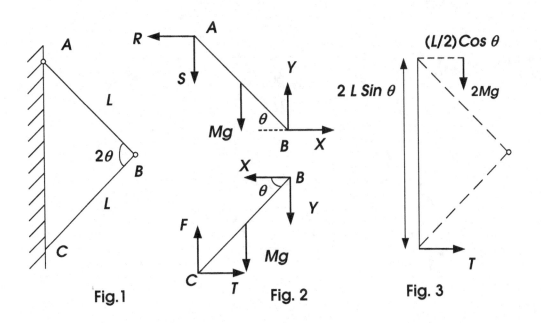

Fig.1 **Fig. 2** **Fig. 3**

Solution

The forces in Newtons on the rods are (See Fig. 2):

R, S = Reactions at hinge A
X, Y = Reactions at hinge B
F = Frictional force
T = Normal reaction component at C
Mg = Weight of rods

139

Take moments about B for BC

$$\left(\frac{L}{2}\right)MgCos\theta + LT\,Sin\theta = FLCos\theta$$

$\Rightarrow MgCos\theta + 2T\,Sin\theta = 2FCos\theta \Rightarrow Mg + 2T\,Tan\theta = 2F$ (i)
Horizontal forces are : $T - X = 0 \Rightarrow T = X$ (ii)
Vertical forces are : $F = Mg + Y$ (iii)
But : $F = \mu T = \frac{1}{3}T \Rightarrow$ equation (i) becomes :

$Mg = \frac{2}{3}T - 2T\,Tan\theta = T\left(\frac{2}{3} - 2\,Tan\theta\right)$ (iv)

Consider the system ABC: the system is shown in Fig. 3:

$$\Rightarrow 2Mg\left(\frac{L}{2}Cos\theta\right) = T2L(Sin\theta)$$

$\Rightarrow Mg = 2T\,Tan\theta$ (v)

Putting this into equation (iv) gives:
$2T\,Tan\theta = T\left(\frac{2}{3} - 2Tan\theta\right)$

$\Rightarrow 4\,Tan\theta = \frac{2}{3} \Rightarrow Tan\theta = \frac{1}{6} = 0.1666 \Rightarrow \theta = 9.45° \Rightarrow 2\theta = 18.9°$

14. Two member frame
A two member frame consists of rods of equal mass M kg, AB (of length L metres) and BC (See Fig. 1). Find the components of the reactions at A and C.
Solution

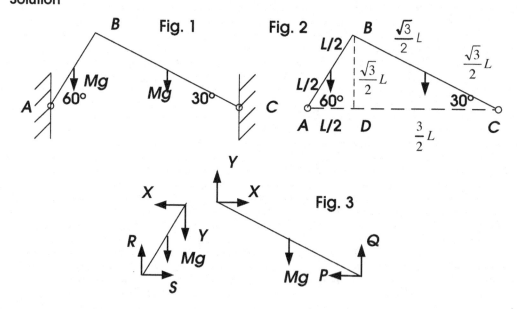

Draw a line between A and C. Draw D so that $BD \perp AC$ (See Fig. 2):

From geometry:
Let $|AB| = L$ It is possible to derive the following :

$$|AD| = \frac{L}{2}, \quad |BD| = \frac{\sqrt{3}}{2}L, \quad |BC| = L\sqrt{3}, \quad |CD| = \frac{3}{2}L, \quad |AC| = 2L$$

The forces in Newtons acting on the rods are shown in Fig. 3:

140

S, R = Horizontal and Vertical components of Reaction at A
P, Q = Horizontal and Vertical components of Reaction at C
X, Y = Horizontal and Vertical components of Reaction at B
Mg = Weight of mass M

For frame ABC: $R + Q = 2Mg$ (i)

Take moments about A for frame ABC:

$$Mg\left(\frac{L}{2} + \frac{3}{4}L\right) + Mg\left(\frac{L}{4}\right) = Q\left(\frac{L}{2} + \frac{3}{2}L\right) = Q(2L)$$

$$\Rightarrow Mg\left(\frac{3}{2}\right)L = Q(2L) \Rightarrow Q = \frac{3}{4}Mg \Rightarrow \text{From equation (i)}: R = \frac{5}{4}Mg \quad \text{(iii)}$$

Take moments about B for AB:

$$Mg\left(\frac{L}{2}\right)Cos60° + SL\,Sin60° = RL\,Sin30°$$

$$\Rightarrow Mg\frac{L}{2}\frac{1}{2} + SL\frac{\sqrt{3}}{2} = RL\frac{1}{2} \Rightarrow Mg + 2S\sqrt{3} = 2R \quad \text{(ii)}$$

$$\Rightarrow \text{From equation (iii)}: 2S\sqrt{3} = 2R - Mg = \frac{3}{2}Mg \Rightarrow S = \frac{\sqrt{3}}{4}Mg$$

Components of the reaction at A are:

Vertically: $R = Y + Mg \Rightarrow Y = R - Mg = \frac{1}{4}Mg$

Horizontally: $S = X \Rightarrow X = \frac{\sqrt{3}}{4}Mg$

Considering BC: Components of the reaction at C are:

$$Q = \frac{3}{4}Mg \text{ and } P = X = \frac{\sqrt{3}}{4}Mg$$

15. Disc and rods
A disc of radius r metres and mass M kg supported by two light rods AB and CD where $|AE| = |ED| = a$, $|CE| = |EB| = b$, hinged at E, which are kept in equilibrium by a string joining A and C (see Fig. 1). If $a = 5r$ and $b = 2r$ show that the tension, T, of the string $= \frac{3}{4}$ Mg Newtons.

Solution
From Fig. 1: Let $|AE| = |DE| = a$ and $|CE| = |BE| = b$
Fig. 2 shows the forces in Newtons acting on the whole system. Fig. 3 shows the relationship between a, b and r. Fig. 4 shows the forces acting on the rods. Fig. 5 shows the forces acting on the disc where:
R = Reaction between rod and horizontal plane
X, Y = Horizontal and vertical components of reaction at hinge at E
S = Normal reaction at C, B (where disc makes contact with supports)
T = Tension in string
Mg = Weight of disc

From Fig. 3: $Tan\alpha = \dfrac{r}{b} \Rightarrow b = \dfrac{r}{Tan\alpha}$ (i)

Fig. 4 shows the forces acting on each rod separately:

For AB: Horizontal: $T + S\,Cos\,\alpha = X$ (ii)   Vertical: $R = Y + S\,Sin\,\alpha$ (iii)

For the overall system: $2R = Mg \Rightarrow R = \frac{1}{2}\,Mg$ (iv)

For the disc: $2S\,Sin\,\alpha = Mg$ (v)

Equations (iii), (iv) and (v) show that: $Y = 0$.

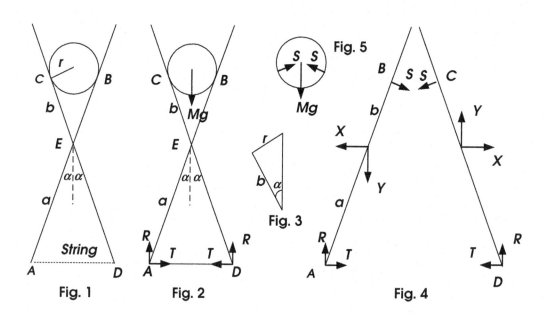

Fig. 5

Fig. 3

Fig. 1   Fig. 2   Fig. 4

Taking moments about B(for AB):

$XCos\alpha(b) + RSin\alpha(a + b) = TCos\alpha(a + b)$ (vi)

$\Rightarrow T = X\left(\dfrac{b}{a+b}\right) + RTan\alpha$ (vii)

Taking moments about A for AC: $XCos\alpha(a) = S(a + b) = 0$

But, from equation (v): $XCos\alpha(a) = \dfrac{Mg}{2Sin\alpha}(a + b)$

$\Rightarrow X = \dfrac{Mg}{2Sin\alpha Cos\alpha}\dfrac{(a + b)}{a}$

$\Rightarrow$ Equation (vii) becomes: $T = \dfrac{Mg}{2Sin\alpha Cos\alpha}\left(\dfrac{a+b}{a}\right)\left(\dfrac{b}{a+b}\right) + \dfrac{Mg}{2}Tan\alpha$

$\Rightarrow T = \left(\dfrac{Mgb}{2aSin\alpha Cos\alpha}\right) + \dfrac{Mg}{2}Tan\alpha.$

But, Fig. 3 shows: $Tan\alpha = \dfrac{r}{b} \Rightarrow b = \dfrac{r}{Tan\alpha}$

$\Rightarrow T = \left(\dfrac{Mgb}{2aSin\alpha Cos\alpha}\right) + \dfrac{Mg}{2}Tan\alpha = \dfrac{Mg}{2}\left(\dfrac{r}{aSin^2\alpha} + Tan\alpha\right)$

If $a = 5r$ and $b = 2r \Rightarrow T = \frac{3}{4}\,Mg$ Newtons

16. Two rods in equilibrium

Two rods, *AB* and *BD*, of equal length *L* metres are freely jointed at *B*. If they are placed standing vertically with ends *A* and *C* on a smooth surface with horizontal forces *P* and *Q* applied to *A* and *C* respectively (See Fig. 1). Find *P* and *Q* for equilibrium conditions if (a) Both rods have equal mass *M* kg, (b) Mass of *CD* is increased to 2*M* kg.

Solution

(a) Both rods have equal mass *M* kg

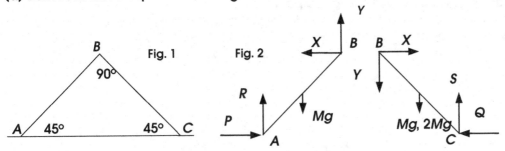

The forces in Newtons acting on each rod are shown in Fig. 2.

X, Y = Horizontal and Vertical Reactions at *B*
R, S = Normal Reactions between rods and floor
P, Q = Horizontal forces being applied, and Mg = Weights of masses

For frame *ABC*: $R + S = 2 Mg$ and $R = S$ (as frame is symmetrical) $\Rightarrow R = S = Mg$
Taking moments about *A* for *AB*:

$$Mg \, Cos\, 45° \left(\frac{L}{2}\right) = L\left(X \, Cos\, 45° + Y \, Cos\, 45°\right) \Rightarrow \frac{Mg}{2} = X + Y \qquad \text{(i)}$$

Taking moments about *C* for *BC*:

$$MgCos\, 45° \left(\frac{L}{2}\right) = L\left(X \, Cos\, 45° - Y \, Cos\, 45°\right) \Rightarrow \frac{Mg}{2} = X - Y \qquad \text{(ii)}$$

$\Rightarrow$ From equations (i) and (ii): $X = \dfrac{Mg}{2}, \quad Y = 0$

Considering each rod separately :
Rod *AB* : Horizontal forces : $X = P \Rightarrow P = 0.5 Mg$
Rod *BC* : Horizontal forces : $X = Q \Rightarrow Q = 0.5 Mg$

(b) The mass of *BD* is increased to 2*M*

For frame *ABC*: $R + S = 3 Mg$

Taking moments about *A* for *ABC* : $SL\sqrt{2} = 2Mg\left(\dfrac{3}{4}\right)L\sqrt{2} + Mg\left(\dfrac{1}{4}\right)L\sqrt{2}$

$$\Rightarrow S = \frac{7}{4} Mg \Rightarrow R = \frac{5}{4} Mg$$

Taking moments about *B* for *AB* : $MgCos45°\left(\dfrac{L}{2}\right) = L\left(RCos45° - PCos45°\right)$

$$\Rightarrow \frac{MgL}{2\sqrt{2}} = L\left(\frac{5}{4\sqrt{2}} Mg - \frac{P}{\sqrt{2}}\right) \Rightarrow P = \frac{3}{4} Mg \Rightarrow X = \frac{3}{4} Mg$$

Considering each rod separately :

Rod AB : Vertical forces : $R + Y = Mg \Rightarrow Y = Mg - R = -\dfrac{1}{4}Mg$

Rod BC : $X = Q \Rightarrow Q = \dfrac{3}{4}Mg$ $\qquad Y = Mg - R = -\dfrac{1}{4}Mg$

17. Two rigidly joined rods hanging from a point

Two identical rods AB, BC each of weight W Newtons are rigidly jointed at B at an angle α and the combination ABC is hinged at A and hangs freely from a fixed point (See Fig. 1). Find the angle θ which AB makes with the vertical where: (a) $\alpha = 90°$ (b) $\alpha = 120°$

Solution
(a) $\alpha = 90°$

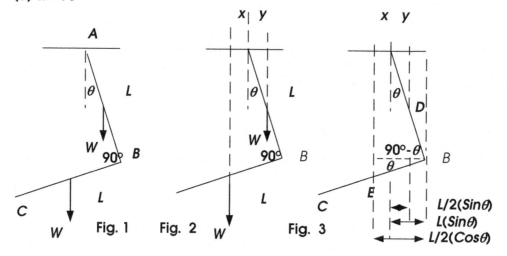

The combination ABC is in equilibrium when the moments created by the weights of the rods are equal and opposite i.e. $Wx = Wy$ (See Fig. 2)
But : if $Wx = Wy \Rightarrow x = y$ and (See Fig.3)

$y = \dfrac{L}{2}Sin\theta, \quad x = \dfrac{L}{2}Cos\theta - LSin\theta$

$\Rightarrow \dfrac{L}{2}Cos\theta - LSin\theta = \dfrac{L}{2}Sin\theta \Rightarrow Cos\theta = 3Sin\theta \Rightarrow \theta = Tan^{-1}\dfrac{1}{3}$

(b) $\alpha = 120°$
But : if $Wx = Wy \Rightarrow x = y$ and (Use Fig.3 again, but angle is $120°$ rather than $90°$)

$y = \dfrac{L}{2}Sin\theta \quad x = \dfrac{L}{2}Cos(\theta + 30) - LSin\theta$

$\Rightarrow \dfrac{L}{2}Cos(\theta + 30) - LSin\theta = \dfrac{L}{2}Sin\theta \Rightarrow Cos(\theta + 30) = 3Sin\theta$

$\Rightarrow Cos\theta \, Cos30° - Sin\theta \, Sin30° = 3Sin\theta \Rightarrow Cos\theta \dfrac{\sqrt{3}}{2} - Sin\theta \dfrac{1}{2} = 3Sin\theta \Rightarrow \theta = Tan^{-1}\dfrac{\sqrt{3}}{7}$

18. Two rigidly joined rods hanging from a point

If two rods AB, BC of equal length L metres and equal mass M kg are freely jointed and supported by a light string at C such that $\angle ABC$ is 120° and the angle AB makes with the vertical is $\theta = \text{Tan}^{-1} \sqrt{3}/7$ find the tension in the string.

Solution

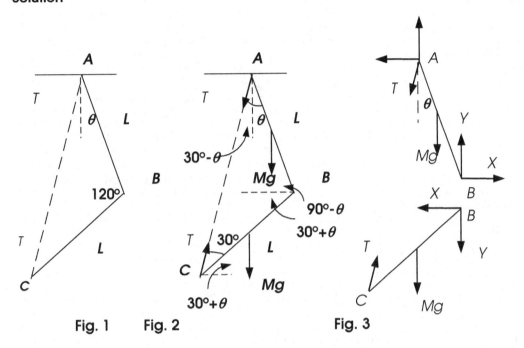

Fig. 1   **Fig. 2**   **Fig. 3**

The relevant angles and forces (in Newtons) are shown in Figs. 2 and 3.
X, Y = Horizontal and Vertical components of the Reaction at B
T = Tension in string
Mg = Weight of rods

Taking moments about B for BC: $Mg\,Cos\left(30° + \theta\right)\left(\dfrac{L}{2}\right) = LTSin30°$

$$\Rightarrow Mg\left(\frac{\sqrt{3}}{2}Cos\theta - \frac{1}{2}Sin\theta\right)\frac{1}{2} = T\frac{1}{2} \qquad \Rightarrow T = Mg\left(\frac{\sqrt{3}}{2}Cos\theta - \frac{1}{2}Sin\theta\right)$$

But: $Tan\theta = \dfrac{\sqrt{3}}{7} \Rightarrow Cos\theta = \dfrac{7}{\sqrt{52}}, \quad Sin\theta = \dfrac{\sqrt{3}}{\sqrt{52}}$

$$\Rightarrow T = Mg\left(\frac{\sqrt{3}}{2}\frac{7}{\sqrt{52}} - \frac{1}{2}\frac{\sqrt{3}}{\sqrt{52}}\right) = Mg\frac{6\sqrt{3}}{2\sqrt{52}} = \frac{3\sqrt{3}}{2\sqrt{13}}Mg = 0.721Mg \quad \text{Newtons}$$

19. Two rigidly joined rods hanging from a point

Two identical rods AB, BC of length L_1 and L_2 metres respectively and weight $M(L_1)$ and $M(L_2)$ Newtons respectively are rigidly jointed at B at an angle of 90° and the combination ABC is hinged at, and hangs freely from, a fixed point at A (See Fig. 1). Find an expression for angle θ in terms of L_1 and L_2.

Solution

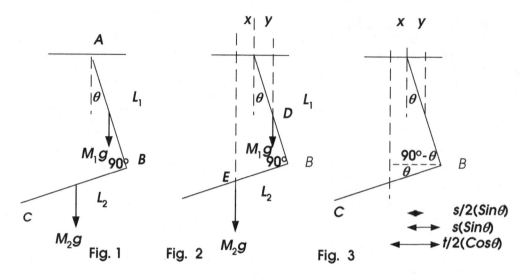

	Fig. 1
	Fig. 2
	Fig. 3

Assume that the midpoints of AB and BC are D and E respectively. The weights of the rods act through these points. The combination ABC is in equilibrium when the moments created by the weights of the rods are equal and opposite i.e. $M(L_1)y = M(L_2)x$ (See Fig. 2). $\Rightarrow (L_1)y = (L_2)x$

$$y = \frac{L_1}{2}Sin\theta, \quad x = \frac{L_2}{2}Cos\theta - L_1 Sin\theta$$

$$\Rightarrow L_2\left(\frac{L_2}{2}Cos\theta - L_1Sin\theta\right) = L_1\left(\frac{L_1}{2}Sin\theta\right) \Rightarrow \frac{L_2^2}{2} - L_1L_2Tan\theta = \frac{L_1^2}{2}Tan\theta$$

$$\Rightarrow \frac{L_1^2}{2}Tan\theta + L_1L_2Tan\theta = \frac{L_2^2}{2} \Rightarrow Tan\theta = \frac{L_2^2}{L_1^2 + 2L_1L_2}$$

20. Rods suspended by strings

A rod AB of mass $2M$ kg and length L metres is suspended by two strings of equal length, D metres.
(a) Find the tension in each strings, T_A and T_B and the ratio $T_A:T_B$
(b) If a second mass M kg is added to AB at a distance $L/4$ metres from end B find the resulting ratio $T_A:T_B$

Solution

(a) Find the tension in each strings, T_A and T_B and the ratio $T_A:T_B$
As the system is system is symmetrical, $T_A = T_B \Rightarrow T_A : T_B = 1$
Let $T_A = T_B = T$. Then, resolving forces vertically:

$$T_A\ Sin\ \alpha + T_B\ Sin\ \alpha = 2\ T\ Sin\ \alpha = 2Mg \Rightarrow T = \frac{2Mg}{2Sin\alpha} = \frac{Mg}{Sin\alpha}$$

(b) If a second mass M kg is added to AB at a distance $L/4$ metres from end B, find the reactions at A, B and C.
Assume that the addition of the second mass causes the rod to tilt at an angle θ to the horizontal.

Taking moments about A for AB: $T_B(L)Sin\alpha = 2Mg\dfrac{L}{2}Cos\theta + Mg\dfrac{3L}{4}Cos\theta = MgL\dfrac{7}{4}Cos\theta$

Taking moments about B for AB: $T_A(L)Sin\alpha = 2Mg\dfrac{L}{2}Cos\theta + Mg\dfrac{L}{4}Cos\theta = MgL\dfrac{5}{4}Cos\theta$

$$\Rightarrow \frac{T_B(L)Sin\alpha}{T_A(L)Sin\alpha} = \frac{MgL\dfrac{7}{4}Cos\theta}{MgL\dfrac{5}{4}Cos\theta} \Rightarrow \frac{T_B}{T_A} = \frac{7}{5}$$

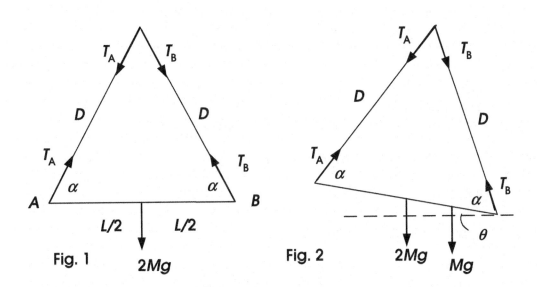

Fig. 1   Fig. 2

2Mg   2Mg   Mg

21. Rods suspended by strings

A light rod AB is suspended by two strings of equal length, L metres from a point O such that OA and OB form a right angle. If two weights W_A, W_B, (where $W_B > W_A$), are placed at A and B respectively, find the resulting angle θ which the rod will make with the horizontal (See Fig. 1).

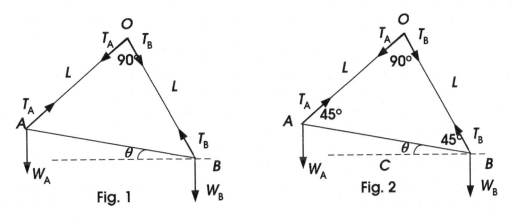

Fig. 1   Fig. 2

Solution

As $\angle AOB$ is a right angle and $|OA| = |OB|$ then $\angle OAB = \angle OBA = 45°$

Taking moments about A for AB: $T_B(L\sqrt{2})Sin45° = W_B L\sqrt{2}Cos\theta \Rightarrow T_B = W_B Cos\theta\sqrt{2}$ (i)

Taking moments about B for AB: $T_A(L\sqrt{2})Sin45° = W_A L\sqrt{2}Cos\theta \Rightarrow T_A = W_A Cos\theta\sqrt{2}$ (ii)

Resolving forces parallel and perpendicular to AB gives:

Parallel: $T_A Cos45° + W_A Sin\,\theta + W_B Sin\,\theta = T_B Cos\,45°$ (iii)

Perpendicular: $T_A Cos45° + T_B Cos45° = W_A Cos\,\theta + W_B Cos\,\theta$ (iv)

From equation (iii): $\dfrac{T_A}{\sqrt{2}} + (W_A + W_B)Sin\theta = \dfrac{T_B}{\sqrt{2}}$

Using values from equations (i), (ii): $W_A Cos\theta + (W_A + W_B)Sin\theta = W_B Cos\theta$

$\Rightarrow (W_A + W_B)Sin\theta = W_B Cos\theta - W_A Cos\theta = (W_B - W_A)Cos\theta \Rightarrow Tan\theta = \dfrac{W_B - W_A}{W_A + W_B}$

22. Three rod symmetrical frame

Find the forces acting along each rod of the frame shown in Fig. 1:

Solution

The forces in Newtons acting on the rods are shown in Fig. 2:

T = Thrust in rod

R, S = Reactions at A, C

Mg = Weight of mass M

Frame ABC: $R + S = Mg$ Point B: $Mg = 2T_1 Cos45° \Rightarrow T_1 = \dfrac{Mg}{\sqrt{2}}$

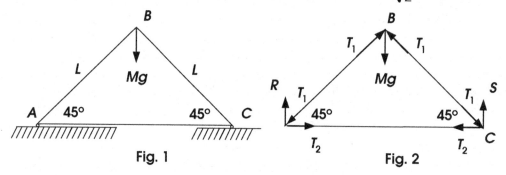

Fig. 1   Fig. 2

Point A:

Vertically : $R = T_1 Cos\,45° = \dfrac{T_1}{\sqrt{2}} = \dfrac{Mg}{2} \Rightarrow S = \dfrac{Mg}{2}$

Horizontally : $T_1 Cos\,45° = T_2 \Rightarrow T_2 = \dfrac{Mg}{2}$

Chapter 8   Equilibrium

1. Diving Board
A person of mass 100 kg is standing on the end of a light diving board as shown below (see Fig. 1). Find the reactions at A and B, if the board is pinned at A and rests on B.

Solution

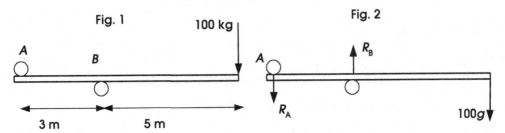

The forces acting in Newtons are:
R_A and R_B = Reactions at A and B, respectively,   100g = Weight of mass

The forces are: Horizontal: Σ Forces = 0   Vertical: $R_B = R_A + 100g$
Taking moments about A: $(100g)(3 + 5) = R_B (3) \Rightarrow R_B = 266.7g$ Newtons
$\Rightarrow R_A = 266.7g - 100g = 166.7g$ Newtons
Also: Taking moments about B: $(100g)(5) = (R_A)(3) \Rightarrow R_A = 166.7g$ Newtons

2. Rod and Sphere
A light rod AB, 5 metres in length supports a sphere of mass M kg. The rod is pivoted at A while B is supported by a cord, fixed to the vertical surface at C as shown in Fig. 1. The sphere touches the surface of the rod AB at D and the vertical surface AC at E. Assume lengths of AD and DB are 2 and 3 metres respectively. $\angle BAC = 45°$. If all contact surfaces are smooth find the tension of the cord, T.

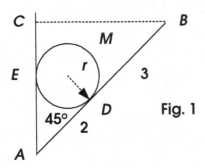

Fig. 1

Solution
From Fig. 1: Angle $\angle ABC = \angle BAC = 45°$ and $\angle ACB = 90°$. The forces on the sphere and rod are shown in separate diagrams. Note: R_V and R_H are the vertical and horizontal reactions respectively at A.
The forces in Newtons acting on the sphere are:

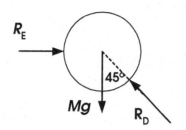 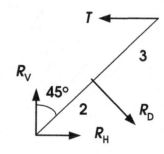

Fig. 2

Horizontally : $R_E = R_D \, Cos \, 45° \Rightarrow R_E = \dfrac{R_D}{\sqrt{2}}$

Vertically : $R_D \, Cos 45° = Mg \Rightarrow R_D = Mg\sqrt{2} \Rightarrow R_E = Mg$ (i)

The forces acting on the rod can be found by taking moments about A :

$5T \, Cos 45° = 2R_D \Rightarrow 5\dfrac{T}{\sqrt{2}} = 2R_D \Rightarrow T = \dfrac{2R_D\sqrt{2}}{5}$ (ii)

But, from equation (i), $R_D = Mg\sqrt{2} \Rightarrow T = \dfrac{(Mg\sqrt{2})(2\sqrt{2})}{5} = \dfrac{4}{5}Mg$ Newtons

3. Sphere and two planes

Two smooth planes are inclined to the horizontal as shown in Fig. 1. A sphere of mass M kg and a radius r metres is placed in contact with the planes. Find the reactions between the planes and the sphere.

Solution

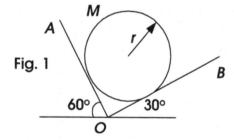 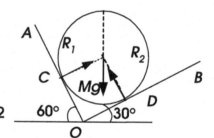

Fig. 1

Fig. 2

Let the planes be OA and OB. Let the sphere be in contact with OA at C and with OB at D. The forces acting in Newtons are:

R_1, R_2 = Reaction between the sphere and planes at C and D respectively.

Mg = Weight of sphere

R_1, R_2 and Mg intersect at the centre of mass of the sphere: this gives a triangle of forces where these forces are balanced. (See Fig. 2). Resolving the forces vertically and horizontally:

Vertically: $R_1 \, Cos \, 60° + R_2 \, Cos \, 30° = Mg$ (i)

Horizontally: $R_1 \, Cos \, 30° - R_2 \, Cos \, 60° = 0$ (ii)

From equation (ii): $R_1 \, Cos \, 30° = R_2 \, Cos 60° \Rightarrow R_1\dfrac{\sqrt{3}}{2} = R_2\dfrac{1}{2} \Rightarrow R_2 = R_1\sqrt{3}$

Putting this in equation (i) $\Rightarrow R_1\left(\dfrac{1}{2}\right) + R_1\sqrt{3}\dfrac{\sqrt{3}}{2} = Mg \Rightarrow R_1\left[\dfrac{1}{2} + \dfrac{3}{2}\right] = Mg$

$\Rightarrow R_1 = \dfrac{1}{2}Mg \Rightarrow R_2 = \dfrac{\sqrt{3}}{2}Mg$ Newtons

4. Two Spheres

Two identical spheres, each of mass m kg and radius r metres are attached by strings each 1 metre long to a fixed point O (See Fig. 1). If $r = 0.5$ metres find the tension in the strings, T, and the reaction between the spheres, R.

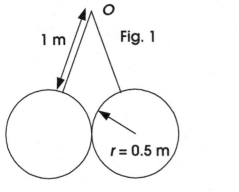

Fig. 1

1 m

$r = 0.5$ m

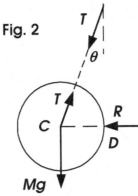

Fig. 2

Solution
The forces in Newtons acting on one sphere is shown in Fig. 2:
R = Reaction between spheres, T = Tension in string, Mg = Weight of sphere

Resolving vertically and horizontally: $R = T \sin\theta$ (i), $T \cos\theta = Mg$ (ii)

But : $\sin\theta = \dfrac{r}{|OC|} = \dfrac{r}{1+r} = \dfrac{0.5}{1.5} = \dfrac{1}{3} \Rightarrow \cos\theta = \dfrac{2\sqrt{2}}{3}$ From equations (i) and (ii) :

$T = \dfrac{3Mg}{2\sqrt{2}} \Rightarrow R = T \sin\theta = \dfrac{3Mg}{2\sqrt{2}} \times \dfrac{1}{3} = \dfrac{Mg}{2\sqrt{2}}$ Newtons

5. Two Spheres

Two steel spheres each of radius r metres and mass m kg are placed in a container of width $7/2\,r$ metres. The line joining their centres makes an angle of θ with the horizontal (See Fig. 1). If the container surfaces are smooth, find the reactions between the spheres and the container R_a, R_b and R_c. (Figs. 1,2)

Solution
The forces acting on the spheres are shown separately in Fig. 2 above:
R_a, R_c = Reactions between the spheres and walls at a and c respectively.
R_1, R_2 = Reactions between the spheres.
mg = Weight of spheres

For equilibrium: (i) $R_1 = R_2$
Resolving the forces on sphere 1:
(ii) Horizontally: $R_c = R_2 \cos\theta$ (iii) Vertically: $mg = R_2 \sin\theta$
Resolving the forces of sphere 2:
(iv) Horizontally: $R_1 \cos\theta = R_a$ (v) Vertically: $R_1 \sin\theta + mg = R_b$

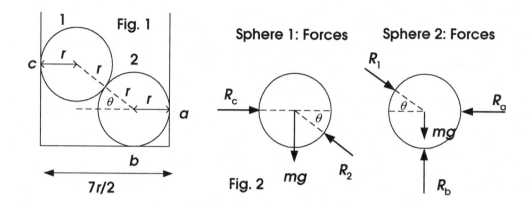

Fig. 1   Sphere 1: Forces   Sphere 2: Forces

Fig. 2

The container width $7r/2 = r + rCos\theta + rCos\theta + r = 2r + 2rCos\theta$ (See Fig. 1)

$$\Rightarrow \frac{7r}{2} = 2r + 2rCos\theta \Rightarrow Cos\theta = \frac{\frac{7r}{2} - 2r}{2r} = \frac{3}{4} \Rightarrow Sin\theta = \frac{\sqrt{7}}{4}$$

From equation (iii): $R_2 Sin\,\theta = mg \Rightarrow R_2 = \frac{mg}{Sin\theta} = \frac{4mg}{\sqrt{7}}$

$\Rightarrow$ From equation (ii): $R_c = R_2 Cos\theta \Rightarrow R_c = \frac{4mg}{\sqrt{7}} \times \frac{3}{4} = \frac{3mg}{\sqrt{7}}$

From equation (iv): $R_a = R_1 Cos\theta = R_2 Cos\theta = \frac{3mg}{\sqrt{7}}$

From equation (v): $R_b = mg + R_1 Sin\theta = mg + R_2 Sin\theta = mg + mg = 2mg$

6. Rod and Cord

A light rod AB of length L metres is pivoted at A and carries a load M kg in mid-span (See Fig. 1). A cord is fastened from C to B. If the cord makes an angle α with the vertical, and the rod makes an angle β with the vertical, express T, the tension of the cord, in terms of M, β and α.

Solution

The forces acting in Newtons are (See Fig. 2): R_H, R_V = Horizontal and vertical components respectively of the reaction at A, T = Tension in cord, Mg = Weight of rod

Draw line $BD \perp AB$ and draw a horizontal line EB through B. See Fig. 3. The angle $\angle CBD = \theta + \beta$ Taking moments about A for AB:

$$Mg\frac{L}{2}Sin\,\beta = TLCos(\theta + \beta) \Rightarrow MgSin\,\beta = 2T(Cos\theta\,Cos\,\beta - Sin\theta\,Sin\,\beta)$$

But since : $\alpha + \theta = 90° \Rightarrow Cos\theta = Sin\alpha$ and $Sin\theta = Cos\alpha$

$$\Rightarrow Mg\,Sin\,\beta = 2T(Sin\,\alpha\,Cos\,\beta - Cos\,\alpha\,Sin\,\beta) = 2T(Sin(\alpha - \beta)) \Rightarrow \quad T = \frac{Mg\,Sin\,\beta}{2\,Sin(\alpha - \beta)}\,\text{Newtons}$$

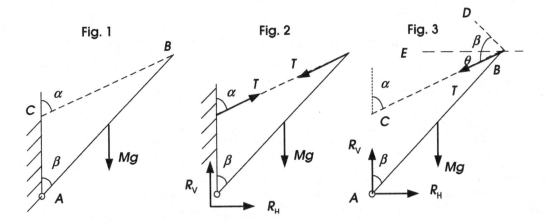

7. Ladder resting against a Wall

A uniform ladder AB of length L metres and mass m kg rests with one end against a smooth vertical wall and the other end on rough ground which slopes away from the wall at an angle α to the horizontal (see Fig. 1). The ladder makes an angle θ with the wall. If the ladder is on the point of slipping when $\theta = 30°$ and $\alpha = 15°$, find the coefficient of friction, μ.

Solution

The forces acting on the ladder in Newtons are:
R_1, R_2 = Reactions between surfaces and ladder at A and B respectively
F = Frictional force at B

Forces on the ladder (See Fig. 2) and relevant angles (See Fig. 3):

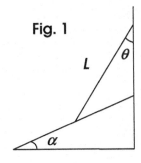
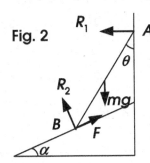
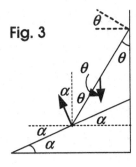

Taking moments about B for AB:

$$R_1 L \cos\theta = mg\sin\theta\frac{L}{2} \Rightarrow R_1 = \frac{mg\tan\theta}{2} \Rightarrow R_1 = \frac{1}{2}mg\tan\theta$$

Resolving forces horizontally : $R_1 + R_2 \sin\alpha = F\cos\alpha$

But : $F = \mu R_2$ (μ = coefficient of friction)

$$\Rightarrow R_1 + R_2\sin\alpha = \mu R_2\cos\alpha \Rightarrow R_1 = R_2(\mu\cos\alpha - \sin\alpha) \Rightarrow \frac{1}{2}mg\tan\theta = R_2(\mu\cos\alpha - \sin\alpha)$$

Resolving forces vertically : $R_2 \, Cos\alpha + F \, Sin\alpha = mg$

$$\Rightarrow R_2 \, Cos\alpha + \mu R_2 Sin\alpha = mg \Rightarrow R_2 \left(Cos\alpha + \mu \, Sin\alpha\right) = mg \Rightarrow R_2 = \frac{mg}{\left(Cos\alpha + \mu \, Sin\alpha\right)}$$

$$\Rightarrow \tfrac{1}{2}mg Tan\theta = mg\left(\frac{\mu \, Cos\alpha - Sin\alpha}{\mu \, Sin\alpha + Cos\alpha}\right)$$

$$\Rightarrow Tan\theta = 2\left(\frac{\mu \, Cos\alpha - Sin\alpha}{\mu \, Sin\alpha + Cos\alpha}\right) \quad \text{If} \quad \theta = 30° \text{ and } \alpha = 15° \text{ then} : \mu = 0.6$$

8. Rod in equilibrium

A rod, AB, of mass M kg rests on two smooth planes as shown in Fig. 1. If the rod carries an additional mass, $3M$ kg at a point a distance x metres from B find the value of x if the rod is to remain horizontal.

Solution

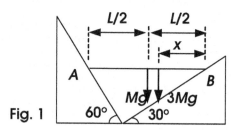

Fig. 1   Fig. 2

The forces acting are shown in Fig. 2:

Horizontally : $R_A Cos30° = R_B Cos60° \Rightarrow R_A\sqrt{3} = R_B$

Vertically : $R_A Cos60° + R_B Cos30° = 4Mg \Rightarrow R_A + R_B\sqrt{3} = 8Mg$

From the above : $R_A = 2Mg$ and $R_B = 2\sqrt{3}\,Mg$

Taking moments about A for AB: $R_B Cos30°(L) = Mg\left(\frac{L}{2}\right) + 3Mg(L - x)$

$$\Rightarrow R_B L\sqrt{3} = MgL + 6Mg(L - x) = 7MgL - 6Mgx$$

$$\Rightarrow 2\sqrt{3}\,MgL\sqrt{3} = 7MgL - 6Mgx \Rightarrow 6L = 7L - 6x \Rightarrow x = \frac{L}{6} \quad \text{metres}$$

9. Rod on two smooth planes

A uniform rod, 2 metres long and of mass $m = 10$ kg, is placed on two smooth planes inclined at 30° and 60° to the horizontal. Find the reactions on each plane and the inclination of the rod to the horizontal when in equilibrium. (See Fig. 1).

Solution

The forces acting on the rod are shown in Fig. 2:

(a) Let R and S be the reactions between the rod and planes at A and B respectively. R and S are acting at 90° to the planes at A and B.

(b) Let Mg be the weight of the rod. Continue AB to C so that angle $\angle BCD = \theta$. The planes intersect at D.

(c) As the rod is uniform its weight will act through its mid-point.

The angles associated with the forces are shown in Fig. 4.

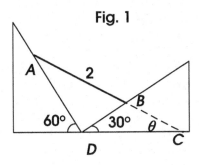

Fig. 1

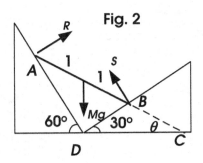

Fig. 2

Resolving forces:

Horizontally: $R \cos 30° = S \cos 60° \Rightarrow R\frac{\sqrt{3}}{2} = S\frac{1}{2} \Rightarrow S = R\sqrt{3}$ (i)

Vertically: $R \sin 30° + S \sin 60° = 10g \Rightarrow R\frac{1}{2} + S\frac{\sqrt{3}}{2} = 10g \Rightarrow R + S\sqrt{3} = 20g$ (ii)

$\Rightarrow 4R = 20g \Rightarrow R = 5g \Rightarrow S = 5g\sqrt{3}$

Taking moments about B for AB (See Fig. 4):

$Mg \cos\theta(1) = R \sin(30° + \theta)(2) \Rightarrow 10g \cos\theta = 10g \sin(30° + \theta)$

$\Rightarrow \cos\theta = \sin(30° + \theta) = (0.5)\cos\theta + \frac{\sqrt{3}}{2} \sin\theta = \cos\theta \Rightarrow \theta = 30°$

$\Rightarrow$ Inclination of the rod to the horizontal when in equilibrium is $\theta = 30°$

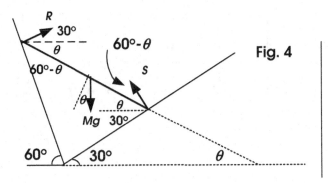

Fig. 4

10. Two rings sliding on a Rod

Two light rings can slide on a rough horizontal rod. The rings are connected by a light inextensible string of length L metres to the mid-point of which is attached a mass M kg. If μ is the coefficient of friction between the rings and the rod, and d is the distance between the rings, find an expression for μ in terms of L and d. The arrangement is shown in Fig 1.

Solution

The forces in Newtons are shown in Fig. 2: R = Reaction between rings and the rod, F = Frictional forces, T = Tension in string, Mg = Weight of mass

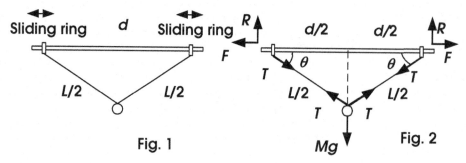

Fig. 1   Fig. 2

Take either ring: We have : $R = T\,Sin\theta$ (i), $F = T\,Cos\theta$ (ii)

But from equations (i), (ii) : $\mu = \dfrac{F}{R} = \dfrac{1}{Tan\theta} = \dfrac{\dfrac{d}{2}}{\sqrt{\left(\dfrac{L}{2}\right)^2 - \left(\dfrac{d}{2}\right)^2}} = \dfrac{d}{\sqrt{L^2 - d^2}}$

11. Mass on a rough surface

A cube of mass M kg and side L metres is placed on a rough surface. The coefficient of friction between the cube and surface is μ. A slowly increasing horizontal force, H Newtons, acts at a point a distance x metres up one side of the cube as shown in Fig. 1. Determine whether the cube slides or topples over for the cases: (a) $x = \frac{1}{2}L$ (b) $x = \frac{3}{4}L$

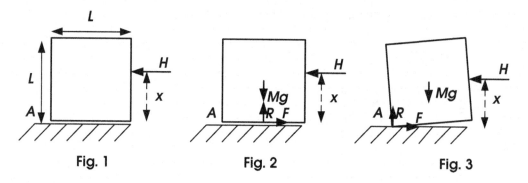

Fig. 1   Fig. 2   Fig. 3

Solution

The forces in Newtons acting on the cube are: R = Normal Reaction, F = Frictional force, Mg = Weight of cube, H = Horizontal force

If the cube slides:

The forces in Newtons are shown in Fig. 2. Resolving the forces:
Horizontally: $H = F$ Vertically: $R = Mg$ But, $F = \mu R = \mu Mg \Rightarrow H = \mu Mg$

If the cube topples over about point A:

The forces in Newtons are shown in Fig. 3 (note: block is shown at a slight angle to the horizontal for illustration only. In equilibrium the cube will remain flat on the plane).

Taking moments about point A: $Hx = Mg\dfrac{L}{2}$

(a) $x = \frac{1}{2}L$

For sliding: $H = \mu Mg$ For toppling: $H\dfrac{L}{2} = Mg\dfrac{L}{2} \Rightarrow H = Mg$

Whichever value of H is reached first will cause an impact on the cube. Therefore, for any value of $\mu < 1$ the cube will slide rather than topple.

(b) $x = ¾ L$

For sliding: $H = \mu Mg$ For toppling: $H\dfrac{3L}{4} = Mg\dfrac{L}{2} \Rightarrow H = \dfrac{2}{3} Mg$

Which value of H is reached first? For any value of $\mu < \dfrac{2}{3}$ the cube will slide rather than

topple. If $\mu > \dfrac{2}{3}$ the block will topple before it slides.

12. Mass on a rough inclined surface
A rectangular block of weight W Newtons, height h metres and width $h/2$ metres is placed on a rough plane hinged at A. (Fig. 1). The coefficient of friction between the block and plane is μ. The plane is slowly swivelled about A (See Fig. 2). Determine whether the block slides or topples over for the cases: (a) $\mu = 0.45$ (b) $\mu = 0.55$

Solution
With the plane inclined at angle θ to the horizontal, the forces in Newtons acting are: R = Normal Reaction, F = Friction, W = Weight of block

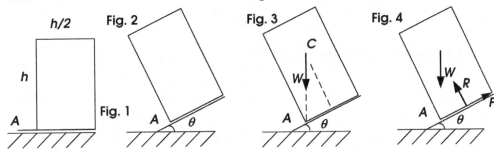

If the cube topples over:
At the moment of toppling over, W acts through the centre of mass, C, through point A.

(Fig. 3). From geometry: $Tan\theta = \dfrac{\dfrac{h}{2}}{h} = 0.5 \Rightarrow \theta = 26.57°$

$\Rightarrow$ The block will topple over when θ reaches a value of 26.57°.

If the cube slides:
The forces in Newtons are shown in Fig. 4. Resolving the forces:
Parallel to plane: $F = WSin\theta$ (i)   Perpendicular to plane: $R = WCos\theta$ (ii)
But : $F = \mu R \Rightarrow \mu = Tan\theta \Rightarrow \theta = Tan^{-1}\mu$
(a) $\mu = 0.45$
The block will begin to slide when: $\theta = Tan^{-1} 0.45 = 24.2°$. In this case, the angle is smaller than that which will cause toppling so the block will slide before toppling.
(b) $\mu = 0.55$
The block begins to slide when: $\theta = Tan^{-1} 0.55 = 28.81°$ so toppling occurs before sliding

13. Rod and Chain
A light rod AB of length 5 metres pivoted at A carries a mass of 500 kg at a distance $x = 2$ metres from A. A chain connects B to C (See Fig. 1). The chain is priced at Euro 0.1/Nm (where: m = chain length in metres and N = Tensile strength in Newtons). Find:
(a)   The tension in the chain in Newtons

(b)   The value of θ to give minimum chain cost
(c)   The minimum chain cost in Euro.

Solution
The forces acting on the rod in Newtons are:
T = Tension in chain, Mg = Weight of mass, R_H, R_V = Horizontal and vertical components respectively of the reaction at A.

(a) The tension in the chain
Taking moments about A (See Fig. 2): $Mg(x) = (T Sin\theta)(L) \Rightarrow T = \dfrac{Mgx}{L Sin\theta} = \dfrac{200g}{Sin\theta}$

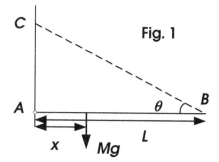

Fig. 1

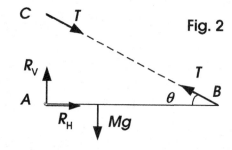

Fig. 2

(b) The value of θ to give minimum chain cost
Chain cost $= (Euro\ 0.1)(T)|BC|$ But $Cos\theta = \dfrac{L}{|BC|} = \dfrac{5}{|BC|} \Rightarrow |BC| = \dfrac{5}{Cos\theta}$

$\Rightarrow$ Chain cost $= (Euro\ 0.1)(\dfrac{200g}{Sin\theta}\dfrac{5}{Cos\theta}) = Euro\left(\dfrac{100g}{Sin\theta\ Cos\theta}\right) = Euro\left(\dfrac{200g}{Sin2\theta}\right)$ (i)

From equation (i): Chain cost is a minimum when $\theta = 45°$
(c) The minimum chain cost
$\Rightarrow$ Minimum chain cost = Euro 1,962

14. Forces on particles
A light string $ABCD$ is attached to fixed points at its ends A and D, where D is vertically below A. Particles of mass 5 kg and 10 kg are attached to the string at B and C, respectively, and a horizontal force H Newtons is applied to particle at B so that the string is in equilibrium as shown in Fig. 1. Find H.

Solution
We have: Angle $\angle BCD = 360 - 45 - 45 - 120 = 150°$
The forces in Newtons acting on B and C are (See Fig. 2):
T_1 = Tension is string AB, T_2 = Tension in string BC, T_3 = Tension is string CD
Resolving forces:
At point B:

Vertically : $T_1 Cos45° = T_2 Cos15° + 5g \Rightarrow T_1 = 1.366T_2 + 5g\sqrt{2}$ (i)

Horizontally : $H = T_1 Cos45° + T_2 Sin15° = 0.7071T_1 + 0.2588\ T_2$ (ii)

$\Rightarrow$ From equation (i): $H = 0.7071(1.366T_2 + 5g\sqrt{2}) + 0.2588T_2 = 1.2247\ T_2 + 49.05$ (iii)

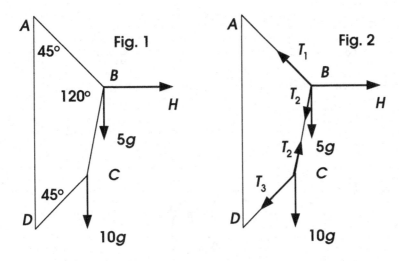

At point C:

Vertically : $T_2 \, Cos15° = T_3 \, Cos45° + 10g \Rightarrow 0.9659T_2 = 0.7071T_3 + 10g$ (iv)

Horizontally : $T_3 \, Cos45° = T_2 \, Sin15° \Rightarrow 0.7071T_3 = 0.2588T_2 \Rightarrow T_3 = 0.366T_2$ (v)

Putting this into equation (iv): $\Rightarrow T_2 = 138.74 \Rightarrow T_3 = 50.78$

Putting values in equation (iii): $H = 1.2247(138.74) + 49.05 \Rightarrow H = 218.96$ Newtons

15. Three Cylinders

Fig. 1 shows three identical cylinders, each of radius r metres and mass m kg. Each cylinder is in contact with the other two cylinders. The coefficient of friction between cylinders, and between cylinders and ground is μ. Find the smallest value of μ for the cylinders to remain in equilibrium.

Solution

The forces in Newtons acting on the cylinders are shown in Fig. 2:

R = Reaction between cylinders, F = Frictional force between cylinders

G = Frictional force between cylinders and ground,

S = Reaction between cylinders and ground, mg = Weight of each cylinder

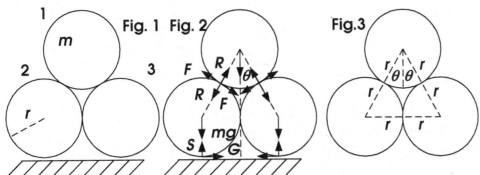

The cylinders' axes are a distance of $2r$ from each other: Thus $2\theta = 60° \Rightarrow \theta = 30°$ (See Fig. 3). Consider the forces on the cylinders (Fig. 4):

Forces acting on cylinder 1:

Resolving vertically: $2R \, Cos \, 30° + 2F \, Cos \, 60° = mg \Rightarrow R\sqrt{3} + F = mg$ (i)

Consider the forces on cylinder 2:

Moments of the frictional forces: $Fr = Gr \Rightarrow F = G$

Resolving horizontally: $R\,Cos60° = F\,Cos\,30° + G = F\,Cos\,30° + F$

$\Rightarrow R = F\sqrt{3} + 2F \Rightarrow R = F(\sqrt{3} + 2) \Rightarrow \dfrac{F}{R} = \dfrac{1}{(\sqrt{3} + 2)}$ But: $\dfrac{F}{R} = \mu \Rightarrow \mu = 0.268$

This is the minimum value of μ for the cylinders to remain in equilibrium.

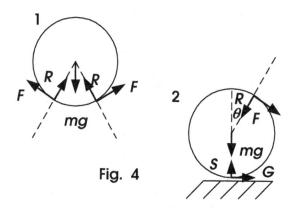

Fig. 4

16. Three Cylinders

Two cylinders of radius r metres and mass m kg lie on a rough table with the centres a distance $3\,r$ apart. A third cylinder is placed in contact with the two cylinders (see Fig.1). If the frictional forces are just enough to keep the cylinders in equilibrium, find:
(a) The least coefficient of friction between upper and lower cylinders, μ_1
(b) The least coefficient of friction between lower cylinders and the ground, μ_2

Solution

The forces in Newtons acting on the cylinders are shown in Fig. 2:

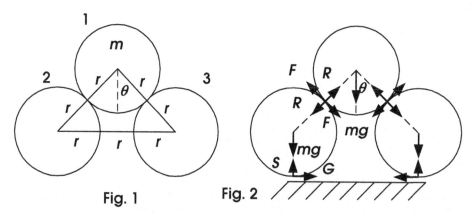

Fig. 1   **Fig. 2**

R = Reaction between cylinders, F = Frictional force between cylinders
G = Frictional force between cylinders and ground,
S = Reaction between cylinders and ground, mg = Weight of each cylinder

Fig. 1 indicates how θ can be calculated:

$Sin\theta = \dfrac{1.5r}{2r} = 0.75, \quad Cos\theta = \dfrac{\sqrt{(2r)^2 - (1.5r)^2}}{2r} = \dfrac{r\sqrt{4 - 2.25}}{2r} = \dfrac{\sqrt{1.75}}{2} = 0.6614$

The forces involved are shown in Figs 2 and 3:

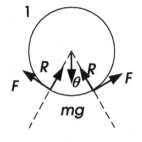

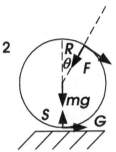

Fig. 3

(a) The least coefficient of friction between upper and lower cylinders, μ_1

From Fig. 2: $2S = 3mg \Rightarrow S = 1.5mg$

Vertical forces acting on cylinder 1: $2R \cos\theta + 2F \sin\theta = mg \Rightarrow 1.328 R + 1.5 F = mg$ (i)

Moments of the frictional forces acting on cylinder 2: $Fr = Gr \Rightarrow F = G$

Resolving horizontally: $F \cos\theta + G = R \sin\theta \Rightarrow 0.6614 F + F = 0.75 R \Rightarrow R = 2.2152 F$ (ii)

Resolving vertically: $R \cos\theta + F \sin\theta + mg = S$ (iii)

$\Rightarrow 0.6614 R + 0.75 F + mg = 1.5 mg$

$\Rightarrow$ (Using equation (ii)): $1.465 F + 0.75 F = 0.5 mg \Rightarrow F = 0.2257 mg = G \Rightarrow R = 0.5 mg$

But: $F = \mu_1 R \Rightarrow$ coefficient of friction between cylinders $= \mu_1 = \dfrac{F}{R} = 0.451$

Thus the least coefficient of friction between the cylinders which can preserve equilibrium is $\mu_1 = 0.451$

(b) The least coefficient of friction between lower cylinders and ground, μ_2

$\mu_2 = \dfrac{G}{S} = \dfrac{F}{S} = \dfrac{0.2257mg}{1.5mg} = 0.15$ Thus the least coefficient of friction between the

lower cylinders and the table which can preserve equilibrium is $\mu_2 = 0.15$

17. Three cylinders in a cylinder

Fig. 1 shows three identical smooth cylinders, each of radius r metres and weight W Newtons resting in equilibrium within a large smooth cylinder of radius R metres (only part-shown). Each small cylinder is in contact with the other two small cylinders. Assuming there is no friction involved, find the largest value of R in terms of r for the small cylinders to remain in equilibrium.

Solution

See Fig. 2. Assume that the large cylinder has centre at O. Draw lines from O to the points A, B where the small cylinders touch the large cylinder. These lines, OA and OB, must pass through the cylinder centres, C and D respectively as the reactions at C and D must pass through the centres of the smaller and larger cylinders. Let the angle $\angle AOE = \theta$.

The small cylinders' axes are a distance of $2r$ from each other: Thus angle $\angle CED =$ angle $\angle ECD =$ angle $\angle CDE = 60^\circ$

The forces in Newtons acting on the cylinders are shown in Fig. 3:

$T =$ Reaction between upper and lower small cylinders

U = Reaction between two small lower cylinders (Note: U acts horizontally)
S = Reaction between large cylinder and small cylinder
W = Weight of each cylinder

Consider the forces on the cylinders:

Forces acting on cylinder 1: Resolving vertically: $2T \cos 30° = W \Rightarrow T\sqrt{3} = W$ (i)

Consider the forces on cylinder 2:

Resolving horizontally: $T \cos 60° + U = S \sin\theta \Rightarrow \dfrac{T}{2} + U = S \sin\theta$

But at the point where equilibrium is lost, $U = 0 \Rightarrow \dfrac{T}{2} \le S \sin\theta$ (ii)

From equation (i) this gives:

$\dfrac{W}{2\sqrt{3}} \le S\sin\theta$ But, considering the whole system : $2S \cos\theta = 3W$ (iii)

$\Rightarrow \dfrac{W}{2\sqrt{3}} \le \dfrac{3W}{2\cos\theta} \sin\theta \Rightarrow \tan\theta = \dfrac{1}{3\sqrt{3}}$

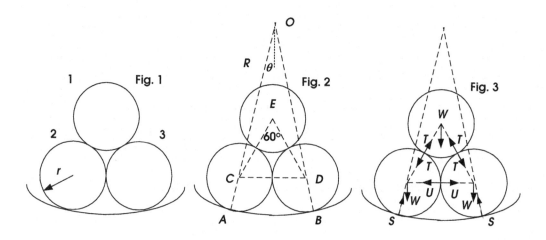

Fig. 1   Fig. 2   Fig. 3

But, from Fig. 2 : $\tan\theta = \dfrac{r}{\sqrt{(R-r)^2 - r^2}} \Rightarrow \dfrac{r}{\sqrt{(R-r)^2 - r^2}} = \dfrac{1}{3\sqrt{3}}$

$\Rightarrow 27r^2 = (R-r)^2 - r^2 \Rightarrow (R-r)^2 = 28r^2 \Rightarrow R-r = 2r\sqrt{7}$

$\Rightarrow R = r + 2r\sqrt{7} = r(1 + 2\sqrt{7})$ metres = maximum value for R for equilibrium to exist.

18. Hemisphere in equilibrium

A solid hemisphere of weight W Newtons and radius r metres is attached by a string to a fixed point A on a smooth vertical wall (See Fig. 1) such that the string makes an angle θ with the wall and the hemisphere (OC) makes an angle of α with the wall . When the system is in equilibrium find the relationship between θ and α.

Solution

Let: AB = String joining hemisphere to A, BC = Diameter of hemisphere = $2r$ metres,
O = Midpoint of BC, G = Centre of gravity of hemisphere

OG = Line joining O, G where $|OG| = \dfrac{3}{8}r$

D = Intersection of continuation of lines AB and OG

162

Let $|DG| = x$ metres $\Rightarrow |OD| = \dfrac{3}{8}r + x$ metres

The forces in Newtons acting are shown in Fig. 2: R = Reaction between wall and hemisphere, T = Tension in string, W = Weight of hemisphere

Fig. 4 (not to same scale as other figures above) is shown for clarity:
As ABF and EBF are right angle triangles then:
Angle $\angle EBF = 90° - \alpha$ Angle $\angle EBA = \beta$
But: $\angle EBA + \angle EBF + \theta = 90° \Rightarrow \beta + 90° - \alpha + \theta = 90° \Rightarrow \beta = \alpha - \theta$
But: $\angle DBO = \beta \Rightarrow \angle DBO = \alpha - \theta$
Equilibrium is achieved when the moments of the forces T and W are equal:
$WCos\alpha|OG| = TCos\beta|OD|$ But :

Resolving forces vertically gives : $W = TCos\theta \Rightarrow WCos\alpha|OG| = W\dfrac{Cos\beta}{Cos\theta}|OD|$

Also : $|OG| = \dfrac{3}{8}r, \quad |OD| = \dfrac{3}{8}r + |DG| = \dfrac{3}{8}r + x, \quad \beta = \alpha - \theta$

$\Rightarrow WCos\alpha\dfrac{3}{8}r = W\dfrac{Cos(\alpha - \theta)}{Cos\theta}\left(\dfrac{3}{8}r + x\right)$ (i)

$Tan\beta = Tan(\alpha - \theta) = \dfrac{|OD|}{|OB|} = \dfrac{\left(\dfrac{3}{8}r + x\right)}{r} \Rightarrow x = rTan(\alpha - \theta) - \dfrac{3}{8}r$

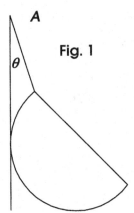

Fig. 1

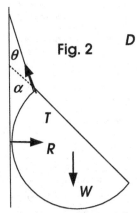

Fig. 2

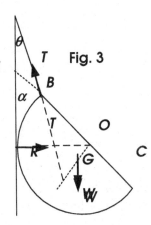

Fig. 3

$\Rightarrow$ Equation (i) can be re – written :

$WCos\alpha\dfrac{3}{8}r = W\dfrac{Cos(\alpha - \theta)}{Cos\theta}\left(\dfrac{3}{8}r + rTan(\alpha - \theta) - \dfrac{3}{8}r\right) = W\dfrac{Cos(\alpha - \theta)}{Cos\theta}rTan(\alpha - \theta)$

$\Rightarrow WCos\alpha\dfrac{3}{8}r = Wr\dfrac{Sin(\alpha - \theta)}{Cos\theta} \Rightarrow \dfrac{3}{8}Cos\alpha Cos\theta = Sin(\alpha - \theta)$

$\Rightarrow \dfrac{3}{8}Cos\alpha Cos\theta = Sin\alpha Cos\theta - Cos\alpha Sin\theta$ Dividing across by $Cos\alpha Cos\theta$:

$\Rightarrow \dfrac{3}{8} = Tan\alpha - Tan\theta \Rightarrow Tan\alpha = Tan\theta + \dfrac{3}{8}$

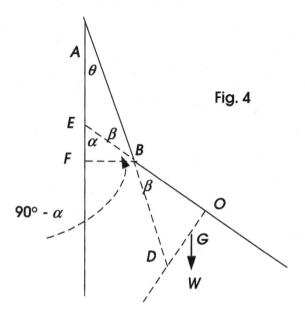

Fig. 4

19. Rod inside Bowl

Two particles of mass m and M kg respectively are carried on the ends of a light rod of length L metres. If the rod is placed inside a smooth hemispherical bowl of radius r metres with m resting at the rim at A, while M rests at B, find the ratio m/M in terms of L and r.

Solution

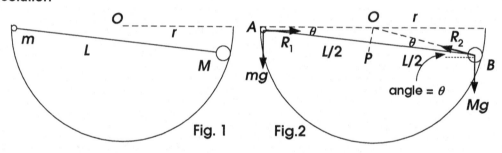

Fig. 1   **Fig.2**

The forces in Newtons acting on the rod are shown in Fig. 2 and are:
R_1 = Reaction at A between m and the bowl, directed towards O.
R_2 = Reaction at B between M and the bowl, directed towards O.
Mg, mg = Weights of each particle

Construct OP so that OP bisects AB i.e. $AP = PB = \dfrac{L}{2}$

$$Cos\theta = \frac{L}{2r}, \quad Sin\theta = \frac{\sqrt{r^2 - \dfrac{L^2}{4}}}{r}, \quad Tan\theta = \frac{2\sqrt{r^2 - \dfrac{L^2}{4}}}{L}$$

Taking moments about B: $mgLCos\theta = R_1 LSin\theta$

Taking moments about A: $MgL\cos\theta = R_2 L\sin\theta$

$$\Rightarrow R_1 = \frac{mg}{\tan\theta} = \frac{mgL}{2\sqrt{r^2 - \frac{L^2}{4}}} \quad \text{and} \quad R_2 = \frac{Mg}{\tan\theta} = \frac{MgL}{2\sqrt{r^2 - \frac{L^2}{4}}}$$

But, resolving forces horizontally:

$\Rightarrow R_1 = R_2\cos 2\theta = R_2\left(\cos^2\theta - \sin^2\theta\right)$ and, replacing R_1, R_2 by the values above

$$\Rightarrow \frac{mgL}{2\sqrt{r^2 - \frac{L^2}{4}}} = \frac{MgL}{2\sqrt{r^2 - \frac{L^2}{4}}}\left(\cos^2\theta - \sin^2\theta\right) = \frac{MgL}{2\sqrt{r^2 - \frac{L^2}{4}}}\left[\frac{(\frac{L^2}{4})}{r^2} - \frac{(r^2 - \frac{L^2}{4})}{r^2}\right]$$

$$= \frac{MgL\left(2\frac{L^2}{4} - r^2\right)}{2\sqrt{r^2 - \frac{L^2}{4}}\ r^2} \Rightarrow m = \frac{M\left(2\frac{L^2}{4} - r^2\right)}{r^2} \Rightarrow \frac{m}{M} = \frac{\left(L^2 - 2r^2\right)}{2r^2}$$

20. Rod, Hemisphere and Floor

A uniform rod of length 3 metres and mass 20 kg is smoothly hinged at one end, A, to a rough horizontal floor. The rod rests on the smooth curved surface of a hemisphere of mass 10 kg and radius 1 metre. The rod is in equilibrium inclined at 45° to the horizontal when the hemisphere is on the point of moving (See Fig. 1). Find:

(a)   The reaction between the rod and the hemisphere, R
(b)   The coefficient of friction between the hemisphere and the floor, μ
(c)   The reaction between the hemisphere and the floor

Solution
(a) The reaction between the rod and the hemisphere
The arrangement of rod and hemisphere are shown in Fig. 2:
$|AB| = 1\,\text{metre}$, $|BC| = 0.5\,\text{metres}$, $|CD| = 1.5\,\text{metres}$

The forces in Newtons acting on the rod and the sphere are shown in Figs. 3 and 4 respectively: $20g$ = Weight of 20kg mass, R = Reaction between rod and sphere, R_1, R_2 = Reaction components at hinge, R_{HEM} = Reaction between hemisphere and floor, F = Frictional force

For the rod:

Vertically: $R_1 + R\cos 45° = 20g \Rightarrow R_1 + R\dfrac{1}{\sqrt{2}} = 20g$ (i)

Horizontally: $R_2 = R\cos 45° \Rightarrow R_2 = R\dfrac{1}{\sqrt{2}}$ (ii)

Taking moments about A:

$$(1)(R) = (1.5)(20g)\cos 45° = \frac{30g}{\sqrt{2}} \Rightarrow R = \frac{30g}{\sqrt{2}} \quad \text{Newtons}$$

$$\Rightarrow \text{From equation (i) above}: \ R_1 = 20g - \frac{30g}{\sqrt{2}}\frac{1}{\sqrt{2}} = 5g \quad \text{Newtons}$$

(b) The coefficient of friction between the hemisphere and the floor

But : $\mu = \dfrac{F}{R_{HEM}} = \dfrac{15g}{25g} \Rightarrow \mu = \dfrac{3}{5}$

For the hemisphere : Horizontally : $F = R\,Cos\,45° = \dfrac{30g}{\sqrt{2}}\dfrac{1}{\sqrt{2}} = 15g$

(c) The reaction between the hemisphere and the floor

Vertically: $R_{HEM} = R\,Cos\,45° + 10g \Rightarrow R_{HEM} = 15g + 10g = 25g$ Newtons

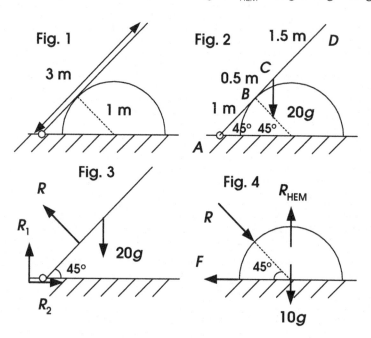

21. Rod inside Sphere

A light rod AB of length 0.8 metres carries a mass M kg at a distance of 0.3 metres from one end and rests in equilibrium inside a smooth hollow sphere of radius r, making an angle θ with the vertical. (See Fig. 1). Find expressions for:

(i) θ in terms of r

(ii) The reaction at A (= R, say) in terms of r, M, g

Solution

In Fig. 2 let R and S be reactions between the rod and sphere at A and B respectively. These are directed towards O. Mg is the weight of mass M.

Draw OC such that $OC \perp AB \Rightarrow OC$ bisects AB, giving $GC = 0.1$, $AC = CB = 0.4$

Consider triangle $OCA : |OC|^2 + 0.4^2 = r^2 \Rightarrow |OC|^2 = r^2 - 0.16$

Consider triangle $OCG : |GC|^2 + |OC|^2 = |OG|^2$

$\Rightarrow |OG|^2 = 0.1^2 + (r^2 - 0.16) = r^2 - 0.15 \Rightarrow |OG| = \sqrt{r^2 - 0.15}$

But from Fig. 3 : $Cos\,\theta = \dfrac{GC}{OG} = \dfrac{0.1}{\sqrt{r^2 - 0.15}}, \Rightarrow \theta = Cos^{-1}\dfrac{0.1}{\sqrt{r^2 - 0.15}} \Rightarrow Sin\theta = \sqrt{\dfrac{r^2 - 0.16}{r^2 - 0.15}}$

Take moments about B:

$$R|AB|Sin(\angle OAB) = Mg|GB|Sin\theta \quad \text{where} \quad Sin(\angle OAB) = \frac{OC}{OA} = \frac{\sqrt{r^2 - 0.16}}{r}$$

$$\Rightarrow R(0.8)\frac{\sqrt{r^2 - 0.16}}{r} = Mg(0.5)\sqrt{\frac{r^2 - 0.16}{r^2 - 0.15}} \Rightarrow R = \frac{0.5Mg\sqrt{\frac{r^2 - 0.16}{r^2 - 0.15}}}{(0.8)\frac{\sqrt{r^2 - 0.16}}{r}} = \frac{0.5rMg}{0.8\sqrt{r^2 - 0.15}}$$

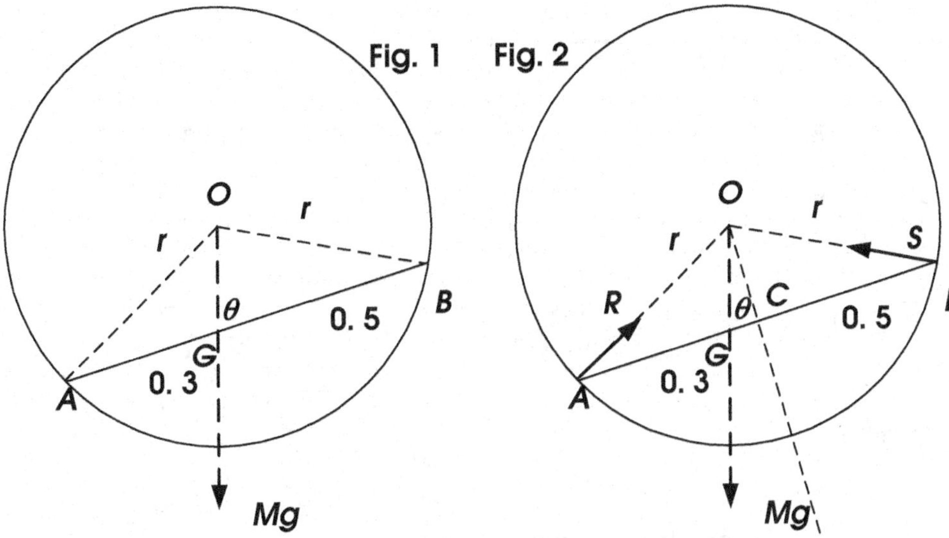

Fig. 1   Fig. 2

22. Column of discs

A column of n discs each of diameter D metres and thickness T metres is placed on a table. If the column is tilted from the vertical such that the centres of gravity of the discs lie along a straight line which makes an angle θ with the vertical, find the maximum value of θ for which the discs will not topple over.

Solution

The tilted column of discs is shown in Fig. 1. Assume that the edge of each disc is displaced a distance x metres from the disc below it. See Fig. 2. Let the centre of gravity of the discs be CG1 (for first disc), CG2 (for second disc), etc. If there are two discs only, the centre of gravity for the two-disc column will be CGc (say) and it will lie along the line joining CG1 and CG2 (and the other disc centres of gravity). CGc will be displaced a horizontal distance $\frac{x}{2}$ metres from the vertical through

CG1 (i.e at O). (See Fig. 3). The horizontal distance from CGc to the edge of the first disc (i.e. at E) will be $\frac{D}{2} - \frac{x}{2}$ metres. If there are n discs, the centre of gravity for the column will lie along the line joining CG1 and CG2, CGc, will be displaced a horizontal distance $(n-1)\frac{x}{2}$ metres from CG1. (See Fig. 3). The horizontal distance from the centre of gravity of the column, CGc, to the edge of the first disc will be $\frac{D}{2} - (n-1)\frac{x}{2}$ metres.

When the horizontal distance between a vertical line through O to a vertical through column's centre of gravity exceeds $\dfrac{D}{2}$ metres the column will topple.

But, when the horizontal distance between a vertical line through O to a vertical column's centre of gravity exceeds $\dfrac{D}{2}$ metres then

$$\frac{D}{2} - (n-1)\frac{x}{2} = 0 \Rightarrow \frac{D}{2} = (n-1)\frac{x}{2} \Rightarrow x = \frac{D}{(n-1)}$$

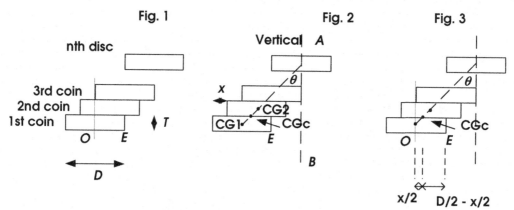

Fig. 1 **Fig. 2** **Fig. 3**

But, the angle θ at this point can be determined by considering the points CG1 and CG2 as:

$$Tan\theta = \frac{x}{\dfrac{T}{2} + \dfrac{T}{2}} = \frac{\dfrac{D}{(n-1)}}{\dfrac{T}{2} + \dfrac{T}{2}} = \frac{D}{(n-1)T}$$

$\Rightarrow$ Angle at maximum inclination to the vertical $= \theta = Tan^{-1}\left[\dfrac{D}{(n-1)T}\right]$

23. Hemisphere on rough slope

A hemisphere of radius r metres and weight W_1 Newtons is kept in equilibrium on a rough inclined slope with its place face horizontal by a weight W_2 (See Fig. 1). Derive an expression for the coefficient of friction μ in terms of W_1 and W_2.

Solution

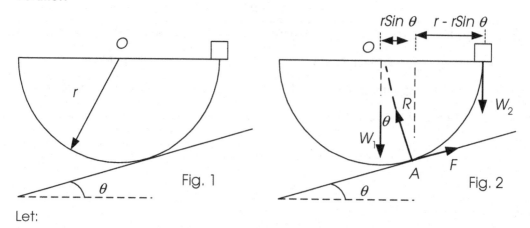

Fig. 1 **Fig. 2**

Let:

The slope be inclined at an angle θ to the horizontal.
The point of contact between slope and hemisphere be at A.
The forces acting are shown in Fig. 2: W_1 = Weight of hemisphere, W_2 = Weight at rim of hemisphere, F = Frictional force, R = Reaction between hemisphere and plane

Resolving these forces horizontally and vertically gives:

Horizontally: $R\,Sin\,\theta = F\,Cos\,\theta \quad \Rightarrow \mu = \dfrac{F}{R} = Tan\theta$ (i)

Vertically: $W_1 + W_2 = R\,Cos\,\theta + F\,Sin\,\theta$ (ii)

Taking moments about A (See Fig. 2): $W_1\,(r\,Sin\,\theta) = W_2\,(r - r\,Sin\,\theta)$ (iii)

From equation (iii) : $(W_1 + W_2)Sin\theta = W_2 \Rightarrow Sin\theta = \dfrac{W_2}{(W_1 + W_2)}$

From Fig. 3: $Tan\theta = \dfrac{W_2}{\sqrt{W_1^2 + 2W_1W_2}} = \mu$

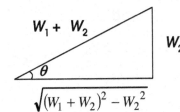

$W_1 + W_2$

W_2

θ

$\sqrt{(W_1 + W_2)^2 - W_2^2}$

Fig. 3

169

Chapter 9   Motion of a Rigid Body

1. Moment of Inertia of a Thin Rod
a. Thin Rod (Axis through centre).
Consider the case of motion of inertia of a thin uniform rod, AB, about an axis (Axis 1) through its centre, O, and perpendicular to its length, 2L metres. Assume the rod has a uniform density λ kg/m. (See Fig. 1):

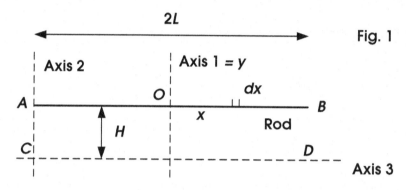

Fig. 1

Solution
Length = AB = 2L, centre at O.
Consider an element dx at a distance x from O. The mass of the element = $\lambda\,dx$. The axis is perpendicular to the rod and passes through the centre of the rod:
$\therefore I$ = Moment of Inertia $(\lambda\,dx)x^2$ of element about y

$\therefore I$ = Moment of Inertia of rod AB about the axis of $y = \int_{-L}^{L} \lambda\, x^2\,dx = \lambda \int_{-L}^{L} x^2\,dx$

$$\Rightarrow I = \lambda \left[\frac{x^3}{3}\right]_{-L}^{L} = \lambda \frac{2L^3}{3}$$

But mass of rod = $\lambda(2L) = M \Rightarrow I = \dfrac{ML^2}{3}$ kgm^2

b. Thin rod (axis through one end)
Find the moment of inertia of rod about an axis perpendicular to AB, passing through A, (Axis 2). (See Fig. 1).
Solution
We get, from the Parallel Axes Theorem:

$I_A = \dfrac{ML^2}{3} + Md^2$ where : d = distance from original axis = L

$\Rightarrow I_A = \dfrac{ML^2}{3} + ML^2 = \dfrac{4}{3}ML^2$ kgm^2

c. Thin rod (axis parallel to rod's length)
Find the moment of inertia of rod about an axis parallel to AB and a distance H metres from it, (Axis 3). (See Fig. 1). $I = MH^2$ kgm^2

2. Moment of Inertia of circular disc
Find the moment of inertia of a disc about an axis perpendicular to its plane, Axis z, passing through its centre O. Assume the rod has a uniform density σ kg/m^2

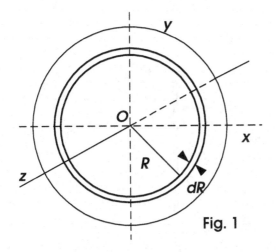

Fig. 1

Solution

Consider a ring whose width is dR (See Fig. 1). Mass of ring = $(2\pi R)(dR)\sigma$.
Its moment of inertia about an axis x through O is $Io = 2\pi R\sigma(dR)(R^2)$
If radius of the disc = r then Io = moment of inertia of disc about its axis, OX

$$\Rightarrow Io = \int_0^r 2\pi\sigma R^3\, dR = 2\pi\sigma \int_0^r R^3\, dR = 2\pi\sigma\left(\frac{R^4}{4}\right)_0^r \Rightarrow Io = 2\pi\sigma\left(\frac{r^4}{4}\right) = \frac{\pi\sigma r^4}{2}$$

But mass of disc = $M = \pi\sigma r^2 \Rightarrow Io = \dfrac{Mr^2}{2}$ kgm^2

3. Moment of Inertia of rectangular lamina
a) Axis through centre

Find the moment of inertia of a rectangular lamina about an axis perpendicular to its plane, Axis y, passing through its centre O. Assume the lamina has uniform density σ kg/m^2. See Fig. 1: $AB = 2b$, $AC = 2a$. Centre = O
Consider a strip of width dx, length $2a$ at a distance x from axis Oy

Mass of strip = $2a\,dx\sigma$ \quad Moment of inertia of strip about $Oy = 2a\,dx\sigma x^2$

$$\Rightarrow \text{Moment of inertia of rectangle about } Oy = \int_{-b}^{b} 2a\,dx\sigma x^2 = I_{Oy}$$

$$\Rightarrow I_{Oy} = 2a\sigma\int_{-b}^{b} x^2\, dx = \frac{2a\sigma}{3}\left(x^3\right)_{-b}^{b} = \frac{4a\sigma b^3}{3}$$

Total mass of lamina = $4ab\sigma = M \Rightarrow I_{Oy} = \dfrac{Mb^2}{3}$

$$\Rightarrow \text{Using the same approach}: Ox = I_{Ox} = \frac{Ma^2}{3} \quad \text{kgm}^2$$

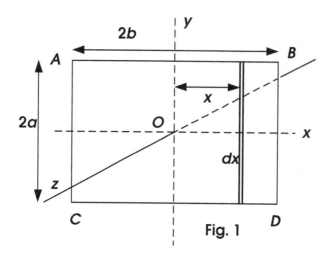

Fig. 1

4. Angular Motion

A particle of mass m kg at point A is acted upon by a force F Newtons at a distance r from its axis of rotation through O and perpendicular to the plane of rotation. The resulting torque τ Newton-metres produces an acceleration α radians/seconds^2, increasing the angular speed of the particle from ω_1 radians/second at A to ω_2 radians/second at point B in a time of t seconds during which time the particle turns through an angle θ. (See Fig. 1).

(a)   Derive expressions for the motion of the particle.
(b)   Derive expressions for the angular velocity, angular acceleration, torque, angular momentum and kinetic energy.
(c)   If a body of mass M kg consists of particles of mass m kg behaving as above find its torque, angular momentum and kinetic energy.

Solution

(a) Derive expressions motion of the particle.

The equations of motion which apply to angular motion are analogous to those for linear motion:

$$\omega_2 = \omega_1 + \alpha t, \quad \theta = \omega_1 t + \tfrac{1}{2}\alpha t^2, \quad \omega_2^{\,2} = \omega_1^{\,2} + 2\alpha\theta$$

(b) Derive expressions for the angular velocity, angular acceleration, torque, angular momentum and kinetic energy.

Angular velocity

If $|AB| = x \Rightarrow x = r\theta \Rightarrow$ Differentiating both sides with respect to

time gives : $\dfrac{dx}{dt} = r\dfrac{d\theta}{dt}$ (i)

But : $\dfrac{dx}{dt}$ = linear velocity = v; $\dfrac{d\theta}{dt}$ = angular velocity = $\omega \Rightarrow v = r\omega$ (ii)

Angular acceleration

⇒ differentiating both sides with respect to time :

$$\frac{d^2x}{dt^2} = \frac{dv}{dt} = r\frac{d^2\theta}{dt^2} = r\frac{d\omega}{dt} \text{ (iii)} \quad \text{But}: \frac{dv}{dt} = \text{linear acceleration} = a$$

$$\frac{d\omega}{dt} = \text{angular acceleration} = \alpha \Rightarrow a = r\alpha \text{ (iv)}$$

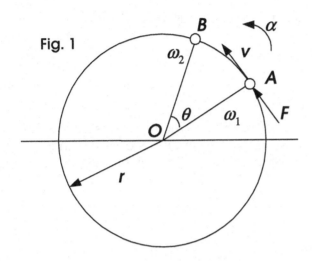

Fig. 1

Torque

Distance moved by force $F = r\theta$ (See Fig. 1)
Work done by force in this time = Force × Distance = $F(r\theta)$ (v)
But : Force × lever arm through which the force acts = Fr = Torque = τ (vi)
⇒ Work done = $\tau\theta$ (vii)

From equation (iv): Linear acceleration = $r\alpha$
⇒ Force required to produce acceleration = $m(r\alpha)$
⇒ From equation (vi), torque required to produce acceleration = $\tau = (m(r\alpha))r$
⇒ $\tau = mr^2\alpha$ (viii)

Angular momentum
Velocity of particle = $v = r\omega \Rightarrow$ Momentum of particle = mass × velocity = $mv = mr\omega$

⇒ Angular momentum = $(mr\omega)r = mr^2\omega$ (ix)

Work Done
Velocity of particle = $v = r\omega$

⇒ Kinetic Energy of particle = $\frac{1}{2}mv^2 = \frac{1}{2}mr^2\omega^2$ (x)

(c) If a body of mass M kg consists of particles of mass m kg behaving as above find its torque, angular momentum and kinetic energy.

Torque
From equation (viii): Total torque to accelerate a body consisting of particles of mass m

kg $= \tau = \alpha \int mr^2 = I\alpha$ (as $I = \int mr^2$) (xi)

Angular Momentum

From equation (ix):

Angular momentum for a body consisting of particles of mass $m = \int mr^2 \omega$

As: $I = \int mr^2 \Rightarrow$ Angular Momentum for a body consisting of particles of mass $m = I\omega$

Kinetic Energy

$\Rightarrow$ Kinetic Energy of body consisting of particles of mass $m = \int \frac{1}{2} mr^2 \omega^2$

$= \frac{1}{2}\omega^2 \int mr^2 = \frac{1}{2} I\omega^2 \quad \left(\text{Since} : I = \int mr^2\right)$

5. Parallel Axes Theorem

Let the moment of inertia of a rigid body about an axis through its centre (and perpendicular to the page) of gravity = I_G kg m². Find the moment of inertia of the body about an axis through O parallel to the axis through its centre of gravity and a distance d metres from this axis.

Solution

Let:
G = Centre of Gravity of the body
d = Distance in metres between axis through G and parallel axis through O
M = Mass of body in kg
m = Mass of one particle of the body in kg
I_G = Moment of Inertia of body about an axis through G in kg m²

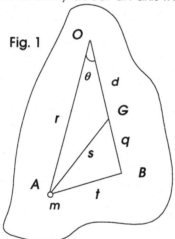

Fig. 1

Consider a particle of the body of mass m at point A. Let the dimensions be as follows:
$|OA| = r$, $|GB| = q$, $|OG| = d$, $|AG| = s$, $|AB| = t$

The moment of inertia of the particle about an axis through G is: $I = m s^2$

Moment of inertia of the body about an axis through G is: $I_G = \int s^2 dm$

The moment of inertia of the particle about the axis at O is: $m r^2$

But : $r^2 = \left(t^2 + (d+q)^2\right) = \left(t^2 + d^2 + 2dq + q^2\right) = \left(s^2 + d^2 + 2dq\right)$

Moment of inertia of whole body about axis at $O = I_o = \int r^2 dm = \int (s^2 + d^2 + 2dq)dm$

$= \int s^2\, dm + \int d^2\, dm + \int 2dq\, dm = \int s^2\, dm + \int d^2\, dm \left[\text{Since} \quad \left(\int 2dq\, dm\right) = 0\right]$

$\Rightarrow I_o = I_G + Md^2$

6. Simple Pendulum and Compound Pendulum

Find the:
(a)   Period of a simple pendulum of length L metres and mass m kg
(b)   Period of a compound pendulum of length L metres and mass m kg
(c)   Length of equivalent simple pendulum with same period as compound pendulum

Solution
(a) Find the period of a simple pendulum of length L metres and mass m kg

A simple pendulum consists of a mass m kg suspended on a light string of length L which is swinging in a vertical plane.
See Fig. 1. Assume the mass, at C, has been displaced through an angle θ where θ is small. The forces are shown in Fig. 2.
The force tending to restore the mass to its position at A is $mgSin\,\theta$
$\Rightarrow$ acceleration of mass towards $A = g\,Sin\,\theta$ ($\approx g\theta$ where θ is small and is measured in

radians). But: $Sin\,\theta = \dfrac{x}{L}$ and $x \approx$ arc $AC \Rightarrow$ acceleration of mass towards $A = \dfrac{g}{L}x$

$\Rightarrow$ acceleration of mass towards A is directly proportion to its displacement from A
$\Rightarrow$ motion is performing approximately simply harmonic motion about A in the form

$$\frac{d^2x}{dt^2} = \omega^2 x \quad \text{where}: \omega = \sqrt{\frac{g}{L}} \Rightarrow \text{Period} = T = 2\pi\sqrt{\frac{L}{g}} \quad \text{(i)}$$

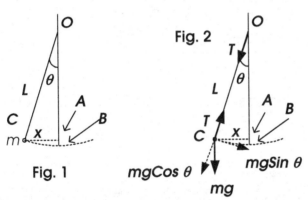

Fig. 2

Fig. 1

$mgCos\,\theta$

$mgSin\,\theta$

mg

(b) Find the period of a compound pendulum of length L metres, mass m kg

Consider a compound pendulum of length L with centre of gravity at G which rotates about a horizontal axis through O (and perpendicular to the page).
Assume the pendulum is displaced through an angle θ where θ is small.
The force tending to restore the mass to its position is: $mgSin\theta$.
$\Rightarrow$ restoring torque $= (mgSin\,\theta)L \approx mg\theta L$ (i) (where θ is small and is measured in radians).
But: Torque $= \tau = I_o\alpha$ (ii) where $I_o =$ moment of inertia of pendulum about O and α is its angular acceleration.
The moment of inertia of the pendulum is, from the Parallel Axes Theorem,

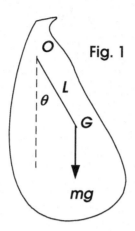

Fig. 1

$I_0 = I_G + mL^2 = ma^2 + mL^2$ (where ma^2 is the moment of inertia about a horizontal axis through centre of gravity at G)

$\Rightarrow$ Using equations (i) and (ii): $\tau = mg\theta L = I_0\alpha = (ma^2 + mL^2)\alpha \Rightarrow g\theta L = (a^2 + L^2)\alpha$

$\Rightarrow \alpha = \dfrac{gL}{(a^2 + L^2)}\theta$ But : $\alpha = \dfrac{d^2\theta}{dt^2} \Rightarrow \dfrac{d^2\theta}{dt^2} = \dfrac{gL}{(a^2 + L^2)}\theta$

This is simple harmonic motion of the form : $\dfrac{d^2\theta}{dt^2} = \omega^2\theta$

$\Rightarrow$ Period $= T = \dfrac{2\pi}{\omega} = \dfrac{2\pi}{\sqrt{\dfrac{gL}{(a^2 + L^2)}}} = 2\pi\sqrt{\dfrac{(a^2 + L^2)}{gL}}$ (ii) $= 2\pi\sqrt{\dfrac{I_0}{mgL}}$

(a)   The length of the equivalent simple pendulum

Comparing equations (i) and (ii) a simple pendulum of length $\dfrac{(L^2 + a^2)}{L}$

will have the same period as a compound pendulum of length L.

$\Rightarrow$ The length of the equivalent simple pendulum : $\dfrac{(L^2 + a^2)}{L}$

7. Solid Sphere

Prove that the moment of inertia of a uniform solid sphere of radius a and of mass m about a diameter is $\dfrac{2ma^2}{5}$

Solution

Let ρ be the density of the sphere in kg/m³

Divide the sphere into circular slices, each of thickness dx, perpendicular to Ox.

Consider a slice a vertical distance x from O:

Volume of slice is : $\pi(a^2 - x^2)dx$ Mass of slice is : $\rho\pi(a^2 - x^2)dx$

The moment of inertia of the slice about Ox is :

$\dfrac{1}{2}(\text{Mass})(a^2 - x^2) = \dfrac{1}{2}(\rho\pi)(a^2 - x^2)dx\,(a^2 - x^2) = \dfrac{1}{2}(\rho\pi)(a^2 - x^2)^2 dx$

Moment of inertia of sphere about Ox is :

$\displaystyle\int_{-a}^{a}\dfrac{1}{2}(\rho\pi)(a^2 - x^2)^2 dx = \dfrac{1}{2}(\rho\pi)\int_{-a}^{a}(a^2 - x^2)^2\,dx = \dfrac{1}{2}(\rho\pi)\int_{-a}^{a}(a^4 - 2a^2x^2 + x^4)dx$

$$= \frac{1}{2}(\rho\pi)\left[a^4x - \frac{2a^2x^3}{3} + \frac{x^5}{5}\right]_{-a}^{a} = \frac{1}{2}(\rho\pi)\frac{16a^5}{15}$$

$$\Rightarrow I = \frac{8\rho\pi a^5}{15} \quad \text{But the mass of the sphere} = m = \frac{4}{3}\pi a^3\rho \Rightarrow I = \frac{2}{5}ma^2 \quad \text{kgm}^2$$

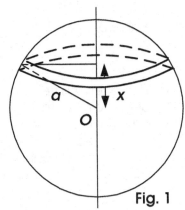

Fig. 1

8. Rotating disc

A uniform circular disc of mass m kg and radius r metres rotates about a fixed horizontal axis through a point A on its rim perpendicular to its plane (See Fig. 1). If the speed of its centre, C, when AB is vertical (i.e. with A above B) is v find the minimum value of v so that the disc will perform complete revolutions.

Solution

Moment of inertia of disc about its centre $= \frac{1}{2}mr^2$

Moment of inertia of disc about $A = \frac{1}{2}mr^2 + mr^2 = \frac{3}{2}mr^2$

Energy when AB is vertical (A above B) = Energy when AB is vertical (B above A)

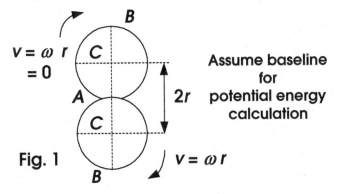

Fig. 1

Assume baseline for potential energy calculation

Energy when AB is vertical (A above B):

The total energy has two components: Potential Energy + Kinetic Energy
Potential Energy: Assume Potential energy in this position is 0
Kinetic Energy: Speed $= v$ m/s But: $v = \omega r$

$$\Rightarrow \text{Kinetic Energy} = \frac{1}{2}I\left(\frac{v}{r}\right)^2 = \frac{1}{2}\left(\frac{3}{2}mr^2\right)\left(\frac{v}{r}\right)^2 = \frac{3}{4}mv^2 \Rightarrow \text{Total Energy} = \frac{3}{4}mv^2 \quad \text{(i)}$$

Energy when AB is vertical (A above B):
Potential Energy: Assume Potential Energy in this position = 0.
Kinetic Energy: Speed = v m/s. But: $v = \omega r$
$\Rightarrow$ Total Energy = $2mgr$ (ii)   But : equation (i) = equation (ii)

$$\Rightarrow \frac{3}{4}mv^2 = 2mgr \Rightarrow v = 2\sqrt{\frac{2}{3}gr} \quad \text{m/s} = \text{minimum value of } v \text{ for disc to continue rotating.}$$

9. Rotating disc

A uniform circular disc has mass m kg and radius r metres. It is free to rotate about a fixed horizontal axis through a point A on its rim perpendicular to its plane (See Fig. 1). A particle of mass $4m$ kg is attached to the disc at a point B on its rim diametrically opposite A. The disc is held with AB horizontal and released from rest.
(a) Find, in terms of r, the angular velocity when B is vertically below A.
(b) If the system were to oscillate as a compound pendulum, find its period.

Solution
(a) Find, in terms of r, the angular velocity when B is vertically below A.

Moment of inertia of disc about $A = \frac{1}{2}mr^2 + mr^2 = \frac{3}{2}mr^2$

Moment of inertia of mass $4m = (4m)(2r)^2 = 16mr^2$

Moment of inertia of system $= 16mr^2 + \frac{3}{2}mr^2 = \frac{35}{2}mr^2 \quad \text{kgm}^2$

The initial and final positions are shown in Fig. 1:

Energy of system in initial position = Potential Energy (P.E.) + Kinetic Energy (K.E.), where:
Potential Energy = $mg(2r) + (4m)g(2r) = 10mgr$,   Kinetic Energy = 0
$\Rightarrow$ Total Energy = $10mgr$ (i)
Energy of system in final position = P.E. + K.E. , where:

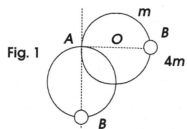

Fig. 1

P.E. = $mg(r)+(4m)g(0)= mgr$,   K.E. $=\frac{1}{2}I\omega^2 \Rightarrow$ Total Energy $= mgr+\frac{1}{2}I\omega^2$ (ii)

But total energy in system remains unchanged $\Rightarrow$ equation (i) = equation (ii) :

$$\Rightarrow 9mgr = \frac{1}{2}I\omega^2 \Rightarrow \omega^2 = \frac{18mgr}{I} = \frac{18mgr}{\frac{35}{2}mr^2} = \frac{36g}{35r} \Rightarrow \omega = \sqrt{\frac{36g}{35r}}$$

(b) If the system were to oscillate as a compound pendulum, find its period.
To find the period of small oscillations: The disc behaves as a compound pendulum:

178

$$T = 2\pi\sqrt{\dfrac{I}{Mgh}} \quad \text{where}: M = m + 4m = 5m$$

$h =$ Distance from centre of gravity to axis $= \dfrac{m(r) + 4m(2r)}{m + 4m} = \dfrac{9mr}{5m} = \dfrac{9}{5}r$

$$\Rightarrow T = 2\pi\sqrt{\dfrac{35mr^2}{2(5m)g\left(\dfrac{9}{5}\right)r}} = 2\pi\sqrt{\dfrac{35r}{18g}}$$

10. Trapdoor

A trapdoor, of dimensions $D \times D$ metres and mass m kg is making small oscillations about a horizontal axis YY (See Fig. 1). Let the trapdoor's central horizontal axis be XX. Find the periodic time.

Moment of inertia of door about $XX = \dfrac{1}{3}m\left(\dfrac{D}{2}\right)^2 = \dfrac{1}{12}mD^2$

Moment of inertia of door about axis $YY = \dfrac{1}{12}mD^2 + m\left(\dfrac{D}{2}\right)^2 = \dfrac{mD^2}{3}$ kgm^2

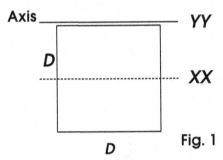

Axis _____ **YY**

D

$- - - - - - - - - - - - - - -$ **XX**

D **Fig. 1**

To find the period of small oscillations: $T = 2\pi\sqrt{\dfrac{I}{Mgh}}$ where:

$M = m, \quad h =$ Distance from centre of gravity to axis $= \dfrac{D}{2}$

$$\Rightarrow T = 2\pi\sqrt{\dfrac{mD^2}{3(m)g\left(\dfrac{D}{2}\right)}} = 2\pi\sqrt{\dfrac{2D}{3g}} \quad \text{seconds}$$

11. Uniform rod and disc

A uniform rod AB of length $3L$ metres is attached to the rim of a uniform disc of diameter L metres with centre at C (see Fig.1). The disc and the rod are both of mass m kg. Find:

(a) The moment of inertia of the compound body about an axis through A perpendicular to the plane of the disc.

(b) The periodic time if the compound body makes small oscillations in a vertical plane about A

(c) The angular speed of C when AB is vertical if the compound body is released from rest with AB horizontal

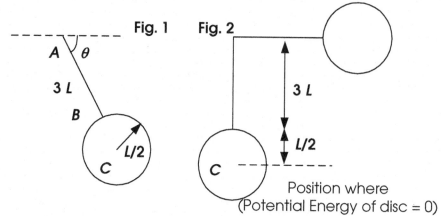

Fig. 1   Fig. 2

Position where
(Potential Energy of disc = 0)

Solution

(a) The moment of inertia

The moment of inertia of the rod about an axis through A perpendicular to

the plane of the disc is: $I_{rod} = \dfrac{4}{3}(m)\left(\dfrac{3L}{2}\right)^2 = 3mL^2 \quad kg\,m^2$

The moment of inertia of disc about axis through A, perpendicular to the plane of the

disc is: $I_{disc} = \dfrac{1}{2}m\left(\dfrac{L}{2}\right)^2 + (m)\left(\dfrac{7L}{2}\right)^2 = \dfrac{99}{8}mL^2 \quad kg\,m^2$

Pendulum: The moment of inertia of the pendulum:

$I_{pendulum} = 3mL^2 + \dfrac{99}{8}mL^2 = \dfrac{123}{8}mL^2 \quad kg\,m^2$

(b) The period of motion of the pendulum

Period of motion of pendulum = $T = 2\pi\sqrt{\dfrac{I_{pendulum}}{Mgh}}$ where :

M = Combined mass = $2m$

h = Centre of gravity of pendulum from Axis at $A = \dfrac{\dfrac{3L}{2}m + \dfrac{7L}{2}m}{2m} = \dfrac{5}{2}L$

$\Rightarrow T = 2\pi\sqrt{\dfrac{\dfrac{123}{8}mL^2}{2mg\dfrac{5}{2}L}} = 2\pi\sqrt{\dfrac{123L}{40g}}$ seconds

(c)   The angular speed of C when AB is vertical if the compound body is released from rest with AB horizontal

Energy of pendulum when AB is horizontal = Energy of pendulum when AB is vertical.
See Fig. 2. (Note that when AB is vertical, Potential Energy of disc = 0).

Energy when AB is horizontal : $mgh_{rod} + mgh_{disc} + \frac{1}{2}I_{pendulum}\omega^2$

$h_{rod} = 3L + \dfrac{L}{2} = \dfrac{7}{2}L$ $h_{disc} = \dfrac{7}{2}L,$ $\omega = 0$

$\Rightarrow$ Energy = $7mgL + 0 = 7mgL$

Energy when AB is vertical : $mgh_{rod} + mgh_{disc} + \frac{1}{2}I_{pendulum}\omega^2$

$$h_{rod} = \frac{3}{2}L + \frac{L}{2} = 2L, \qquad h_{disc} = 0$$

$\Rightarrow$ Energy $= 2mgL + 0 + \frac{1}{2}I_{pendulum}\omega^2 = 2mgL + \frac{1}{2}I_{pendulum}\omega^2$

$\Rightarrow 7mgL = 2mgL + \frac{1}{2}I_{pendulum}\omega^2 \Rightarrow I_{pendulum}\omega^2 = 10mgL$

$$\Rightarrow \omega^2 = \frac{10mgL}{I_{pendulum}} = \frac{10mgL}{\frac{123}{8}mL^2} = \frac{80g}{123L} \Rightarrow \omega = \sqrt{\frac{80g}{123L}} \quad \text{radian/s}$$

12. Pendulum

A pendulum clock consists of a thin uniform rod AC of mass M kg and length $6L$ metres where $|AB| = 4L$ and $|BC| = 2L$ (See Fig. 1). A uniform circular disc of mass $4M$ kg and radius L is rigidly attached at B and C. Find:

(a)   The moment of inertia of the pendulum about an axis through A perpendicular to the plane of the disc.

(b)   The angular velocity of B when AB is vertical if the pendulum is released from rest with AB horizontal.

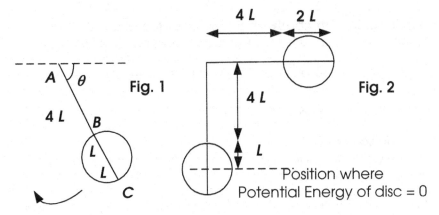

(a) Moment of inertia

(See Fig. 2)

Disc:

The moment of inertia of disc about axis through A perpendicular to the plane of the disc is: $I_{disc} = \frac{1}{2}(4M)L^2 + (4M)(5L)^2 = 102ML^2 \quad$ kg m^2

Rod:

The moment of inertia of rod about axis through A, perpendicular to the plane of the rod is: $I_{rod} = \frac{1}{3}M(6L)^2 = 12ML^2$

$\Rightarrow$ The moment of inertia of pendulum about this axis $= 114ML^2 \quad$ kgm^2

(b) The angular velocity of B when AB is vertical

Let h = vertical height above position of zero potential energy.

Energy when AC is horizontal : $Mgh_{rod} + 4Mgh_{disc} + \frac{1}{2}I_{pendulum}\omega^2$

$h_{rod} = 5L, h_{disc} = 5L, \omega = 0 \Rightarrow$ Energy $= 5MgL + 20MgL + 0 = 25MgL$

Energy when AC is vertical : $Mgh_{rod} + 4Mgh_{disc} + \frac{1}{2}I_{pendulum}\omega^2$

$h_{rod} = 2L, h_{disc} = 0 \Rightarrow$ Energy $= 2MgL + 0 + \frac{1}{2}I_{pendulum}\omega^2$

$\Rightarrow 25MgL = 2MgL + \frac{1}{2}I_{pendulum}\omega^2$

$\Rightarrow \omega^2 = \dfrac{46MgL}{I_{pendulum}} = \dfrac{46MgL}{114ML^2} = \dfrac{23g}{57L} \Rightarrow \omega = \sqrt{\dfrac{23g}{57L}}$ radians/sec

13. Dumbbell

A symmetrical dumbbell consists of two solid spheres each of mass m kg and radius r metres joined by a narrow uniform rigid bar of mass m kg and length $2r$ metres so that the centres of the spheres are a distance $4r$ metres apart (See Fig. 1). If the dumbbell is

freely-pivoted about an axis through point D perpendicular to BE a distance $\dfrac{r}{2}$ from its

centre so that it can perform small oscillations in a vertical plane,

(a) Find the moment of inertia of the body about the axis of rotation
(b) Find the period of oscillation
Solution
(a) Find the moment of inertia of the body about the axis of rotation
Consider the dumb-bell as being made up of four components (See Figs 2,3)

Sphere (iv): About B: $\dfrac{2}{5}mr^2$ About D: $\dfrac{2}{5}mr^2 + m\left(\dfrac{5r}{2}\right)^2 = \dfrac{133}{20}mr^2$

Total moment of inertia = sum of above = $\dfrac{593}{60}mr^2$ kgm^2

(b) Find the period of oscillation.
Consider the dumb-bell when it is making an angle θ with the vertical. The moment of the restoring force about D is:

$\dfrac{3mg}{4}\dfrac{3r}{4}Sin\theta + mg\dfrac{5r}{2}Sin\theta - \dfrac{mg}{4}\dfrac{r}{4}Sin\theta - mg\dfrac{3r}{2}Sin\theta = \dfrac{3}{2}mgr\,Sin\theta$

$\dfrac{d^2\theta}{dt^2} = \left(\dfrac{\dfrac{3}{2}mgrSin\theta}{I}\right)$ But : θ is small $\Rightarrow Sin\theta \approx \theta$

$\Rightarrow \dfrac{d^2\theta}{dt^2} = \left(\dfrac{3mgr}{2I}\right)\theta \Rightarrow$ This is in the form : $\dfrac{d^2\theta}{dt^2} = \left(\dfrac{3mgr}{2I}\right)\theta \Rightarrow \omega = \sqrt{\dfrac{3mgr}{2I}}$

$\Rightarrow$ Period $= T = \dfrac{2\pi}{\omega} = 2\pi\sqrt{\dfrac{2I}{3mgr}}$

Using the value previously calculated for I: $T = 2\pi\sqrt{\dfrac{(2)(593mr^2)}{3mgr(60)}} = 2\pi\sqrt{\dfrac{593r}{90g}}$ seconds

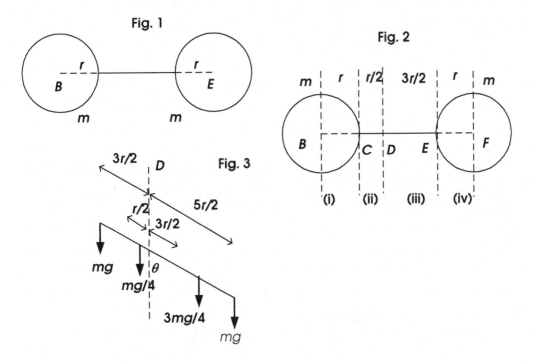

Fig. 1

Fig. 2

Fig. 3

14. Rotating rod

A rod has mass M kg and length $2L$ metres. The rod is free to rotate in a vertical plane about a fixed horizontal axis at A with a particle of mass $4M$ kg attached to its other end (See Fig. 2). The system is released from rest with the rod vertical. Find the angular speed of the rod when the particle reaches point B (its lowest point).

Solution

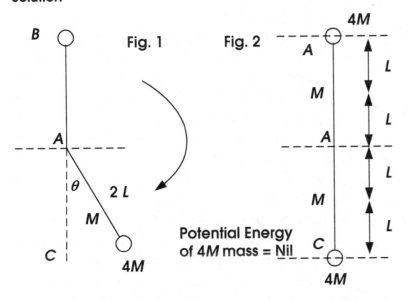

Fig. 1

Fig. 2

Potential Energy of $4M$ mass = Nil

The moment of inertia of the system

$$I_{rod} \text{ about } A = \frac{1}{3} M(2L)^2 = \frac{4}{3} ML^2 \qquad I_{particle} \text{ about } A = (4M)(2L)^2 = 16ML^2$$

$$\Rightarrow I_{system} \text{ about } A = \frac{4}{3} ML^2 + 16ML^2 = \frac{52}{3} ML^2 \quad \text{kg} \, \text{m}^2$$

Energy at any point (See Fig. 2)= Potential Energy + Kinetic Energy
Energy at A: = $(M)g(L + 2L) + (4M)g(2L + 2L) + \frac{1}{2} I \omega^2 = 19MgL$ (Since $\omega = 0$)
Energy at B: = $(M)g(L) + (4M)g(0) + \frac{1}{2} I \omega^2 = MgL + \frac{1}{2} I \omega^2$
But the energy is unchanged $\Rightarrow 19MgL = MgL + \frac{1}{2} I \omega^2 \Rightarrow 18MgL = \frac{1}{2} I \omega^2$

$$\omega^2 = \frac{36MgL}{I} = \frac{36MgL}{\frac{52}{3}ML^2} = \frac{108g}{52L} \Rightarrow \omega = \sqrt{\frac{27g}{13L}} \quad \text{radians/second}$$

15. Rotating Rod

A rod AB is free to rotate in a vertical plane about a fixed horizontal axis at C. The mass and length components are shown in Fig. 1. The system is released from rest with the rod horizontal. Find the angular speed and kinetic energy of the rod when B reaches its lowest point.

Solution

The moment of inertia of the system is

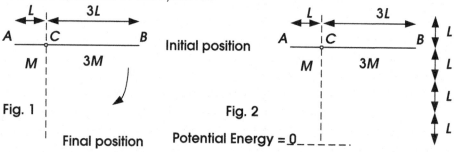

$$I_{rod}(AC)\text{ about } C = \frac{4}{3}M\left(\frac{L}{2}\right)^2 = \frac{ML^2}{3}$$

$$I_{rod}(CB)\text{ about } C = \frac{4}{3}(3M)\left(\frac{3L}{2}\right)^2 = \left(\frac{4}{3}\right)\frac{27ML^2}{4} = 9ML^2 \Rightarrow I_{system}\text{ about } C = \frac{28}{3}ML^2 \quad kgm^2$$

Energy at any point (See Fig. 2) = Potential Energy + Kinetic Energy
Energy at Initial Position = $(4M)g(3L) + \frac{1}{2} I \omega^2 = 12MgL$ (Since $\omega = 0$)

Energy at Final Position = $(M)g\left(\frac{L}{2} + 3L\right) + (3M)g\left(\frac{3L}{2}\right) + \frac{1}{2}I\omega^2$

$$\Rightarrow 12MgL = 8MgL + \frac{1}{2}I\omega^2 \Rightarrow \omega^2 = \frac{8MgL}{I} = \frac{8MgL(3)}{28ML^2} = \frac{24g}{28L} \Rightarrow \omega = \sqrt{\frac{6g}{7L}}$$

and Kinetic Energy when B is at its lowest point: $\frac{1}{2}I\omega^2 = \frac{1}{2}\left(\frac{28ML^2}{3}\right)\left(\frac{6g}{7L}\right) = 4MLg$ Joules

16. Hollow Sphere

Sphere 1 of radius R metres and density ρ kg/m³ has a spherical cavity (Sphere 2) of radius r metres. See Fig. 1. Find the moment of inertia of the body about its symmetrical axis oz.

Solution

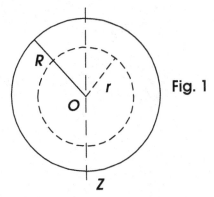

Fig. 1

The moment of inertia of a body about OZ

= moment of inertia of Sphere 1 about OZ – moment of inertia of Sphere 2 about OZ

Mass of Sphere 1 = $M_1 = \frac{4}{3}\pi\rho R^3$ kg   Mass of Sphere 2 = $M_2 = \frac{4}{3}\pi\rho r^3$ kg

$\Rightarrow$ Moment of inertia of body $= \frac{2}{5}\left(\frac{4}{3}\pi\rho R^3\right)R^2 - \frac{2}{5}\left(\frac{4}{3}\pi\rho r^3\right)r^2 = \frac{8}{15}\pi\rho\left(R^5 - r^5\right)$

But : (Volume)$\times \rho$ = Mass

$\Rightarrow \rho = \dfrac{\text{Mass of Sphere 1}}{\text{Volume of Sphere 1}} = \dfrac{\text{Mass of Sphere 2}}{\text{Volume of Sphere 2}} = \dfrac{\text{Mass of Body (i.e. } M_1 - M_2 = M)}{\text{Volume of Body (i.e. Vol}_1 - \text{Vol}_2)}$

$\rho = \dfrac{M_1}{\frac{4}{3}\pi R^3} = \dfrac{M_2}{\frac{4}{3}\pi r^3} = \dfrac{M}{\frac{4}{3}\pi R^3 - \frac{4}{3}\pi r^3} = \dfrac{M}{\frac{4}{3}\pi\left(R^3 - r^3\right)}$

$\Rightarrow$ Moment of Inertia $= \dfrac{8}{15}\pi\left(\dfrac{M}{\frac{4}{3}\pi\left(R^3 - r^3\right)}\right)\left(R^5 - r^5\right) = \dfrac{2}{5}M\dfrac{\left(R^5 - r^5\right)}{\left(R^3 - r^3\right)}$

17. String wound on rim of disc

A light string is wound around the rim of a uniform disc of radius r and mass m kg and one end is attached to a fixed point above the disc (See Fig. 1). When the disc is released at time $t = 0$ seconds from rest it falls vertically as the string unwinds. After time t seconds the centre of gravity is a distance x metres below its original position. Find the vertical acceleration of the disc and the tension in the string.

Solution

The forces in Newtons acting on the system are (See Fig. 2):

T = Tension in the string, Mg = Weight of disc

At time t, the centre of gravity, G, is a distance x below O. The distance x can be thought of as a segment of the circumference of the disc. The segment has a length $r\theta \Rightarrow x = r\theta$. (See Fig. 3)

Approach 1: Linear and Angular acceleration

Linear acceleration of disc $= \dfrac{d^2x}{dt^2}$ where : $m\dfrac{d^2x}{dt^2} = mg - T$ (i) and

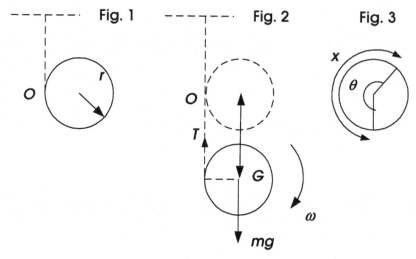

Fig. 1   **Fig. 2**   **Fig. 3**

(note : I = moment of inertia of disc $= \dfrac{1}{2}mr^2$ (ii))

From equation (i): $T = mg - m\left(\dfrac{d^2x}{dt^2}\right)$ But : Torque $= Tr = mgr - mr\left(\dfrac{d^2x}{dt^2}\right)$ (iii)

But : Torque $= I\dfrac{d\omega}{dt}$ where :

I = moment of inertia and $\dfrac{d\omega}{dt}$ = angular acceleration   Also : $\dfrac{d^2x}{dt^2} = r\dfrac{d\omega}{dt}$

$\Rightarrow$ Torque $= I\dfrac{d\omega}{dt} = \dfrac{I}{r}\left(\dfrac{d^2x}{dt^2}\right)$ (iv)   Using equations (iii) and (iv) :

$\Rightarrow mgr - mr\left(\dfrac{d^2x}{dt^2}\right) = \dfrac{I}{r}\left(\dfrac{d^2x}{dt^2}\right) \Rightarrow \dfrac{d^2x}{dt^2} = \dfrac{mgr^2}{I + mr^2} = \dfrac{2mgr^2}{3mr^2}$ using equation (ii) $= \dfrac{2}{3}g$ m/s^2

From equation (i): $T = m\left(g - \dfrac{d^2x}{dt^2}\right) = m\left(g - \dfrac{2}{3}g\right) = \dfrac{mg}{3}$ Newtons

Approach 2: Energy approach

Total energy of disc at any point = Potential Energy + Rotational Kinetic Energy + Linear Kinetic Energy

Energy at Initial Position at point A = $(m)g(x) + ½\,I\,\omega^2 + ½\,mv^2 = mgx$ (Since v, $\omega = 0$)

Assume that after falling a distance x (i.e. to point B) the Potential Energy = 0, then (See Fig. 2):

Energy at point B = Rotational Kinetic Energy + Linear Kinetic Energy

Angular velocity of the disc = ω radians/second

Speed of disc = v m/s

Energy at point $B = \dfrac{1}{2}I\omega^2 + \dfrac{1}{2}mv^2$

But $v = \omega r$ and $I = \dfrac{1}{2}mr^2$

Since: initial energy of system = final energy of system

$$\Rightarrow mgx = \frac{1}{2}mv^2 + \frac{1}{2}\left(\frac{1}{2}mr^2\right)\left(\frac{v}{r}\right)^2 = \frac{3}{4}mv^2 \Rightarrow v^2 = \frac{4gx}{3} \quad m/s$$

But, using the standard equation: $v^2 = u^2 + 2ax$, where: $u = 0$

$$\Rightarrow \frac{4gx}{3} = 2ax \Rightarrow a = \frac{2}{3}g \quad m/s^2 \text{ (As in Approach 1 above)}$$

$\Rightarrow$ Tension, T, will be as in Approach 1 above.

18. String wound on rim of disc

A light inextensible string is wound around the rim of a uniform disc of radius r and mass M_d which can rotate freely about its fixed horizontal axis oq. One end of the string is attached to the rim of the disc and the other end is attached to a particle P of mass M_p which hangs vertically (See Fig. 1). When the system is released from rest P falls vertically and the string unwinds. Find the velocity of P after it has fallen a distance x.

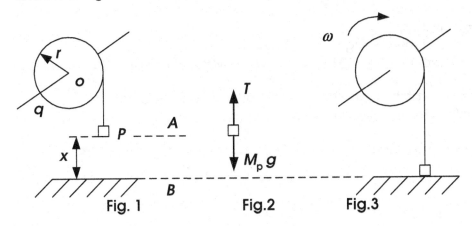

Fig. 1   Fig.2   Fig.3

Solution
Consider the system at time $t = 0$
The forces in Newtons acting on the particle are:
T = Tension in string, $M_p g$ = Weight of the particle

Assume the disc falls a distance x while it turns through an angle θ.

Approach 1: Linear and Angular acceleration

Linear acceleration of particle $= \dfrac{d^2x}{dt^2}$ where : $M_p\dfrac{d^2x}{dt^2} = M_p g - T$ (i)

(note : I = moment of inertia of disc $= \dfrac{1}{2}M_d r^2$ (ii))

From equation (i): $T = M_p g - M_p\left(\dfrac{d^2x}{dt^2}\right)$

But : Torque $= Tr \Rightarrow$ Torque $= M_p gr - M_p r\left(\dfrac{d^2x}{dt^2}\right)$ (iii)

But : Torque $= I\dfrac{d\omega}{dt}$ where : I = moment of inertia of disc and

$$\frac{d\omega}{dt} = \text{angular acceleration} \qquad \text{Also}: \frac{d^2x}{dt^2} = r\frac{d\omega}{dt}$$

$$\Rightarrow \text{Torque} = I\frac{d\omega}{dt} = \frac{I}{r}\left(\frac{d^2x}{dt^2}\right) \quad \text{(iv)} \quad \text{Using equations (iii) and (iv)}:$$

$$\Rightarrow M_p gr - M_p r\left(\frac{d^2x}{dt^2}\right) = \frac{I}{r}\left(\frac{d^2x}{dt^2}\right) \Rightarrow \frac{d^2x}{dt^2} = \frac{M_p gr^2}{I + M_p r^2} = \frac{M_p gr^2}{\frac{1}{2}M_d r^2 + M_p r^2} = \frac{2M_p g}{M_d + 2M_p}$$

$(= \text{acceleration } a\, \text{m/s}^2, \text{say}) \Rightarrow \dfrac{d^2x}{dt^2}$ is constant

Using the standard equation : $v^2 = u^2 + 2ax$ (note : $u = 0$)

$$\Rightarrow v^2 = 2\left(\frac{2M_p gr^2}{2M_p r^2 + M_d r^2}\right)x \Rightarrow v = 2\sqrt{\frac{M_p gx}{2M_p + M_d}} \quad \text{m/s}$$

Approach 2: Energy approach
Energy at any point = Potential Energy + Kinetic Energy
Energy at Initial Position at point $A = (M_p)g(x) + \frac{1}{2}I\omega^2 = M_p gx$ (Since $\omega = 0$)
Assume that after falling a distance x (See Fig. 2):
The energy of the system at this point = energy of particle + energy of disc
Angular velocity of the disc = ω radians/second
Speed of particle $P = v$ m/s
Energy at Final Position at point $B = (M_p)g(0) + \frac{1}{2}I\omega^2 + \frac{1}{2}M_p v^2 = \frac{1}{2}I\omega^2 + \frac{1}{2}M_p v^2$

But $v = \omega r$ and $I = \dfrac{1}{2}M_d r^2$

Since: initial energy of system = final energy of system

19. Solids rolling up an inclined plane
A wheel, hoop, solid cylinder and solid sphere each of mass m kg and radius r metres are rolling separately along a plane with a horizontal speed of v m/s (See Fig. 1). If they mount an inclined plane at an angle of θ to the horizontal, then, assuming no slipping takes place, find how far up the plane each mass will travel before coming to rest.

Solution
When the object is moving at v m/s assume it has an angular velocity of ω radians/second.
Its energy will consist of the following components:
Potential Energy + Kinetic Energy (Rotational) + Kinetic Energy (linear)
Assume that, along the horizontal plane, the Potential Energy = 0.

$$\Rightarrow M_p gx = \frac{1}{2}M_p v^2 + \frac{1}{2}\left(\frac{1}{2}M_d r^2\right)\left(\frac{v}{r}\right)^2 = \frac{1}{2}M_p v^2 + \frac{1}{4}M_d v^2 \Rightarrow v = 2\sqrt{\frac{M_p gx}{2M_p + M_d}} \quad \text{m/s}$$

Initial Energy $= \frac{1}{2}I\omega^2 + \frac{1}{2}mv^2$

Final Energy = Potential Energy only (as object has come to rest) $= mgh$

$$\Rightarrow \frac{1}{2}I\omega^2 + \frac{1}{2}mv^2 = mgh \quad \text{(i)}$$

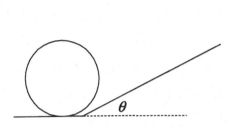

Fig. 1   **Fig. 2**

But : $R = \dfrac{h}{Sin\theta} \Rightarrow R = \dfrac{\frac{1}{2}I\omega^2 + \frac{1}{2}mv^2}{mgSin\theta}$ (ii) $v = \omega r \Rightarrow R = \dfrac{\frac{1}{2}I\frac{v^2}{r^2} + \frac{1}{2}mv^2}{mgSin\theta} = \dfrac{I\frac{v^2}{r^2} + mv^2}{2mgSin\theta}$ (iii)

Using equation (iii) and the moments of inertia in kg.m² as given below the values of R are as follows:

Wheel (i.e. disc of radius r): $I = \dfrac{mr^2}{2}$ $R = \dfrac{3v^2}{4gSin\theta}$

Hoop (of radius r): $I = mr^2$ $R = \dfrac{v^2}{gSin\theta}$

Solid cylinder (of radius r): $I = \dfrac{mr^2}{2}$ $R = \dfrac{3v^2}{4gSin\theta}$

Solid sphere (of radius r): $I = \dfrac{2mr^2}{5}$ $R = \dfrac{7v^2}{10gSin\theta}$

20. Objects rolling without slipping

A circular disc, hoop and solid sphere each of mass M kg and radius r metres are rolling down a rough plane inclined at an angle a to the horizontal with a speed of v m/s parallel to the plane (See Fig. 1). Find the minimum value of μ for each object in order that no slipping takes place (i.e. object keeps rolling instead of slipping).

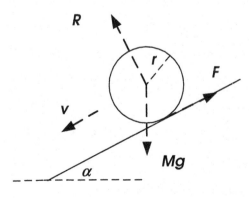

Fig. 1

Solution

When the object is moving at v m/s assume it has an angular velocity of ω radians/second.

The forces in acting are shown in Fig. 1: Mg = Weight of object, R = Reaction between object and plane, F = Frictional force

Equation of motion parallel to inclined plane:

$$M\frac{d^2x}{dt^2} = MgSin\alpha - F \Rightarrow F = MgSin\alpha - M\frac{d^2x}{dt^2} \quad \text{(i)}$$

Equation of motion perpendicular to inclined plane:
$R = MgCos\alpha$ (ii) (Since there is no motion in this direction)

But: Torque $= I\alpha = Fr$ (iii) where α = angular acceleration of object

Assume that object turns through an angle θ in time t, then:

$$x = r\theta \quad \text{and} \quad \frac{d^2x}{dt^2} = r\frac{d^2\theta}{dt^2} \quad \text{From equation (iii)}: F = \frac{I\alpha}{r} = \frac{I\frac{d^2\theta}{dt^2}}{r} = \frac{I\frac{d^2x}{dt^2}}{r^2} \quad \text{(iv)}$$

But, using equation (i): $F = MgSin\alpha - M\dfrac{d^2x}{dt^2} = \dfrac{I\frac{d^2x}{dt^2}}{r^2}$

$$\Rightarrow MgSin\alpha = \frac{d^2x}{dt^2}\left(M + \frac{I}{r^2}\right) \Rightarrow \frac{d^2x}{dt^2} = \frac{MgSin\alpha}{\left(M + \frac{I}{r^2}\right)} \quad \text{From equation (iv)}$$

$$\Rightarrow F = \frac{I\alpha}{r} = \frac{I\frac{d^2\theta}{dt^2}}{r} = \frac{I\frac{d^2x}{dt^2}}{r^2} = \frac{\frac{IMgSin\alpha}{\left(M + \frac{I}{r^2}\right)}}{r^2} = \frac{IMgSin\alpha}{I + Mr^2}$$

But : $F = \mu R \Rightarrow \mu = \dfrac{F}{R} = \dfrac{\frac{IMgSin\alpha}{I + Mr^2}}{MgCos\alpha} = \left(\dfrac{I}{I + Mr^2}\right)Tan\alpha$

The minimum values for coefficient of friction which will allow rolling without slipping are:

Circular disc: $I = \dfrac{1}{2}Mr^2 \Rightarrow \mu = \dfrac{1}{3}Tan\alpha$

Hoop: $I = Mr^2 \Rightarrow \mu = \dfrac{1}{2}Tan\alpha$

Solid Sphere: $I = \dfrac{2}{5}Mr^2 \Rightarrow \mu = \dfrac{2}{7}Tan\alpha$

Chapter 10   Hydrostatics

1. Relative density
Define relative density (also called specific gravity).

Solution
Relative density of a substance = ratio of the density/mass/weight of a given volume of the substance to the density/mass/weight of an equivalent volume of water (at 4°C)

2. Metal alloys
A metal A has a density $\rho_A = 2$ kg/m³ and metal B has a density $\rho_B = 3$ kg/m³
What proportion of A and B by volume will give an alloy whose density is $\rho = 2.8$ kg/m³ ?

Solution
Let masses and volumes of A and B used be m_A, V_A and m_B, V_B respectively. Let the mass and volume of the alloy produced be m kg and V m³.
$\Rightarrow V_A + V_B = V$ (i)   and $m_A + m_B = m$
But: $m_A = \rho_A V_A$ and $m_B = \rho_B V_B$ while $m = \rho V \Rightarrow \rho_A V_A + \rho_B V_B = \rho V$ (ii)

Subtracting:
Equation (ii)-(ρ_A)(equation (i)): i.e. $(\rho_A V_A + \rho_B V_B = \rho V) - (\rho_A V_A + \rho_A V_B = \rho_A V)$
This gives: $(\rho_B - \rho_A)V_B = (\rho - \rho_A) V$
Inserting the values above gives: $(3 - 2)V_B = (2.8 - 2)V \Rightarrow V_B = 0.8V \Rightarrow V_A = 0.2V \Rightarrow$
proportion by volumes is $V_A : V_B = 1:4$

3. Mix of two liquids
If equal volumes (V m³) of two liquids, A and B, are mixed the relative density of the mixture is 4. If equal masses (M kg) of the two liquids are mixed the relative density of the mixture is 3.5. Find the densities of the liquids.

Solution
Let density of liquids A and $B = \rho_1$, ρ_2 respectively.

Equal volumes
Use the information given to derive the quantities (i) – (viii) below:

Liquid	Mass kg		Volume m³	Density kg/m³	Relative Density
A	$\rho_1 V$	(i)	V	ρ_1	$\rho_1/1{,}000$ (vi)
B	$\rho_2 V$	(ii)	V	ρ_2	$\rho_2/1{,}000$ (vii)
Mixture	$\rho_1 V + \rho_2 V$ (iii)		2V (iv)	$(\rho_1 V + \rho_2 V)/2V$ (v)	$(\rho_1 V + \rho_2 V)/2V /(1{,}000) = 4 \Rightarrow$ $\rho_1 + \rho_2 = 8{,}000$ (viii)

Equal masses
Use the information given to derive the quantities (i) – (viii) below:

Liquid	Mass		Volume	Density	Relative Density
A	M	(i)	M/ρ_1	ρ_1	$\rho_1/1{,}000$ (vi)
B	M	(ii)	M/ρ_2	ρ_2	$\rho_2/1{,}000$ (vii)
Mixture	2M	(iii)	$M(1/\rho_1 + 1/\rho_2)$ (iv)	$2M/M(1/\rho_1 + 1/\rho_2)$ $= 2/(1/\rho_1 + 1/\rho_2)$ (v)	$2/(1/\rho_1 + 1/\rho_2)1{,}000 = 3.5$ $\Rightarrow 1/\rho_1 + 1/\rho_2 = 1/1{,}750$ (viii)

From above : $\rho_1 + \rho_2 = 8{,}000$ Equation (i)

Also : $\dfrac{1}{\rho_1} + \dfrac{1}{\rho_2} = \dfrac{1}{1{,}750} \Rightarrow \dfrac{\rho_1 + \rho_2}{\rho_1 \rho_2} = \dfrac{1}{1.750} \Rightarrow 1{,}750(\rho_1 + \rho_2) = \rho_1 \rho_2$

$\Rightarrow 1{,}750(8{,}000) = \rho_1 \rho_2 \Rightarrow \rho_1 \rho_2 = 14{,}000{,}000$ Equation (ii)

Using these equations : $\rho_1, \rho_2 = 2{,}585.6, 5{,}414.2 \ \dfrac{kg}{m^3}$ (in any order)

4. Mix of three liquids

Three liquids A, B, C have relative densities $R_A = 1.2$, $R_B = 1.4$, $R_C = 1.6$ respectively. If 1 m^3 of liquid A, 2 m^3, of liquid B and 3 m^3 of liquid C are fully mixed in a vat what is the relative density of the resulting mix.

Solution
Let densities of liquids A, B, $C = \rho_1, \rho_2, \rho_3$ respectively.

Use the information given to derive the quantities (i) – (vii) below:

Liquid	Mass kg	Volume m³	Density kg/m³	Relative Density
A	$(1)(\rho_1)$ (i)	1	$\rho_1 = 1{,}200$	$R_1 = 1.2$
B	$(2)(\rho_2)$ (ii)	2	$\rho_2 = 1{,}400$	$R_2 = 1.4$
C	$(3)(\rho_3)$ (iii)	3	$\rho_3 = 1{,}700$	$R_3 = 1.7$
Mixture	$1\rho_1 + 2\rho_2 + 3\rho_3 = 9{,}100$ (iv)	6 (v)	$9{,}100/6 = 1{,}516.7$ (vi)	Relative density $=1{,}516.7/1{,}000$ $= 1.517$ (vii)

5. Mix of three liquids

Three vats (Vats 1, 2 and 3), each of volume V, are half filled with liquids L_1, L_2, L_3 respectively. The liquids L_1, L_2, L_3 have densities ρ_1, ρ_2, ρ_3 respectively

If liquid L_1 is poured into Vat 2 until it is filled. Then this mixture (Mixture 1) is poured into Vat 3 until it is filled. What is the density of the final mix (Mixture 2) in Vat 3.

Solution
Characteristics of resulting mixture in Vat 2
Use the information given to derive the quantities (i) – (viii) below:

Liquid	Mass kg	Volume m³	Density kg/m³
L_1	$\frac{1}{2}\rho_1 V$ (i)	☐ V	ρ_1
L_1	$\frac{1}{2}\rho_2 V$ (ii)	$\frac{1}{2}$ V	ρ_2
Mixture 1	$\frac{1}{2} V(\rho_1 + \rho_2)$ (iii)	V (iv)	$\frac{1}{2} V(\rho_1 + \rho_2)/ V = \frac{1}{2}(\rho_1 + \rho_2)$ (v)

Characteristics of resulting mixture in Vat 3
Use the information given to derive the quantities (i) – (viii) below:

Liquid	Mass kg	Volume m³	Density kg/m³
Mixture 1	$\frac{1}{4} V(\rho_1 + \rho_2)$ (i)	☐ V	$\frac{1}{2}(\rho_1 + \rho_2)$
L_3	$\frac{1}{2}\rho_3 V$ (ii)	$\frac{1}{2}$ V	ρ_3
Mixture 2	$\frac{1}{4} V(\rho_1 + \rho_2 + 2\rho_3)$ (iii)	V (iv)	$\frac{1}{4} V(\rho_1 + \rho_2 + 2\rho_3)/V = \frac{1}{4}(\rho_1 + \rho_2 + 2\rho_3)$ (v)

The density of the final mix (Mixture 2) in Vat 3 is $\frac{1}{4}(\rho_1 + \rho_2 + 2\rho_3)$.

6. Pressure in a vessel

A vessel has a cross section of 0.005 m². It is partially filled with water. What is the pressure at this depth? If a solid block of mass 0.05 kg of material of the same relative density is placed on the surface find the change in pressure?

Solution
If relative density = 1 ⇒ density = ρ = 1,000 kg/m³.
Initial pressure at depth H metres = ρgH = 1,000(9.81)H = 9,810H Newtons/m².
Density = mass/volume. Therefore, a block of mass 0.05 kg of material of the same relative density will have a volume of 0.05/1000 = 0.00005 m³. If this is regarded as having a cross sectional area of 0.005 m² then it will have a height of 0.01 metres. Thus the surface of the liquid will be increased to a height of H + 0.008 metres above the bottom of the vessel. The pressure at the bottom of the vessel will be increased by ρg(increase in height) = 1,000(9.81)(0.01) = 98.1 Newtons/m².

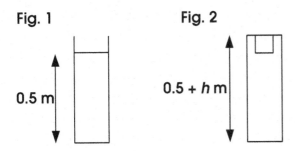

Fig. 1   0.5 m

Fig. 2   0.5 + h m

7. Lamina floating in tank
A rectangular tank, 1 metre long and 0.5 metres wide contains water to a depth of 0.4 metres.
(a)   Find the pressure on the base of the tank and the total thrust on the base.
(b)   If a plane lamina of length 0.2 metres and width 0.2 metres (Area = 0.04 m²) is placed horizontally in the tank 0.2 metres below the surface, find the pressure at, and the thrust on, the lamina.

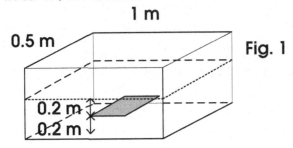

1 m
0.5 m
0.2 m
0.2 m
Fig. 1

Solution

(a) Find the pressure on the base of the tank and the total thrust on the base
See Fig. 1
Area of base of tank = 1 × 0.5 = 0.5 m²   Depth of water = 0.4 metres (= h, say)
Pressure at base = ρgh, where ρ = 1,000 kg/m³, g = 9.81 m/s²
⇒ Pressure at base = (1,000)(9.81)(0.4) = 3,924 N/m²
The centre of gravity of the base is also 0.4 metres below the surface.
Thrust on base = (Surface area of base)(Pressure at the centre of gravity)
= 0.5 m² × 3,924 N/m² = 1,962 Newtons

(b) Find the pressure at, and the thrust on, the horizontal lamina.

If a plane lamina with area = 0.04 m² is placed 0.2 metres below the surface then:
Pressure at centre of gravity = pressure on lamina =
$\rho g h = (1,000)(9.81)(0.2) = 1,962$ N/m²
The thrust on lamina = $(0.04)(1,962) = 78.48$ Newtons

8. Trough filled with water

A trough, 2 metres long, 1 metres deep and width tapering from 0.8 metres (*AB*) to 0.4 metres (*EF*) is full of water. (Assume density = $\rho = 1,000$ kg/m³). (See Fig. 1). Find:

(a)　　The total weight of water.
(b)　　The total force on the base.

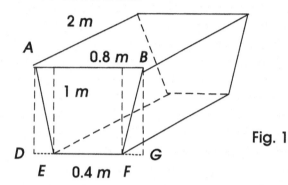

Fig. 1

Solution
(a) Total weight of water
Volume of trough, V = (Length) (area *ABEF*)

$$V = (2)\left(\frac{0.8 + 0.4}{2}\right)(1) = 1.2 \text{ m}^3 \Rightarrow \text{Weight of water} = V \times \rho \times g = 11,772 \text{ Newtons}$$

(b) Force on base:
See Fig. 1.
Pressure (gauge) on base = $\rho g h = (1,000)(9.81)(1) = 9,810$ N/m²
Force on base = Pressure at base × Area of base = $(9,810)(0.4 \times 2) = 7,848$ Newtons

9. Hemispheres filled with water

A hemisphere of radius *r* metres is placed on a horizontal plane. The hemisphere is then filled to the top with a liquid of weight W_W Newtons through a hole in its top. (See Fig. 1). If the minimum weight of the hemisphere to prevent it from lifting is W_H, find W_H in terms of W_W

Solution
The forces in Newtons are shown in Fig. 2:

W_W = Weight of water = (Volume)(density)$(g) = \frac{2}{3}\pi r^3 (1000)(g)$　Newtons

W_H = Weight of hemisphere
U = Up-thrust of liquid
And:　$U = W_W + W_H$　　But:　$U = (\rho g r)$(Area of base) = $1,000 g \pi r^3$

$$\Rightarrow 1000 g\pi\, r^3 = \frac{2}{3}\pi\, r^3(1000g) + W_H \Rightarrow W_H = \frac{1}{3}\pi\, r^3(1000g) \quad \text{Newtons}$$

$$\Rightarrow \frac{W_H}{W_W} = \frac{\frac{1}{3}\pi\, r^3(1000g)}{\frac{2}{3}\pi\, r^3(1000g)} = \frac{1}{2} \Rightarrow W_H = \frac{1}{2}W_W$$

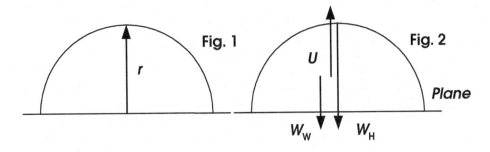

Fig. 1

Fig. 2

Plane

10. Masses floating in water

A wooden cube of volume V m³ and mass M_1 kg floats in water (Assume $\rho = 1,000$ kg/m³) with one quarter above the surface. A mass M_2 kg is placed on top of the wooden cube, causing to sink until its top is level with the surface of the water (See Figs. 1 and 2). Find M_2 in terms of M_1.

Solution

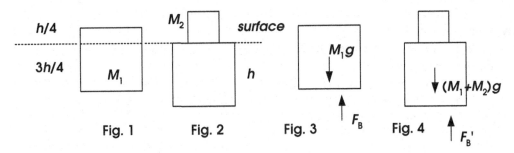

Fig. 1   **Fig. 2**   **Fig. 3**   **Fig. 4**

Mass M_1:
The forces acting are shown in Figs. 3, 4: Mg = Weights of masses, F_B = Buoyancy forces
Buoyancy force = $F_B = M_1 g$
But : F_B = (volume of immersed solid)(density of liquid)(g)
$$\Rightarrow F_B = \frac{3}{4}V\rho g \Rightarrow \frac{3}{4}V\rho g = M_1 g \Rightarrow V = \frac{4M_1 g}{3\rho g} = \frac{M_1}{750} \quad \text{m}^3$$
Masses M_1, M_2 : $F_B' = (M_1 + M_2)g$ But : F_B' = (volume of immersed solid)(density of liquid)(g)
$$\Rightarrow F_B' = V\rho g \Rightarrow V\rho g = (M_1 + M_2)g \Rightarrow \frac{1000 g M_1}{750} = \frac{4g M_1}{3} = M_1 + M_2 \Rightarrow M_2 = \frac{M_1}{3} \quad \text{kg}$$

11. Connected cubes submerged in liquid

Two cubes, identical in dimensions with volume 1 m³ but of mass M_1 and M_2 kg respectively (where $M_1 > M_2$), are connected together by a light string and submerged in water. Find the sum $M_1 + M_2$ if only half of the cube M_2 is submerged. (See Fig. 1).

Solution
The forces acting on the masses are:

T = Tension in the string Mg = Weights of masses F_B = Buoyancy forces

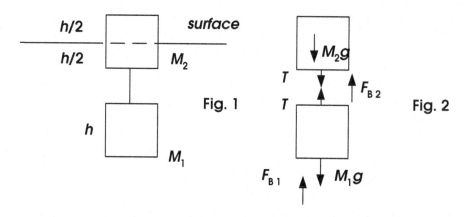

surface

Fig. 1

Fig. 2

Mass M_1 : $M_1g = T + F_{B1}$ But : $F_{B1} = V\rho g \Rightarrow M_1g = T + V\rho g \Rightarrow T = M_1g - V\rho g$ (i)

Mass M_2 : $F_{B2} = M_2g + T$ But : $F_{B2} = \dfrac{V}{2}\rho g \Rightarrow \dfrac{V}{2}\rho g = M_2g + T \Rightarrow T = \dfrac{V}{2}\rho g - M_2g$ (ii)

Using equations (i) and (ii) : $T = M_1g - V\rho g = \dfrac{V}{2}\rho g - M_2g \Rightarrow M_1 + M_2 = \dfrac{3V}{2}\rho$

But : $V = 1 \text{m}^3$ (given) and $\rho = 1,000$ kg/m$^3 \Rightarrow M_1 + M_2 = 1,500$ kg

12. Mass lowered into tank
A cylindrical tank of diameter D metres is filled with a liquid of density ρ kg/m³ to a depth h metres. See Fig. 1.
(a)   Find the thrust on the base of the tank.
(b)   If a solid of mass M kg and density ρ_s kg/m³ is lowered on a wire until immersed in the tank, find the thrust on the base (See Fig. 2).

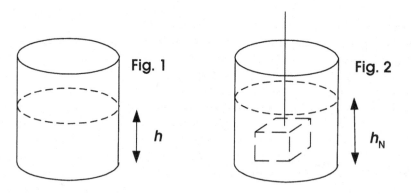

Fig. 1

Fig. 2

Solution

(a) Find the thrust on the base of the tank.
Initial thrust on base = Pressure at base × Area of base

$$= (\rho_1 \, gh)\left(\dfrac{\pi D^2}{4}\right) = \dfrac{\pi \rho_1 \, ghD^2}{4} \quad \text{Newtons}$$

(b) If a solid of mass M kg and density ρ_S kg/m³ is lowered on a wire until immersed in the tank, find the thrust on the base

See Fig. 2. The new depth of liquid in the tank = h_N

Volume of immersed solid $= \dfrac{M}{\rho_S}$ m³

Total volume $= \dfrac{\pi D^2 h}{4}(liquid) + \dfrac{M}{\rho_S}(solid) = \dfrac{\pi D^2 (h_N)}{4}$ m³

$$\Rightarrow h_N = \left(\dfrac{\pi D^2 h}{4} + \dfrac{M}{\rho_S}\right)\left(\dfrac{4}{\pi D^2}\right) \quad \text{metres} \Rightarrow \text{New thrust on base}$$

$$= (\rho_1 \, g)\left(\dfrac{\pi D^2 h}{4} + \dfrac{M}{\rho_S}\right)\left(\dfrac{4}{\pi D^2}\right)\left(\dfrac{\pi D^2}{4}\right) = (\rho_1 \, g)\left(\dfrac{\pi D^2 h}{4} + \dfrac{M}{\rho_S}\right) \quad \text{Newtons}$$

13. Rod partially immersed in water

A thin uniform rod ab of length L metres and weight W kg can turn freely about the end a, which is pivoted at a height $L/2$ metres above the surface of water into which the other end dips. See Fig. 1. If the rod is in equilibrium when inclined at 45° to the vertical, find the magnitude of the reaction at a and the relative density of the rod.

Solution
(a) The magnitude of the reaction at a

Length of rod = $|ab| = L$ (See Fig. 2). But: $Cos\ 45° = \dfrac{\dfrac{L}{2}}{|ad|} = \dfrac{1}{\sqrt{2}}$

$\Rightarrow$ Length of rod above water $= |ac| = \dfrac{L}{\sqrt{2}}$

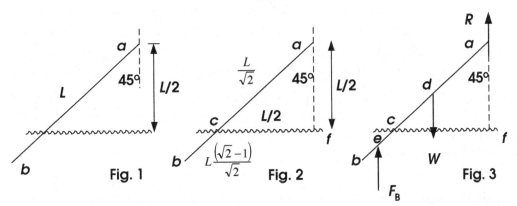

Fig. 1   **Fig. 2**   **Fig. 3**

$\Rightarrow$ Length of rod under water $= |bc| = L\left(\dfrac{\sqrt{2}-1}{\sqrt{2}}\right)$

Let: d = Mid-point of ab and e = Mid-point of bc

The forces in Newtons acting on the rod are shown in Fig. 3:
R = Vertical reaction at hinge at a
W = Weight of rod (acting through d, midpoint of ab)

F_B = Buoyancy force on rod (acting through e, midpoint of bc).

$R + F_B = W$ and, taking moments about a, gives:

$$W \sin 45° \times ad = F_B \sin 45° \times ae = F_B \sin 45° \, (ac + ce)$$

$$\Rightarrow W\left(\frac{1}{\sqrt{2}}\right)\frac{L}{2\sqrt{2}} = F_B\left(\frac{1}{\sqrt{2}}\right)\left[\frac{L}{\sqrt{2}} + \frac{L}{2}\left(\frac{\sqrt{2}-1}{\sqrt{2}}\right)\right] = F_B L\left(\frac{1+\sqrt{2}}{4}\right)$$

$$\Rightarrow W = F_B\sqrt{2}\left(\frac{1+\sqrt{2}}{\sqrt{2}}\right) \Rightarrow F_B = W\left(\frac{1}{1+\sqrt{2}}\right) \Rightarrow R = W - F_B = W - W\left(\frac{1}{1+\sqrt{2}}\right) = W\frac{\sqrt{2}}{1+\sqrt{2}}$$

(b) The relative density of the rod

But : F_B = Volume of water displaced by rod $\times$ density of water $\times$ g
$\Rightarrow F_B$ = (Volume of rod under water)$\rho_{water}g$
The rod is uniform $\Rightarrow$ If L weighs W then

$$bc \, (= \text{length of rod under water}) = L\left(\frac{\sqrt{2}-1}{\sqrt{2}}\right) \text{ weighs } W\left(\frac{\sqrt{2}-1}{\sqrt{2}}\right)$$

But this is equal to:
(Volume of rod under water, V) $\times$ (density of rod) $\times$ $g = V\rho_{rod}g$

$$\Rightarrow V\rho_{rod}g = W\left(\frac{\sqrt{2}-1}{\sqrt{2}}\right) \text{ and from above : } F_B = V\rho_{water}g = W\left(\frac{1}{1+\sqrt{2}}\right)$$

$$\text{Relative density of rod} = \frac{\text{Weight of rod}}{\text{Weight of equal volume of water}}$$

From above:

$$\frac{V\rho_{rod}g}{V\rho_{water}g} = \frac{\rho_{rod}}{\rho_{water}} = \text{relative density of rod} = \frac{W\left(\frac{\sqrt{2}-1}{\sqrt{2}}\right)}{W\left(\frac{1}{1+\sqrt{2}}\right)} = \frac{1}{\sqrt{2}}$$

14. Rod partially immersed in water

A uniform rod ab of length L metres and weight W N in equilibrium is inclined to the vertical with one quarter of its length immersed under water and its upper end, a, supported by a vertical force F. (Assume density of water $= \rho = 1,000$ kg/m³). (See Fig. 1). Find:

(a)   The magnitude of F
(b)   The relative density of the rod

Solution
(a)   Find the magnitude of F
The forces in Newtons acting on the rod are (see Fig. 2):
W = Weight of rod, acting through e, the midpoint of ab
F_B = Buoyancy force on rod, acting through d, the midpoint of bc
F = Vertical force on rod, acting at a

Taking moments about a: $W\dfrac{L}{2}\sin\theta = F_B\dfrac{7L}{8}\sin\theta \Rightarrow \dfrac{W}{2} = \dfrac{7F_B}{8} \Rightarrow W = \dfrac{7}{4}F_B$

$$\Rightarrow F_B = \frac{4}{7}W \quad \text{Equation (i)}$$

Resolving forces vertically: $F + F_B = W \Rightarrow F + \frac{4}{7}W = W \Rightarrow F = \frac{3}{7}W$

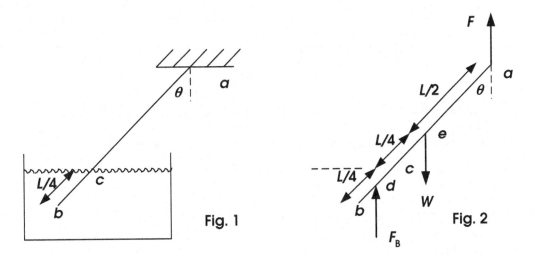

Fig. 1   Fig. 2

(b) The relative density of the rod

But, F_B = (Volume of water displaced by rod) × (density of water) × g

$\Rightarrow F_B$ = (Volume of rod under water)$(\rho_w)(g) = V\rho_w g$

But: $bc = \frac{1}{4}$ of rod length (= immersed length) $\Rightarrow$ it weighs $\frac{W}{4}$

i.e. V(density of rod)$g = \frac{W}{4} \Rightarrow V\rho_r g = \frac{W}{4}$ Equation (ii)

From above equations (i) and (ii): $\dfrac{V\rho_r g}{V\rho_w g} \Rightarrow \dfrac{\rho_r}{\rho_w} = \dfrac{\dfrac{W}{4}}{W\dfrac{4}{7}} = \dfrac{7}{16}$

$\Rightarrow$ The relative density of the rod $= \dfrac{7}{16}$

16. Cylinder floating in water

A small uniform cylinder of density ρ kg/m³, mass m kg, total length L metres and uniform cross section A m² floats in a liquid of density 2ρ kg/m³ with its axis vertical (see Fig. 1). Find:

(a) The thrust on the cylinder when displaced vertically, without being completely immersed, through a distance x from the equilibrium position

(b) The period of oscillation when it is released in this position

Solution

(a) **The thrust on the cylinder when displaced vertically, without being completely immersed, through a distance x from the equilibrium position.**

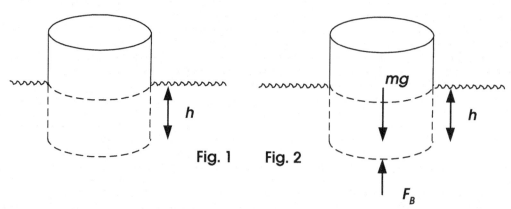

Fig. 1   **Fig. 2**

Let the length of cylinder beneath the surface of the liquid at equilibrium = h
Then, for equilibrium:
Weight of cylinder = Weight of liquid which is contained in a cylinder of cross- sectional area A and height h

i.e. $(A)(L)(\rho)(g) = (A)(h)(2\rho)g \Rightarrow L = 2h \Rightarrow h = \dfrac{L}{2}$

$\Rightarrow$ the cylinder floats with half its length under the liquid. If the cylinder is displaced downwards through a distance x then the buoyancy force will be greater than the weight of the cylinder (See Fig. 2):

mg = Downward force = Weight of cylinder = $AL\rho g$
F_B = Buoyancy force = Weight of liquid of density 2ρ contained in cylinder of cross sectional area A and height $h + x = (A)(h + x)2\rho g$

$\Rightarrow$ Resultant Buoyancy force = $F_B - mg = (A)(h + x)2\rho g - AL\rho g$
$\Rightarrow F_B - mg = (A)(h + x)2\rho g - Ah2\rho g = (2A\rho g) x$

$\Rightarrow$ Resultant Buoyancy force = $(2A\rho g) x$ i.e. the upward force $\propto x$

$\Rightarrow m\dfrac{d^2x}{dt^2} = -2A\rho gx$ = thrust on the cylinder   (i)

(b) The period of oscillation when it is released in this position

Equation (i) is an equation for simple harmonic motion of the form:
$\dfrac{d^2x}{dt^2} = -\omega^2 x \Rightarrow \dfrac{d^2x}{dt^2} = -\dfrac{2A\rho g}{m}x = -\dfrac{2A\rho g}{AL\rho}x = -\dfrac{2g}{L}x \Rightarrow \omega^2 = \dfrac{2g}{L} \Rightarrow \omega = \sqrt{\dfrac{2g}{L}}$

But the periodic time : $T = \dfrac{2\pi}{\omega} \Rightarrow T = 2\pi\sqrt{\dfrac{L}{2g}}$ seconds

17. Rod partially immersed in water
A uniform thin rod AB is of length L metres and weight W Newtons. The rod is partly immersed in water with its lower end freely pivoted to a fixed point at a depth L/3 metres below the surface of the water. (See Fig. 1). (Assume density of water = ρ = 1,000 kg/m³). If the rod rests in equilibrium at an angle of 60° to the vertical, find the relative density of the rod.

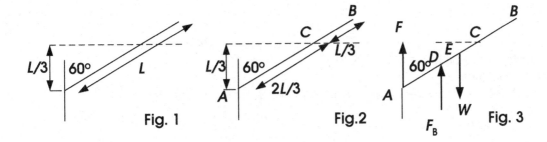

Fig. 1   **Fig.2**   **Fig. 3**

Solution

The inclination of rod to the vertical = 60° (See Fig. 2):

$$\Rightarrow \cos 60° = \frac{\frac{L}{3}}{\text{Length of rod under water}} = \frac{1}{2} \Rightarrow$$

Length of rod under water = $AC = \frac{2L}{3} \Rightarrow$ Length of rod above water = $CB = \frac{L}{3}$

The forces in Newtons on the rod are (See Fig. 3):
F = Upwards force at support A
W = Weight of rod, acting through E the midpoint of AB
F_B = Buoyancy force acting on rod acting through D the midpoint of AC

$$|AD| = \frac{L}{3}, |AE| = \frac{L}{2}$$

Equating the forces for equilibrium gives: $F + F_B = W$
Taking moment about A gives:

$$F_B \sin 60° |AD| = W \sin 60° |AE| \Rightarrow F_B \frac{L}{3} = W \frac{L}{2} \Rightarrow F_B = W \frac{3}{2}$$

But: F_B = (Volume of water displace by rod) × density of water × g

= Volume of rod under water × ρ_{water} × g = $V \rho_{water} g \Rightarrow F_B = W \frac{3}{2} = V \rho_{water} g$ (i)

For the rod: Length L weighs $W \Rightarrow$ length $\frac{2}{3} L$ weighs $\frac{2}{3} W$

$$\Rightarrow \frac{2}{3} W = \text{(volume of rod under water)} \times \text{(density of rod)} \times g$$

$$\Rightarrow \frac{2}{3} W = V \rho_{rod} g \quad \text{(ii)} \quad \text{But}: \quad \frac{V \rho_{rod} g}{V \rho_{water} g} = \frac{\rho_{rod}}{\rho_{water}} = \frac{\frac{2}{3} W}{\frac{3}{2} W}$$

$$\Rightarrow \text{Relative density of the rod} = \frac{4}{9}$$

18. Timber block floating in oil/water layers

A cube of timber (density of 600 kg/m²) of side 0.1 metre floats in water. Oil of density 900 kg/m³ is poured on the water until the top of the oil layer is 0.035 metres below the top of the block. (Assume density of water = ρ = 1,000 kg/m³). Find:
(a) The depth of the oil layer, L

(b) The gauge pressure at the lower face of the block

Solution
(a) The depth of the oil layer (See Fig. 2)
Let depth of oil layer = L
Assume that, below the layer of oil, the remaining depth of timber is T
Thus: $L + T = 0.065$ (i)

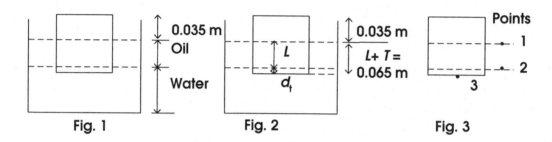

Fig. 1   Fig. 2   Fig. 3

In equilibrium, the downward force must be equivalent to the upward force:
Downward force = Weight of block = (Volume of block)(Density of timber)g
$= (0.1)^3 600g$ Newtons

Upward force = Buoyancy forces for oil and water
Buoyancy force (for oil) = (Area of block)(Depth of block in oil)(Density of oil)g
$= (0.1)^2 L900g$ Newtons

Buoyancy force (for water) = (Area of block)(Depth of block in water)(Density of water)g
$= (0.1)^2 (T)1,000\, g$ Newtons $\Rightarrow (0.1)^3 600\, g = (0.1)^2 L\, 900\, g + (0.1)^2 (T)\, 1,000\, g$
$\Rightarrow 9L + 10T = 0.6$ (ii)
Now: subtracting equation (ii) from $10 \times$ equation (i): $L = 0.05$ and $T = 0.015$
Thus, the layer of oil is 0.05 metres deep.
(b) The gauge pressure at the lower face of the block (See Fig. 3)
To get gauge pressure: (i.e. neglect atmospheric pressure Pa)
P_1 = pressure at surface of oil = 0
$P_2 = (0.05)\rho_{oil} g$ $P_3 = P_2 + (0.015)\rho_{water}\, g$
$\Rightarrow P_3 = (0.05)900\, g + (0.015)1,000\, g = 60\, g = 588.6\ \text{N/m}^2$

19. Board partially immersed in water
A uniform rectangular board abcd of weight W Newtons floats in a tank of water with the diagonal ac on the surface of the water, the lowest corner b being attached to the bottom of the tank by a light inelastic string as shown in Fig. 1. (Assume density of water $= \rho = 1,000$ kg/m³). Find:
(a)   The relative density of the board.
(b)   The tension in the string.

Solution
(a)   The relative density of the board
The forces in Newtons acting on the board are shown in Figs. 2 and 3:

T = Tension of string, W = Weight of board, F_B = Buoyancy force on board

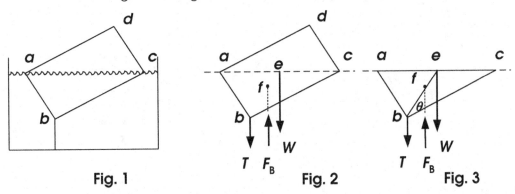

Fig. 1   **Fig. 2**   **Fig. 3**

Let e be the midpoint of diagonal ac and f be the centroid of triangle abc

Let $be = L \Rightarrow bf = \dfrac{2}{3}L$

Resolving forces vertically for equilibrium: $T + F_B = W$ (i)

Taking moments about b:

We get: $F_B \ Sin\ \theta |bf| = W\ Sin\ \theta |be| \Rightarrow F_B = W \left| \dfrac{L}{\frac{2}{3}L} \right| = \dfrac{3W}{2}$

But: F_B = (Volume of water displaced by board) × (density of water) × g

   = (Volume of board below water) $\rho_w g \Rightarrow F_B = V\rho_w g = \dfrac{3W}{2}$ (ii)

But, the weight of the board below water is:

(Volume of board below water) × (density of board) × g $\Rightarrow V\rho_b g = \dfrac{W}{2}$ (iii)

Thus : $\dfrac{V\rho_b g}{V\rho_w g} = \dfrac{\text{weight of board}}{\text{weight of equal volume of water}}$

$\Rightarrow \dfrac{\rho_b}{\rho_w} = \dfrac{1}{3}$ (from equations (i) and (ii)) $\Rightarrow$ relative density $= \dfrac{1}{3}$

(b)   The tension in the string

Since $F_B = \dfrac{3W}{2}$, $T + W = F_B \Rightarrow T = \dfrac{W}{2}$ Newtons

20. Cones submerged in liquid

Two right angle cones, A and B, each of diameter and vertical height d, h metres respectively and mass m kg are held submerged in a tank of water to a depth of x metres by light strings connected to the base of the tank (See Fig. 1). Find the tension in each string. (ρ = Density of water = 1,000 kg/m³)

Solution

The forces in Newtons acting in both cases are:

mg = Weight of cone, T = Tension in string, F_B = Buoyancy force

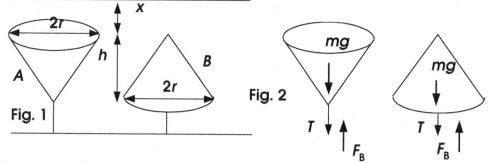

Fig. 1

Fig. 2

Cone A: $F_B = mg + T \Rightarrow T = F_B - mg$

But : $F_B = $ (Volume of cone)$(\rho)(g) = \dfrac{1}{3}\pi r^2 h\rho\, g \Rightarrow T = \dfrac{1}{3}\pi r^2 h\rho\, g - mg$ Newtons

Cone B: $F_B = mg + T \Rightarrow T = F_B - mg$

$F_B = $ (Volume of cone)$(\rho)(g) = \dfrac{1}{3}\pi r^2 h\rho\, g \Rightarrow T = \dfrac{1}{3}\pi r^2 h\rho\, g - mg$ Newtons

$\Rightarrow$ Tension in string is the same as that for cone A.

21. Submerged hemispheres

Two solid hemispheres each of mass m kg and radius a metres are held submerged in a liquid of density ρ kg/m³. They are held at a distance $2a$ below the free surface by light strings as shown in Figs. 1 and 2. For each hemisphere, calculate the:

(a)   Buoyancy force
(b)   Force exerted by the liquid on the plane face
(c)   Force exerted by the liquid on the curved surface
(d)   Tension in the string, T

Solution

The forces in Newtons acting are:
mg = Weight of hemisphere, F_B = Buoyancy force, T = Tension in string

Hemisphere in Fig. 1. See Forces in Fig. 3.

(a) Buoyancy force = (volume)(density of liquid)g $= F_B = \dfrac{2}{3}\pi\rho\, g\, a^3$ (i)

(b) Force exerted by liquid on place face area = Downward Force =
(Pressure on plane face)(Area of plane face) $= (\rho g(2a))(\pi a^2) = 2\pi\rho g a^3$ (ii)

(c) As: Force exerted by liquid on curved surface area = Upward force and
      Buoyancy = Upward force – Downward force
$\Rightarrow$ Force on curved surface = Buoyancy + Downward force
$\Rightarrow$ Using equations (i) and (ii):

Force on curved surface $= \dfrac{2}{3}\pi\rho\, g\, a^3 + 2\pi\rho\, g\, a^3 = \dfrac{8}{3}\pi\rho\, g\, a^3$

(d) Tension in string:

$T + mg = F_B = \dfrac{2}{3}\pi\rho\, g a^3 \Rightarrow T = \dfrac{2}{3}\pi\rho\, g a^3 - mg$

Hemisphere in Fig. 2. See Forces in Fig. 4.

204

(a) Buoyancy force = (volume)(density of liquid)$g \Rightarrow F_B = \frac{2}{3}\pi a^3 \rho g$

(b) Force exerted by liquid on place face area = Upward Force
= (Pressure on plane face)(Area of plane face) = $(\rho g(3a))(\pi a^2) = 3\pi\rho g a^3$

(c)   As:
Force exerted by liquid on curved surface area = Downward force   and
Buoyancy = Upward force − Downward force
$\Rightarrow$ Force on curved surface = Downward force = Upward force - Buoyancy

$\Rightarrow$ Force on curved surface = $3\pi\rho\, g\, a^3 - \frac{2}{3}\pi\rho\, g\, a^3 = \frac{7}{3}\pi\rho\, g\, a^3$

(d) Tension in string:

$T + mg = F_B = \frac{2}{3}\pi\rho\, g a^3 \Rightarrow T = \frac{2}{3}\pi\cdot\rho\, g a^3 - mg$

Note: Tension is the same for both strings

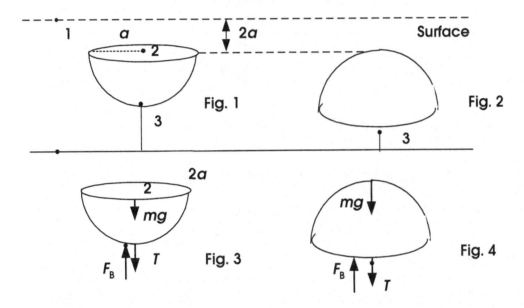

Fig. 1   Fig. 2   Fig. 3   Fig. 4

22. Floating sphere

A light hollow sphere of mass = M = 10 kg and V = volume = 1 m³ is connected to the bottom of a tank of water by a light string (See Fig. 1). If the submerged portion represents 90% of the sphere's total volume find the tension in the string, T Newtons. (Assume water density of ρ_w = 1,000 kg/m³)

Solution

The forces in Newtons acting on the sphere are:
Mg = Weight of sphere, T = Tension in string, F_B = Buoyancy force
These forces are shown in Fig. 2: $F_B = T + Mg = T + 10g \Rightarrow T = F_B - 10g$ (i)

But: F_B = (Density of liquid)(Volume of submerged portion of object)(g)
= $(\rho_w)(0.9V)g = (1,000)(0.9)g = 900g \Rightarrow$ From equation (i): $T = 890g$ Newtons

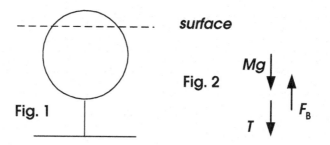

Fig. 1

Fig. 2

Mg

T

F_B

23. Balloon
A light balloon of volume, V, ($V = 1$ m³) is filled with a light gas of density 0.1 kg/m³. The balloon is connected to the ground by a light string (See Fig. 1). If the density of the surrounding atmosphere is $\rho_A = 1$ kg/m³ find the tension in the string, T Newtons.

Solution
The forces in Newtons acting on the balloon are:
Mg = Weight of balloon, T = Tension in string, F_B = Buoyancy force

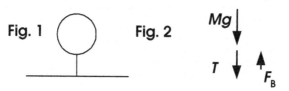

Fig. 1   **Fig. 2** Mg

T F_B

These forces are shown in Fig. 2: $Mg + T = F_B \Rightarrow T = F_B - Mg$ (i)

Mass, M, can be found from:
M = (Density)(Volume) = (0.1 kg/m³)(1m³) = 0.1 kg $\Rightarrow Mg = 0.1g$
The buoyancy force can be found from:
F_B = (Density of atmosphere)(Volume of submerged object)(g)
= $(\rho_A)(V)g = (1)(1)g = 1g = g \Rightarrow$ From equation (i): $T = 0.9g$ Newtons

24. Object immersed in water
A metal object is suspended from a spring balance in air and the balance reads 10 Newtons (See Fig. 1). When the mass is suspended in water (density = $\rho_W = 1,000$ kg/m³) the balance reading is 8 Newtons (See Fig. 3). Find the buoyancy force and the mass and volume of the object.

Solution
The forces in Newtons acting are shown in Fig. 2 and Fig. 4:
Mg = Weight of object (in air), T_1, T_2 = Tension in spring balance (= balance reading)
F_B = Buoyancy force

From Fig. 2: $T_1 = Mg = 10$ Newtons   (i) (given)
From Fig. 4: $T_2 + F_B = Mg$ Newtons   But spring balance reading = $T_2 = 8$ Newtons
(given) $\Rightarrow 8 + F_B = Mg$ (ii)
Using equations (i) and (ii): $F_B = 2$ Newtons
But: F_B = (Density of liquid)(Volume of submerged object)(g)
$\Rightarrow F_B = (\rho_W)(V)g = (1,000)(V)g = 2$ Newtons
$\Rightarrow V = 0.0002$ m³ approx. = volume of block

Mass of block = Weight/g = 10/g kg

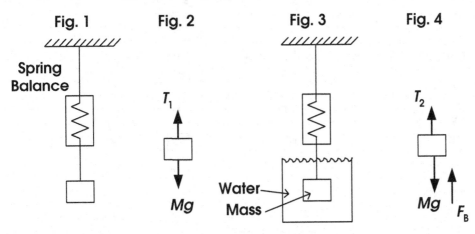

Fig. 1   Fig. 2   Fig. 3   Fig. 4

25. Suspension in oil

A solid metal ball (of volume $V = 0.001$ m³ and density $\rho_M = 8,000$ kg/m³) is suspended from a spring balance in oil (density = 900 kg/m³) and half of the ball is submerged. (See Fig.1). Find the reading on the balance (T) in Newtons.

Solution

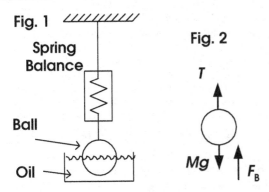

The forces in Newtons acting are shown in Fig. 2:
Mg = Weight of ball (in air), T = Tension in spring balance, F_B = Buoyancy force
From Fig. 2: $T + F_B = Mg \Rightarrow T = Mg - F_B$ (i)
But:
$M = (\rho_M)(V) = 8,000V$ kg $\Rightarrow Mg = 8,000Vg$
F_B = (Density of liquid)(Volume of submerged object)(g)
$\Rightarrow F_B = (900)(0.5V)g = 450Vg \Rightarrow T = Mg - F_B \Rightarrow Vg(8,000 - 450) = 7,550Vg = 74.07$ Newtons

26. Suspended block submerged in water and another liquid

A block weighing 10 Newtons is suspended from a spring balance and immersed (in separate operations) in a beaker of water and a beaker of liquid. When the block is immersed in the water the spring balance reads 7.5 Newtons. When immersed in the liquid the spring balance reads 5 Newtons. Find the relative densities of: (a) The block, (b) The liquid.

Solution

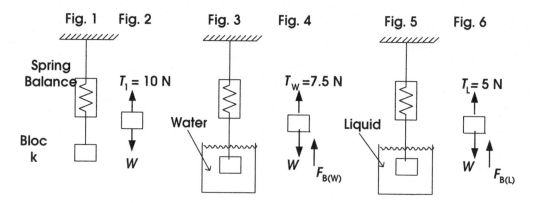

Fig. 1 Fig. 2 Fig. 3 Fig. 4 Fig. 5 Fig. 6

The forces acting in Newtons are:
W = Weight of block, $F_{B(W)}$ = Buoyancy force in water, $F_{B(L)}$ = Buoyancy force in liquid
T_1 = Reading on Spring Balance when Mass is not immersed
T_W = Reading on Spring Balance when Mass is immersed in water
T_L = Reading on Spring Balance when Mass is immersed in liquid

(a) Relative density of the block
Block not immersed: Spring Balance reads 10 Newtons
See Figs. 3 and 4:
Block immersed in water $\Rightarrow$ Spring Balance reads 7.5 N (i.e. T_W = 7.5 Newtons)
But the forces acting are: $W = T_W + F_{B(W)} \Rightarrow 10 = 7.5 + F_{B(W)} \Rightarrow F_{B(W)} = 2.5$ Newtons
Thus, the weight of water displaced by the volume of the block = 2.5 Newtons
$\Rightarrow$ Relative density $= \dfrac{10}{2.5} = 4$

(b) Relative density of the liquid
Block not immersed: Spring Balance reads 10 Newtons
See Figs. 5 and 6:
Block immersed in liquid $\Rightarrow$ Spring Balance reads 5 Newtons (i.e. T_L = 5 Newtons)
But the forces acting are: $W = T_W + F_{B(L)} \Rightarrow 10 = 5 + F_{B(L)} \Rightarrow F_{B(L)} = 5$ Newtons
Thus, the weight of liquid displaced by the volume of the block = 5 Newtons
$\Rightarrow$ Relative density $= \dfrac{\text{weight of liquid displaced by volume of block}}{\text{weight of water displaced by volume of block}} = \dfrac{5}{2.5} = 2$

27. Suspended block submerged in liquid
A mass of m kg and volume 1 litre (0.001 m³) hangs by a string from a spring balance and is submerged in a liquid of density ρ kg/m³, contained in a beaker. The beaker has a weight of 3 Newtons and weight of the liquid is 4 Newtons. The spring balance reads 5 Newtons and the pan balance A reads 16 Newtons. Find:
(a) The density of the liquid.
(b) The reading on each balance if the Mass is removed from the liquid.

Solution
The arrangement of the balances is shown in Fig. 1.
The forces acting in Newtons are: W_M = Weight of Mass, W_W = Weight of water and beaker, F_B = Buoyancy force
S_1 = Reading on Spring Balance when Mass is submerged in liquid
S_2 = Reading on Spring Balance when Mass is removed in liquid
P_1 = Reading on Pan Balance when Mass is submerged in liquid

P_2 = Reading on Pan Balance when Mass is removed from liquid

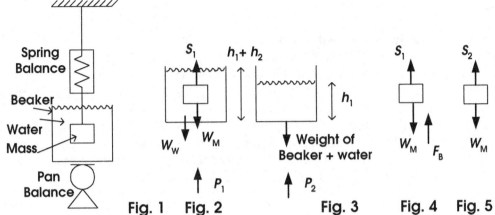

Fig. 1   Fig. 2   Fig. 3   Fig. 4   Fig. 5

(a) Find the density of the liquid.

Mass submerged in liquid $\Rightarrow$ Pan Balance reads 16 (i.e. P_1 = 16 Newtons) See Fig. 2.

Mass removed from liquid $\Rightarrow$ Pan Balance measures only the weights of the beaker and liquid (i.e. P_2 = 3 Newtons + 4 Newtons = 7 Newtons. (See Fig. 3)

Thus, the weight of liquid displaced by the Mass = 16 – 7 = 9 Newtons $\Rightarrow F_B$ = 9 Newtons

But : weight of liquid displaced by the Mass = (Volume of Mass)(density of liquid)(g)

$$\Rightarrow (0.001)\rho\,g = 9\,\text{Newtons} \Rightarrow \rho = \frac{9}{0.001g} = 917.4\,\text{kg/m}^3$$

Alternative approach:

Assume base area of beaker = $A\,\text{m}^2$.

Mass removed from liquid $\Rightarrow$ Pan Balance measures only the weights of the beaker and liquid (i.e. P_2 = 3 Newtons + 4 Newtons = 7 Newtons. (See Fig. 3)

$\Rightarrow$ The liquid causes a thrust of 4 Newtons on the base $\Rightarrow (\rho\,gh_1)(A) = 4$ Newtons

Mass submerged in liquid $\Rightarrow$ the new volume in the beaker

= volume of liquid + volume of Mass = $\Rightarrow (h_1)(A) + 0.001\ \text{m}^3$

New height of liquid in beaker = $h_1 + h_2$ where $h_1 + h_2 = \dfrac{h_1 A + 0.001}{A}$

But : New thrust on base = 16 Newtons (given)

Of this amount, 3 Newtons is contributed by the beaker's own weight

$\Rightarrow$ 13 Newtons is contributed by the weight of the liquid and Mass

But the new thrust on the base can also be expressed as :

$$(h_1 + h_2)Ag\rho\ \text{Newtons} \Rightarrow \left(\frac{h_1 A + 0.001}{A}\right)Ag\rho = 13 \Rightarrow (h_1 A + 0.001)g\rho = 13$$

But, from above : $\rho\,gh_1\,A = 4$ Newtons $\Rightarrow 0.001g\rho = 9$ Newtons $\Rightarrow \rho = 917.4\,\text{kg/m}^3$

(b) Find the reading on each balance if the Mass is removed from the liquid.

Spring balance

When the Mass is submerged in the liquid then (See Fig. 4): $S_1 + F_B = W_M$

But: $S_1 = 5$ N (given) $\Rightarrow 5 + F_B = W_M$

But, from above, $F_B = 9$ Newtons $\Rightarrow W_M = 14$ Newtons

$\Rightarrow$ Spring balance will read 14 Newtons when the Mass is removed from the liquid (i.e. Fig. 5 will show : $S_2 = 14$ Newtons)

Pan Balance:

Pan Balance will read: $P_2 = 3$ Newtons + 4 Newtons = 7 Newtons when Mass is removed from the liquid.

28. Object suspended in tank filled with a liquid
A truncated cone of weight W Newtons is suspended by a rope in an open tank of water. (Assume density of water = $\rho = 1,000$ kg/m³). (See Fig. 1).
(a) Find the total upward force exerted by the liquid on the base of the cone (of area 1 m²) (i.e. at point 3)
(b) Find the total downward force exerted by the liquid on the top of the truncated cone (of area 0.5 m²) (i.e. at point 2) and on curved surface of truncated cone
(c) Find the tension, T, in Newtons in the rope supporting the truncated cone.
(d) Find the minimum mass of the truncated cone in kg for $T > 0$?

Solution
The forces in Newtons acting on the truncated cone are shown in Fig. 2:
F_B = Buoyancy force, T = Tension in rope, W = Weight of truncated cone

$T + F_B = W$
F_B = Weight of fluid displaced by the truncated cone
 = (volume of truncated cone) × (density of fluid) × g

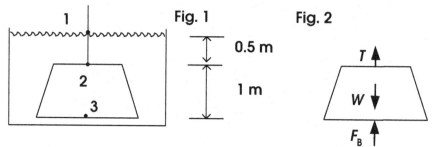

But, the volume of a truncated cone = $V = \frac{1}{3}\pi h(r^2 + R^2 + rR) = \frac{1}{3}h(\pi r^2 + \pi R^2 + \pi rR)$

and: $\pi r^2 = 0.5$ m²(given), $\pi R^2 = 1$m²(given), $h = 1$ metre(given), $\pi rR = \dfrac{1}{\sqrt{2}}$

$\Rightarrow V = \frac{1}{3}(1)\left(0.5 + 1 + \dfrac{1}{\sqrt{2}}\right)$ m³ $\Rightarrow F_B = \frac{1}{3}\left(1.5 + \dfrac{1}{\sqrt{2}}\right)\rho g = 7,217.24$ Newtons

(a) Total upward force exerted by the liquid on the base of the cone
Upward force on lower base = 1 m² × Pressure at 3 = $1.5\rho g = 14,715$ Newtons

(b) Total downward force exerted by liquid on the top of the truncated cone
Downward force on top surface = 0.5 m² × Pressure at 2 = $0.25\rho g = 2,452.5$ Newtons.
But: F_B = Upward force - Downward force
Total downward force = Downward forces on flat and curved surfaces
                    = 2,452.5 + Downward force on curved surface
$\Rightarrow F_B = 7,217.24 = 14,715 - 2,452.5 -$ Downward force on curved surface
$\Rightarrow$ Downward force on curved surface = 5,045.3 Newtons

(c) Tension in rope
$T = W - F_B = W - 7,217.24$ Newtons

(d) Minimum mass of truncated cone
If $T = 0 \Rightarrow W = 7,217.24$ Newtons $\Rightarrow m = 735.7$ kg

Chapter 11   Differentiation and Integration

A. Notes: Indices
The power to which a number is raised is indicated by its index. The rules for indices are:

$$\left(a^m\right)\left(a^n\right)=a^{m+n} \qquad \frac{a^m}{a^n}=a^{m-n} \qquad a^{-n}=\frac{1}{a^n} \qquad a^0=1 \qquad \left(a^m\right)^n=a^{mn}$$

$$(ab)^n=a^nb^n \qquad \left(a^{\frac{1}{n}}\right)^n=a^{\frac{n}{n}}=a^1=a \qquad a^{\frac{m}{n}}=\sqrt[n]{a^m} \qquad \left(a^P\right)^{\frac{1}{Q}}=a^{\frac{P}{Q}}=\left(a^{\frac{1}{Q}}\right)^P$$

B. Notes: Natural logarithms (Note : $\log_e y = I_n y$)

$$\log_e(ab)=\log_e(a)+\log_e(b) \quad \log_e\left(\frac{a}{b}\right)=\log_e(a)-\log_e(b) \qquad x=I_n y \ \Rightarrow y=e^x \Rightarrow y=e^{In y}$$

$$y=e^{ax} \ \Rightarrow (I_n y)=ax(I_n e)\Rightarrow I_n y=ax \qquad I_n x^n=nI_n x \qquad I_n x=I_n\left(\frac{x}{a}.a\right) \ =I_n\left(\frac{x}{a}\right)+I_n(a)$$

C. Differentiation
$$y=kx^n \Rightarrow \frac{dy}{dx}=knx^{n-1} \qquad y=Sin\theta \Rightarrow \frac{dy}{dx}=Cos\theta \quad y=Cos\theta \Rightarrow \frac{dy}{dx}=-Sin\theta$$

$$y=uv \ \Rightarrow \frac{dy}{dx}=v\frac{du}{dx}+u\frac{dv}{dx} \qquad y=\frac{u}{v} \ \Rightarrow \frac{dy}{dx}=\frac{v\frac{du}{dx}-u\frac{dv}{dx}}{v^2}$$

$$y=I_n x \ \Rightarrow \frac{dy}{dx}=\frac{1}{x} \qquad y=I_n ax \ \Rightarrow \frac{dy}{dx}=\frac{a}{ax}=\frac{1}{x} \qquad y=I_n x^n \ \Rightarrow \frac{dy}{dx}=\frac{nx^{n-1}}{x^n}=\frac{n}{x}$$

$$y=e^x \Rightarrow \frac{dy}{dx}=e^x \qquad y=e^{ax} \ \Rightarrow (I_n y)=ax(I_n e)\Rightarrow I_n y=ax\Rightarrow \frac{dy}{dx}=ae^x$$

D. Integral calculus (Note: Constants of integration are not shown)

1. Solve $\displaystyle\int x^n dx=\frac{x^{n+1}}{n+1}$ 2. Solve $\displaystyle\int \frac{1}{x}dx=I_n x$ 3. Solve $\displaystyle\int e^{ax}dx=\frac{e^{ax}}{a}$

4. Solve $\displaystyle\int\left(\frac{1}{a+bx}\right)dx$: Using substitution : Let $u=a+bx \Rightarrow \frac{du}{dx}=b\Rightarrow dx=\frac{du}{b}$

Re – write : $\displaystyle\int\left(\frac{1}{a+bx}\right)dx$ as $\displaystyle\int\left(\frac{1}{u}\right)\frac{du}{b}=\frac{1}{b}\int\left(\frac{du}{u}\right)=\frac{1}{b}I_n u=\frac{1}{b}I_n(a+bx)$

5. Solve $\displaystyle\int\left(\frac{1}{\sqrt{a+bx}}\right)dx$: Using substitution : Let $u=a+bx \Rightarrow \frac{du}{dx}=b\Rightarrow dx=\frac{du}{b}$

Re – write : $\displaystyle\int\left(\frac{1}{\sqrt{a+bx}}\right)dx$ as $\displaystyle\int\left(\frac{1}{\sqrt{u}}\right)\frac{du}{b}=\frac{1}{b}\int\left(u^{-\frac{1}{2}}du\right)=\frac{2}{b}u^{\frac{1}{2}}=\frac{2}{b}\sqrt{u}=\frac{2}{b}\sqrt{a+bx}$

6. Solve $\int\left(\dfrac{x}{a+bx}\right)dx$ Using substitution : Let $u = a+bx \Rightarrow \dfrac{du}{dx} = b \Rightarrow dx = \dfrac{du}{b}$

Re – write : $\int\left(\dfrac{x}{a+bx}\right)dx = \dfrac{1}{b}\int\left(\dfrac{x}{u}\right)du = \dfrac{1}{b^2}\int\left(\dfrac{u-a}{u}\right)du = \dfrac{1}{b^2}\int\left(1-\dfrac{a}{u}\right)du$

$= \dfrac{1}{b^2}\left[u - al_n u\right] = \dfrac{1}{b^2}\left[a+bx - al_n(a+bx)\right]$

7. Solve $\int\left(\dfrac{1}{a+bx^2}\right)dx$

Re – write as : $\dfrac{1}{b}\int\left(\dfrac{1}{\dfrac{a}{b}+x^2}\right)dx$

Using substitution, let : $x = \sqrt{\dfrac{a}{b}}\,Tan\theta \Rightarrow \dfrac{dx}{d\theta} = \sqrt{\dfrac{a}{b}}\left(\dfrac{1}{Cos^2\theta}\right)$

But : $\dfrac{a}{b}+x^2 = \dfrac{a}{b}\left(\dfrac{1}{Cos^2\theta}\right) \Rightarrow \dfrac{1}{b}\int\left(\dfrac{1}{\dfrac{a}{b}+x^2}\right)dx$ becomes $\dfrac{1}{b}\int\dfrac{1}{\dfrac{a}{b}\left(\dfrac{1}{Cos^2\theta}\right)}\sqrt{\dfrac{a}{b}}\left(\dfrac{1}{Cos^2\theta}\right)d\theta$

$\dfrac{1}{\sqrt{ab}}\int d\theta = \dfrac{1}{\sqrt{ab}}\theta$ But : $\theta = Tan^{-1}\sqrt{\dfrac{b}{a}}x$

$\Rightarrow \int\left(\dfrac{1}{a+bx^2}\right)dx = \dfrac{1}{\sqrt{ab}}Tan^{-1}\left(x\sqrt{\dfrac{b}{a}}\right) = \dfrac{1}{\sqrt{ab}}Tan^{-1}\dfrac{x\sqrt{ab}}{a}$

8. Solve $\int\left(\dfrac{1}{a-bx^2}\right)dx$

Re – write as : $\int\left(\dfrac{1}{a-bx^2}\right)dx = \dfrac{1}{2\sqrt{a}}\int\left(\dfrac{1}{\sqrt{a}+\sqrt{b}x} + \dfrac{1}{\sqrt{a}-\sqrt{b}x}\right)dx =$

$\dfrac{1}{2\sqrt{a}}\left[\dfrac{1}{\sqrt{b}}l_n\left(\sqrt{a}+\sqrt{b}x\right) - \dfrac{1}{\sqrt{b}}l_n\left(\sqrt{a}-\sqrt{b}x\right)\right] = \dfrac{1}{2\sqrt{ab}}\left[l_n\left(\dfrac{\sqrt{a}+\sqrt{b}x}{\sqrt{a}-\sqrt{b}x}\right)\right] = \dfrac{1}{2\sqrt{ab}}l_n\left(\dfrac{\sqrt{\dfrac{a}{b}}+x}{\sqrt{\dfrac{a}{b}}-x}\right)$

9. Solve $\int\left(\dfrac{x}{a+bx^2}\right)dx$

Using substitution, let : $u = a+bx^2 \Rightarrow \dfrac{du}{dx} = 2bx$

$\Rightarrow x\,dx = \dfrac{du}{2b}$ Re - write : $\int\left(\dfrac{x}{a+bx^2}\right)dx = \dfrac{1}{2b}\int\left(\dfrac{du}{u}\right) = \dfrac{1}{2b}l_n(u) = \dfrac{1}{2b}l_n\left(a+bx^2\right)$

10. Solve $\int \left(\dfrac{1}{\sqrt{x^2 + a^2}} \right) dx$

Using substitution, let : $x = aTan\theta \Rightarrow \dfrac{dx}{d\theta} = aSec^2\theta \Rightarrow dx = aSec^2\theta \, d\theta$

But : $\sqrt{x^2 + a^2} = \sqrt{a^2 Tan^2\theta + a^2} = a\sqrt{Tan^2\theta + 1} = aSec\theta$

Re – write $\int \left(\dfrac{1}{\sqrt{x^2 + a^2}} \right) dx$ as $\int \left(\dfrac{1}{aSec\theta} \right) aSec^2\theta \, d\theta = \int Sec\theta \, d\theta$

$= I_n \left| Sec\theta + Tan\theta \right| = I_n \left| \dfrac{\sqrt{x^2 + a^2}}{a} + \dfrac{x}{a} \right| = I_n \left| \dfrac{x + \sqrt{x^2 + a^2}}{a} \right|$

11: Solve $\int \left(\dfrac{1}{\sqrt{x^2 - a^2}} \right) dx$

Using substitution, let : $x = aSec\theta \Rightarrow \dfrac{dx}{d\theta} = aSec\theta \, Tan\theta \Rightarrow dx = aSec\theta \, Tan\theta \, d\theta$

But : $\sqrt{x^2 - a^2} = \sqrt{a^2 Sec^2\theta - a^2} = aTan\theta$

Re – write $\int \left(\dfrac{1}{\sqrt{x^2 + a^2}} \right) dx$ as $\int \left(\dfrac{1}{aTan\theta} \right) aSec\theta \, Tan\theta d\theta = \int Sec\theta \, d\theta$

$= I_n \left| Sec\theta + Tan\theta \right| = I_n \left| \dfrac{\sqrt{x^2 - a^2}}{a} + \dfrac{x}{a} \right| = I_n \left| \dfrac{x + \sqrt{x^2 - a^2}}{a} \right|$

12: Solve $\int \left(\dfrac{1}{\sqrt{a^2 - x^2}} \right) dx$

Using substitution, let : $x = aSin\theta \Rightarrow \dfrac{dx}{d\theta} = aCos\theta.$

Also : $Cos\theta = \dfrac{\sqrt{a^2 - x^2}}{a} \Rightarrow \sqrt{a^2 - x^2} = aCos\theta$

$\Rightarrow$ Re – write : $\int \left(\dfrac{1}{\sqrt{a^2 - x^2}} \right) dx$ as $\int \left(\dfrac{1}{aCos\theta} \right) aCos\theta \, d\theta = \int d\theta = \theta = Sin^{-1} \dfrac{x}{a}$

E. General Questions

13. Distance, Velocity and Acceleration

If a body starts from rest with an acceleration, $a = 5 - \dfrac{x}{2}$ m/s² where x = distance travelled, in metres express the velocity, v m/s, as a function of x.

Solution

Acceleration $= \dfrac{d^2x}{dt^2} = v\dfrac{dv}{dx}$ But $\dfrac{dx}{dt} = v \Rightarrow \dfrac{d^2x}{dt^2}$ can be expressed

as $v\dfrac{dv}{dx} = 5 - \dfrac{x}{2} \Rightarrow v\,dv = \left(5 - \dfrac{x}{2}\right)dx$

The limits are : $x = 0 \Rightarrow v = 0$; $x = x \Rightarrow v = v$; $\Rightarrow \displaystyle\int_0^v v\,dv = \int_0^x \left(5 - \dfrac{x}{2}\right)dx$

$\Rightarrow \left[\dfrac{1}{2}v^2\right]_0^v = \left[\left(5x - \dfrac{x^2}{4}\right)\right]_0^x \Rightarrow v^2 = 10x - \dfrac{x^2}{2} \Rightarrow v = \sqrt{10x - \dfrac{x^2}{2}}$

14. Distance, Velocity and Acceleration

If a body starts moving with initial velocity u m/s and constant acceleration a m/s² and s is the distance travelled in metres in time t seconds show that:

(a) $v^2 = u^2 + 2as$ (b) $s = ut + \frac{1}{2}at^2$

Solution

(a) $v^2 = u^2 + 2as$

Acceleration $= \dfrac{d^2x}{dt^2} = v\dfrac{dv}{dx} = a \Rightarrow v\,dv = a\,dx \Rightarrow \displaystyle\int v\,dv = \int a\,dx$

The limits are : $x = 0 \Rightarrow v = u$; $x = s \Rightarrow v = v$

$\Rightarrow \displaystyle\int_u^v v\,dv = \int_0^s a\,dx \Rightarrow \dfrac{1}{2}\left(v^2 - u^2\right) = as \Rightarrow v^2 = u^2 + 2as$

(b) $s = ut + \frac{1}{2}at^2$

From above : $v = \sqrt{u^2 + 2ax} \Rightarrow \dfrac{dx}{dt} = \sqrt{u^2 + 2ax} \Rightarrow \displaystyle\int \dfrac{dx}{\sqrt{u^2 + 2ax}} = \int dt$

The limits are : $t = 0 \Rightarrow x = 0$; $t = t \Rightarrow x = s$

$\Rightarrow \displaystyle\int_0^s \dfrac{dx}{\sqrt{u^2 + 2ax}} = \int_0^t dt \Rightarrow \left[\dfrac{2\sqrt{u^2 + 2ax}}{2a}\right]_0^s = t \Rightarrow \left[\sqrt{u^2 + 2as} - \sqrt{u^2 + 0}\right] = at$

$\Rightarrow \sqrt{u^2 + 2as} = u + at \Rightarrow u^2 + 2as = \left(u + at\right)^2 = u^2 + 2uat + a^2t^2 \Rightarrow s = ut + \frac{1}{2}at^2$

15. Rate of decay of radioactive substance

At a time $t = 0$ a radioactive substance has a mass M kg. The rate of decay of the substance at any time is proportional to the mass at that time. After 200 days 75% of the mass remains. How many days will it take for the mass to be reduced to 50% of the original amount.

Solution

Let: M = Initial mass (i.e. at time $t = 0$), m = Mass at time $t = t$, e^{-dt} = Rate of decay of radioactive substance

We can write : $m = Me^{-dt}$ At $t = 0, m = M(e^0) = M(1) = M$
At $t = 200, m = 0.75M$ (given) $\Rightarrow 0.75M = Me^{-d200} \Rightarrow d = 0.0014384$
At $t = t, m = 0.5M \Rightarrow 0.5M = Me^{-0.0014384t} \Rightarrow t = 481.9$ days
$\Rightarrow$ After 481.9 days the remaining mass is 50% of the original mass M

16. Population growth

Assume that world population in 1950 and 1976 respectively was 1.5 and 3.5 billion
people and that growth at any time was proportional to the population at that time.
Using this data, estimate the world population in 2002.

Solution
Let: P = Population in billions in 1950, p = Population in billions after t years,
e^{kt} = Rate of growth of population
 We can write : $p = Pe^{kt}$ For 1950, $t = 0 \Rightarrow p = P(e^0) = P(1) \Rightarrow P = 1.5$
For 1976 : $t = 26$ years $\Rightarrow p = Pe^{26k} = 1.5e^{26k} = 3.5$ billion $\Rightarrow k = 0.032588$
For 2002 : $p = 1.5e^{52k}$ where : $k = 0.032588 \Rightarrow p = 1.5 \times 5.444 = 8.17$ billion

17. Ball travelling against resistive force

A ball of mass $m = 1$ kg is struck so that its initial velocity along the ground is u m/s. If it
experiences a resistive force of kv Newtons where v is its velocity at any point find its
velocity after travelling a distance of x metres.

Solution

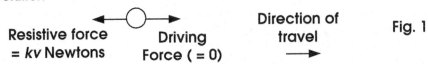

			Direction of	Fig. 1
Resistive force	Driving		travel	
= kv Newtons	Force (= 0)			

The forces acting on the ball after it has been struck are shown in Fig. 1.

Equation of Motion : $m\dfrac{d^2x}{dt^2} = -kv$ but $\dfrac{d^2x}{dt^2} = v\dfrac{dv}{dx} \Rightarrow mv\dfrac{dv}{dx} = -kv$

$\Rightarrow m\int dv = -k\int dx$ But : $m = 1$ kg

$\Rightarrow \int dv = -k\int dx$ The limits are : $v = u \Rightarrow x = 0; v = v \Rightarrow x = x$

$\Rightarrow \int_{u}^{v} dv = -k\int_{0}^{x} dx = -kx \Rightarrow [v]_{u}^{v} = -kx \Rightarrow (u - v) = kx$ $\Rightarrow v = u - kx$ m/s

18. Engine pulling train in resistive medium

An engine of mass m kg pulls a train along a track while working at a constant power of
10 kmu^2 Watts where k and u are constants. The resistance is k times the momentum.
Find the equation of motion and the time to increase speed from: (a) 0 to u m/s (b) u to
$3u$ m/s

Solution

Resistance   Driving Force   Direction of travel   Fig. 1

Power = Driving Force × Velocity = $Fv = 10kmu^2$ Watts

$\Rightarrow$ Driving Force (imparted by engine) = $\dfrac{10kmu^2}{v}$

$\Rightarrow m\dfrac{dv}{dt}$ = Driving Force – Resistance = $\dfrac{10kmu^2}{v} - k(\text{momentum})$

But : momentum = $mv \Rightarrow$ Equation of Motion is :$\Rightarrow m\dfrac{dv}{dt} = \dfrac{10kmu^2}{v} - kmv = km\left(\dfrac{10u^2 - v^2}{v}\right)$

$\Rightarrow v\dfrac{dv}{dt} = k(10u^2 - v^2) \Rightarrow \int \dfrac{v}{10u^2 - v^2}dv = \int k\,dt$

(a) 0 to u m/s
The limits are : $t = 0 \Rightarrow v = 0$; $t = T \Rightarrow v = u$

$\displaystyle\int_0^u \dfrac{v}{10u^2 - v^2}dv = \int_0^T k\,dt = kT \Rightarrow \dfrac{1}{-2}\int_0^u \dfrac{-2v}{10u^2 - v^2}dv = \left(\dfrac{1}{-2}\right)\left[l_n(10u^2 - v^2)\right]_0^u = kT$

$\Rightarrow l_n(10u^2 - u^2) - l_n(10u^2) = -2kT \Rightarrow l_n(10u^2) - l_n(9u^2) = 2kT \Rightarrow l_n\left(\dfrac{10}{9}\right) = 2kT \Rightarrow T = \left(\dfrac{1}{2k}\right)l_n\left(\dfrac{10}{9}\right)$ s

= time taken to increase speed from 0 to u m/s

(b) u to $3u$ m/s
The limits are : $t = 0 \Rightarrow v = u$; $t = T \Rightarrow v = 3u$

$\displaystyle\int_u^{3u} \dfrac{v}{10u^2 - v^2}dv = \int_0^T k\,dt = kT \Rightarrow \dfrac{1}{-2}\int_u^{3u} \dfrac{-2v}{10u^2 - v^2}dv = \left(\dfrac{1}{-2}\right)\left[l_n(10u^2 - v^2)\right]_u^{3u} = kT$

$\Rightarrow l_n(10u^2 - 9u^2) - l_n(10u^2 - u^2) = -2kT \Rightarrow l_n(9u^2) - l_n(u^2) = 2kT \Rightarrow l_n\left(\dfrac{9}{1}\right) = 2kT \Rightarrow T = \left(\dfrac{1}{2k}\right)l_n(9)$ s

= time taken to increase speed from u to $3u$ m/s

19. Hydrofoil travelling on water
A hydrofoil of mass 100 Tonnes is travelling at a constant speed of 20 m/s and the resistance to its motion, R, is 50,000 Newtons. If R is proportional to the square of the speed how far does the hydrofoil travel when increasing speed from 20 m/s to 30 m/s at constant driving power of 5,000 kW.

Solution
Let D = Driving force   The equation of motion is :

$mv\dfrac{dv}{dx}$ = Driving force – Resistance = $D - R$ Let $R = kv^2$

(Given) : $m = 100$ Tonnes = 10^5 kg

$R = 50,000$ Newtons   (when : $v = 20$ m/s) $\Rightarrow 50,000 = k(20)^2 = 400k$

$\Rightarrow k = 125 \Rightarrow$ Equation of motion = $10^5 v\dfrac{dv}{dx} = D - 125v^2$

But: Power = Force × Velocity = Dv (in this case)

$\Rightarrow 5 \times 10^6$ Watts $= Dv = D \times 20 \Rightarrow D = 2.5 \times 10^5$ Newtons

$\Rightarrow 10^5 v\dfrac{dv}{dx} = 2.5 \times 10^5 - 125v^2 \Rightarrow 800v\dfrac{dv}{dx} = 2{,}000 - v^2$

$\Rightarrow \dfrac{800v}{(2{,}000 - v^2)}dv = dx \Rightarrow 800\displaystyle\int\dfrac{v}{(2{,}000 - v^2)}dv = \int dx$

The limits are: At $v = 20$, $x = 0$; At $v = 30$, $x = x$

$\Rightarrow 800\displaystyle\int_{20}^{30}\dfrac{v}{(2{,}000 - v^2)}dv = \int_{0}^{x}dx \Rightarrow 800\left(\dfrac{1}{-2}\right)l_n\big[2{,}000 - v^2\big]_{20}^{30} = x$

$\Rightarrow -400[l_n(2{,}000 - 900) - l_n(2{,}000 - 400)] = x$

$\Rightarrow -400[l_n(1{,}100) - l_n(1{,}600)] = x \Rightarrow 400[l_n(1{,}600) - l_n(1{,}100)] = x \Rightarrow x = 400l_n\left(\dfrac{16}{11}\right) = 149.9$ metres

$\Rightarrow$ Distance travelled when increasing speed from 20 to 30 m/s = 149.9 metres

20. Ship travelling on water

A 30,000 Tonne ship can be driven at a constant speed of 20 m/s by a power of 20,000 kW. The resistance to motion, R, is proportional to the square of the speed. If the ship starts from rest, find its speed after travelling 200 metres.

Solution

Let D = Driving force   The equation of motion is :

$mv\dfrac{dv}{dx} = D - R = D - kv^2$ where k is a constant

$m = 30{,}000$ Tonnes $= 3 \times 10^7$ kg

But : Power $=$ Force $\times$ Velocity $= Dv \Rightarrow 2 \times 10^7$ Watts $= D(20) \Rightarrow D = 10^6$ Newtons

$\Rightarrow 12{,}000v\dfrac{dv}{dx} = 400 - v^2 \Rightarrow \dfrac{12{,}000v}{(400 - v^2)}dv = dx$

But, when $v = 20$ m/s the speed is constant $\Rightarrow v\dfrac{dv}{dx} = 0$

$\Rightarrow D = R \Rightarrow 10^6 = 400k \Rightarrow k = 2{,}500 \Rightarrow \left(3 \times 10^7\right)v\dfrac{dv}{dx} = 10^6 - 2{,}500v^2$

The limits are : At $x = 0$, $v = 0$; At $x = 200$, $v = v$

$\Rightarrow 12{,}000\displaystyle\int_{0}^{v}\dfrac{v}{(400 - v^2)}dv = \int_{0}^{200}dx \Rightarrow 12{,}000\left(-\dfrac{1}{2}\right)l_n\big[400 - v^2\big]_{0}^{v} = 200$

$\Rightarrow -6{,}000 l_n\big[400 - v^2\big]_{0}^{v} = 200 \Rightarrow 30 l_n\big[400 - v^2\big]_{v}^{0} = 1 \Rightarrow l_n\big[400 - v^2\big]_{v}^{0} = 0.0333$

$\Rightarrow \dfrac{400}{400 - v^2} = e^{0.0333} = 1.034 \Rightarrow v = 3.63$ m/s

21. Particle moving with variable acceleration

A particle starts moving from rest at time $t = 0$ with an acceleration of $A + Bv$ where $v =$ velocity at time t. Find expressions for:

(a)   Find an expression for velocity at time t

(b)   If $A > 0$ and $B < 0$ show that as t increases v tends towards a maximum value and find this maximum value

(c)   Find an expression for x, the distance travelled in time t

Solution
(a) Find an expression for velocity at time t

Equation of Motion : $\dfrac{d^2x}{dt^2} = \dfrac{dv}{dt} = A + Bv \Rightarrow \dfrac{dv}{A + Bv} = dt$

$\Rightarrow \displaystyle\int \dfrac{dv}{A + Bv} = \int dt$ The limits are : $t = 0; v = 0 \Rightarrow t = t \Rightarrow v = v$

$\Rightarrow \displaystyle\int_0^v \dfrac{dv}{A + Bv} = \int_0^t dt \Rightarrow \dfrac{1}{B}[l_n(A + Bv)]_0^v = t \Rightarrow \dfrac{1}{B}l_n\left(\dfrac{A + Bv}{A}\right) = t$ (i)

$\Rightarrow \dfrac{A + Bv}{A} = e^{Bt} \Rightarrow v = \dfrac{A}{B}\left(e^{Bt} - 1\right)$

(b) If $A > 0$ and $B < 0$ show that as t increases v tends towards a maximum value and find this maximum value

Since : $v = \dfrac{A}{B}\left(e^{Bt} - 1\right)$ (Note : If $B = -P \Rightarrow e^{Bt} = \dfrac{1}{e^{Pt}} \Rightarrow$ as t increases the value

of e^{Pt} increases $\Rightarrow$ the value of e^{Bt} decreases and tends towards 0)

$\Rightarrow v$ tends towards : $\dfrac{A}{B}(-1)$. But : $\dfrac{A}{B} < 0 \Rightarrow |v| \to \left|\dfrac{A}{B}\right|$

(c) Find an expression for x, the distance travelled in time t

Equation of Motion : $\dfrac{d^2x}{dt^2} = v\dfrac{dv}{dx} = A + Bv \Rightarrow \dfrac{vdv}{A + Bv} = dx$

$\Rightarrow \displaystyle\int \dfrac{v\,dv}{A + Bv} = \int dx$ The limits are : $x = 0; v = 0 \Rightarrow x = x \Rightarrow v = v$

$\Rightarrow \displaystyle\int_0^v \dfrac{v\,dv}{A + Bv} = \int_0^x dx \Rightarrow \dfrac{1}{B^2}[A + Bv - Al_n(A + Bv)]_0^v = x$

$\Rightarrow \dfrac{1}{B^2}[A + Bv - Al_n(A + Bv) - A + Al_nA] = x$

$\Rightarrow x = \dfrac{1}{B^2}[Bv - Al_n(A + Bv) + Al_nA] = \dfrac{1}{B^2}\left[Bv - Al_n\left(\dfrac{A + Bv}{A}\right)\right]$

22. Train slowing down
The brakes of a train of mass M kg are applied when the train is travelling at u m/s at a distance D metres from the buffers at the end of the line. If the brakes produce a resistive force of $A + Bv$ (where v is the speed of the train) and the train just reaches the buffers, find D.

Solution

Equation of Motion : $M\dfrac{d^2x}{dt^2} = Mv\dfrac{dv}{dx} = -(A + Bv) \Rightarrow \dfrac{Mvdv}{A + Bv} = -dx$

$\Rightarrow \displaystyle\int \dfrac{Mv\,dv}{A + Bv} = -\int dx$ The limits are : $x = 0; v = u \Rightarrow x = D \Rightarrow v = 0$

$$\Rightarrow M\int_u^0 \frac{v\,dv}{A+Bv} = -\int_0^D dx \Rightarrow \frac{M}{B^2}[A+Bv-Al_n(A+Bv)]_u^0 = -D$$

$$\Rightarrow \frac{M}{B^2}[A-Al_n(A)-(A+Bu-Al_n(A+Bu))]=-D$$

$$\Rightarrow D = -\frac{M}{B^2}[-Bu+Al_n(A+Bu)-Al_nA] = -\frac{M}{B^2}\left[-Bu+Al_n\left(\frac{A+Bu}{A}\right)\right] = \frac{M}{B^2}\left[Bu-Al_n\left(\frac{A+Bu}{A}\right)\right]$$

23. Boats travelling on water

Three boats of mass m kg travelling in parallel at u m/s stop their engines simultaneously. After they have moved a further distance of d metres each has a speed of $u/2$ m/s. Find the speed of each boat after it has travelled a distance of $2d$ metres if their resistance to the motion is kv, kv^2 and kv^3 respectively where v is the speed and k is a constant.

Solution

(a)   Boat 1: Resistance = kv

Equation of Motion : $m\dfrac{d^2x}{dt^2} = -kv$ but $\dfrac{d^2x}{dt^2} = v\dfrac{dv}{dx} \Rightarrow mv\dfrac{dv}{dx} = -kv$

$\Rightarrow m\int dv = -k\int dx$ The limits are : $v = u \Rightarrow x = 0$; $v = v \Rightarrow x = x$

$\Rightarrow m\int_u^v dv = -k\int_0^x dx = -kx \Rightarrow m[v]_u^v = -kx \Rightarrow m(u-v) = kx$ (i)

But, if : $x = d \Rightarrow v = \frac{u}{2}$ (given) $\Rightarrow m\left(\frac{u}{2}\right) = kd \Rightarrow k = \dfrac{mu}{2d}$

If : $x = 2d \Rightarrow$ from equation (i): $m(u-v) = 2dk = \dfrac{2dmu}{2d} = mu \Rightarrow v = 0$

(b)   Boat 2: Resistance = kv^2

Equation of Motion : $m\dfrac{d^2x}{dt^2} = -kv^2$ but $\dfrac{d^2x}{dt^2} = v\dfrac{dv}{dx} \Rightarrow mv\dfrac{dv}{dx} = -kv^2$

$\Rightarrow m\int \dfrac{dv}{v} = -k\int dx$ The limits are : $v = u \Rightarrow x = 0$; $v = v \Rightarrow x = x$

$\Rightarrow m\int_u^v \dfrac{dv}{v} = -k\int_0^x dx = -kx \Rightarrow m[l_n v]_u^v = -kx \Rightarrow m\left[l_n\dfrac{u}{v}\right] = kx$ (ii)

But, if : $x = d \Rightarrow v = \frac{u}{2} \Rightarrow m[l_n 2] = kd \Rightarrow k = \dfrac{ml_n 2}{d}$

If $x = 2d \Rightarrow$ from equation (ii): $m\left[l_n\dfrac{u}{v}\right] = 2dk = \dfrac{2dml_n 2}{d} \Rightarrow \left[l_n\dfrac{u}{v}\right] = 2l_n 2 = 1.386$

$\Rightarrow l_n\dfrac{u}{v} = 1.386 \Rightarrow v = ue^{-1.386} = \dfrac{u}{4}$

(c) Resistance = kv^3

Equation of Motion : $m\dfrac{d^2x}{dt^2} = -kv^3$ but $\dfrac{d^2x}{dt^2} = v\dfrac{dv}{dx} \Rightarrow mv\dfrac{dv}{dx} = -kv^3$

$\Rightarrow m\int \dfrac{v}{v^3}dv = -k\int dx \Rightarrow m\int \dfrac{dv}{v^2} = -k\int dx$

The limits are : $v = u \Rightarrow x = 0$; $v = v \Rightarrow x = x$

$$\Rightarrow m\int_{u}^{v}\frac{dv}{v^2} = -k\int_{0}^{x}dx = -kx \Rightarrow m\left[-\frac{1}{v}\right]_{u}^{v} = -kx \Rightarrow m\left[\frac{1}{v}-\frac{1}{u}\right] = kx \quad \text{(iii)}$$

But, if : $x = d \Rightarrow v = \frac{u}{2} \Rightarrow m\left[\frac{2}{u}-\frac{1}{u}\right] = kd \Rightarrow k = \frac{m}{du}$

If : $x = 2d \Rightarrow$ from equation (iii) : $m\left[\frac{1}{v}-\frac{1}{u}\right] = 2dk = \frac{2dm}{du} = \frac{2m}{u} \Rightarrow \frac{1}{v} = \frac{3}{u} \Rightarrow v = \frac{u}{3}$

24. Boat travelling on water

A boat of mass m kg stops its engines at a speed of u m/s. The resistance to motion is kv, where v is the speed and k is a constant. Find
(a)   The distance travelled before the boat comes to rest.
(b)   The time elapsed before the boat comes to rest.

Solution
(a) The distance travelled before the boat comes to rest.

Equation of Motion : $m\dfrac{d^2x}{dt^2} = -kv$ (i) but $\dfrac{d^2x}{dt^2} = \dfrac{dv}{dt} = v\dfrac{dv}{dx} \Rightarrow mv\dfrac{dv}{dx} = -kv$

$\Rightarrow m\int dv = -k\int dx$ The limits are : $v = u, \Rightarrow x = 0; v = 0 \Rightarrow x = x$

$\Rightarrow m\int_{u}^{0}dv = -k\int_{0}^{x}dx \Rightarrow m[v]_{u}^{0} = -kx$

$\Rightarrow -mu = -kx \Rightarrow x = \dfrac{mu}{k} =$ distance travelled before boat comes to a stop

(b) The time elapsed before the boat comes to rest

Also, from equation (i) we can write : $m\dfrac{dv}{dt} = -kv \Rightarrow m\int\dfrac{dv}{v} = -k\int dt$

The limits are : $v = u, \Rightarrow t = 0; v = 0 \Rightarrow t = t \Rightarrow m\int_{u}^{0}\dfrac{dv}{v} = -k\int_{0}^{t}dt \Rightarrow m[l_n v]_{u}^{0} = -kt$

$\Rightarrow m[l_n v]_{0}^{u} = kt \Rightarrow t = \dfrac{m}{k}(l_n u)$ seconds = time elapsed before boat comes to rest

25. Particle falling in resistive medium

A particle of mass 0.2 kg falls vertically from rest in a medium which exerts a resistive force of $0.03v$ Newtons where v is the speed of the particle. Find expressions for:
(a)   The speed of the particle at any time t
(b)   The displacement of the particle at time t

Solution
The forces acting on the mass are shown in Fig. 1:

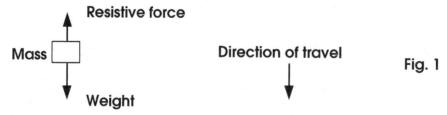

Resistive force

Mass

Direction of travel

Weight

Fig. 1

(a) The speed of the particle at any time t

Equation of motion : $m\dfrac{dv}{dt} = mg - 0.03v \Rightarrow 0.2\dfrac{dv}{dt} = 0.2g - 0.03v \Rightarrow \dfrac{dv}{dt} = g - 0.15v$

$\Rightarrow \displaystyle\int \dfrac{dv}{g - 0.15v} = \dfrac{1}{0.15}\int \dfrac{dv}{\left(\dfrac{g}{0.15} - v\right)} = \int dt$

The limits are : $t = 0 \Rightarrow v = 0;\quad t = t \Rightarrow v = v \Rightarrow \dfrac{1}{0.15}\displaystyle\int_0^v \dfrac{dv}{\left(\dfrac{g}{0.15} - v\right)} = \int_0^t dt = t$

$\Rightarrow -\dfrac{1}{0.15}\left[l_n\left(\dfrac{g}{0.15} - v\right)\right]_0^v = t \quad \Rightarrow \dfrac{1}{0.15}\left[l_n\left(\dfrac{g}{0.15} - v\right)\right]_v^0 = t$

$\Rightarrow \left[l_n\left(\dfrac{g}{0.15}\right) - l_n\left(\dfrac{g}{0.15} - v\right)\right] = 0.15t \quad \Rightarrow l_n\left(\dfrac{\dfrac{g}{0.15}}{\dfrac{g}{0.15} - v}\right) = 0.15t$

$\Rightarrow \dfrac{\dfrac{g}{0.15}}{\dfrac{g}{0.15} - v} = e^{0.15t} \Rightarrow v = \dfrac{g}{0.15}\left(1 - e^{-0.15t}\right)$

(b) The displacement of the particle at time t

But : $v = \dfrac{dx}{dt} \Rightarrow \dfrac{dx}{dt} = \dfrac{g}{0.15}\left(1 - e^{-0.15t}\right)$

The limits are : $t = 0 \Rightarrow x = 0;\quad t = t \Rightarrow x = x \Rightarrow \displaystyle\int_0^x dx = \dfrac{g}{0.15}\int_0^t\left(1 - e^{-0.15t}\right)dt$

$\Rightarrow x = \dfrac{g}{0.15}\left[t + \dfrac{e^{-0.15t}}{0.15}\right]_0^t = \dfrac{g}{0.15}\left[t + \dfrac{e^{-0.15t}}{0.15} - \dfrac{1}{0.15}\right]$

26. Stone falling through resistive medium

A stone with mass $m = 1$ kg falls from rest a height. If the wind resistance has a resistive force of kv^2 where v is the speed of the stone in m/s and k is a constant, find:
(a)   The equation of motion of the stone.
(b)   The terminal velocity of the stone, V_T.
(c)   The velocity, v, after time t.

Solution
(a) The equation of motion of the stone

Equation of motion $= m\dfrac{d^2x}{dt^2} =$ Downward forces – Upward forces

$\Rightarrow m\dfrac{d^2x}{dt^2} = mg - kv^2 \quad$ But $\quad m = 1 \Rightarrow \dfrac{d^2x}{dt^2} = g - kv^2$

(b) The terminal velocity of the stone, V_T

Terminal velocity, V_T occurs when : $\dfrac{dv}{dt} = \dfrac{d^2x}{dt^2} = 0 \Rightarrow g - kv^2 = 0 \Rightarrow v = V_T = \sqrt{\dfrac{g}{k}}$

(c) The velocity, v, after time t

To find velocity after time t:

$$\frac{d^2x}{dt^2} = \frac{dv}{dt} = g - kv^2 \Rightarrow \int \frac{dv}{g-kv^2} = \int dt \Rightarrow \frac{1}{k}\int \frac{dv}{\frac{g}{k} - v^2} = \int dt$$

The standard solution for an integration problem of this type is:

$$= \frac{1}{k}\frac{1}{2}\sqrt{\frac{k}{g}}\left[l_n\left(\frac{\sqrt{\frac{g}{k}}+v}{\sqrt{\frac{g}{k}}-v}\right)\right] \Rightarrow \frac{1}{2\sqrt{gk}}\left[l_n\left(\frac{\sqrt{\frac{g}{k}}+v}{\sqrt{\frac{g}{k}}-v}\right)\right] = t$$

The limits are: $t = 0 \Rightarrow v = 0$; and $t = t \Rightarrow v = v$, giving:

$$\Rightarrow \frac{1}{2\sqrt{gk}}\left[l_n\left(\frac{\sqrt{\frac{g}{k}}+v}{\sqrt{\frac{g}{k}}-v}\right)\right]_0^v = [t]_0^t \Rightarrow \frac{1}{2\sqrt{gk}}l_n\left(\frac{\sqrt{\frac{g}{k}}+v}{\sqrt{\frac{g}{k}}-v}\right) = t \Rightarrow \frac{\sqrt{\frac{g}{k}}+v}{\sqrt{\frac{g}{k}}-v} = e^{2t\sqrt{gk}} \Rightarrow v = \sqrt{\frac{g}{k}}\left(\frac{e^{2t\sqrt{gk}}-1}{e^{2t\sqrt{gk}}+1}\right)$$

27. Descent by parachute

A probe of mass 100 kg is dropped from a helicopter hovering (i.e. at rest) at a distance D above ground. The probe falls a distance D_1 against a resistive force of $10v$ where v is its velocity, reaching a speed of $5g$ m/s. The parachute opens and the probe falls the remaining distance D_2 to the ground against a resistive force of $100v^2/g$. Show that if the velocity of the probe on impact is not to exceed $1.2g$ m/s then $D > 209.1$ metres.

Solution

From above: $D = D_1 + D_2$

For D_1:

Equation of Motion $= mv\frac{dv}{dx} = mg - 10v$

$$100v\frac{dv}{dx} = 100g - 10v \Rightarrow 10v\frac{dv}{dx} = 10g - v \Rightarrow 10\int\frac{v}{10g-v}dv = \int dx$$

The limits are: $v = 0$ at $x = 0$; $v = 6g$ at $x = D_1 \Rightarrow 10\int_0^{5g}\frac{v}{10g-v}dv = \int_0^D dx = D_1$

$$\Rightarrow 10[10g - v - 10gl_n(10g-v)]_0^{5g} = 100gl_n2 - 50g = D_1 \Rightarrow D_1 = 100gl_n2 - 50g$$

For D_2:

Equation of Motion $= 100v\frac{dv}{dx} = 100g - \frac{100}{g}v^2$

$$\Rightarrow v\frac{dv}{dx} = g - \frac{v^2}{g} = \frac{g^2 - v^2}{g} \Rightarrow g\int\frac{v}{(g^2-v^2)}dv = \int dx$$

The limits are: $v = 5g$ at $x = 0$; $v = 1.2g$ at $x = D_2$

$$\Rightarrow g\int_{5g}^{1.2g}\frac{v}{(g^2-v^2)}dv = \int_0^{D_2}dx \Rightarrow g\left(\frac{1}{-2}\right)[l_n(g^2-v^2)]_{5g}^{1.2g} = D_2$$

$$\Rightarrow D_2 = \frac{g}{2}[l_n(g^2-v^2)]_{1.2g}^{5g} = \frac{g}{2}[l_n(g^2 - 25g^2) - l_n(g^2 - 1.44g^2)]$$

$$\Rightarrow D_2 = \frac{g}{2}[l_n(-24g^2) - l_n(0.44g^2)] = \frac{g}{2}\left[l_n\left(\frac{-24g^2}{-0.44g^2}\right)\right] = \frac{g}{2}l_n(54.55)$$

$$\Rightarrow D = D_1 + D_2 = 100gl_n2 - 50g + \frac{g}{2}l_n54.5 = 209.1 \text{ metres}$$

Thus, the minimum dropping height is 209.1 metres

28. Mass suspended from elastic string

A mass m kg is attached to one end of an elastic string of natural length L and modulus $\lambda = mg/k$ N/m. The other end is attached to a fixed point, A. If the mass is allowed to fall from A find its maximum velocity during the subsequent motion.

Solution
The travel of the mass can be split into two components:

Component 1: Mass falls under gravity (i.e. string is slack)
When the mass falls through a distance L from A the only force acting is gravity
$\Rightarrow v^2 = u^2 + 2gL$; But: $u = 0 \Rightarrow v^2 = 2gL \Rightarrow v = \sqrt{2gL}$ is the maximum velocity reached.

Component 2: Mass falls under gravity against the resistance of the elastic string
After falling a distance $> L$, the mass is acted on by gravity and the tension of the string. The equation of motion of the mass is given by:

$$mv\frac{dv}{dx} = mg - \frac{\lambda}{L}x = mg - \frac{mg}{kL}x \Rightarrow v\frac{dv}{dx} = g - \frac{g}{kL}x \Rightarrow \int v\,dv = \int\left(g - \frac{g}{kL}x\right)dx$$

The limits are : $v = \sqrt{2gL}$ at $x = 0$; $v = v$ at $x = x$

$$\Rightarrow \int_{\sqrt{2gL}}^{v} v\,dv = \int_{0}^{x}\left(g - \frac{g}{kL}x\right)dx \Rightarrow \left[\frac{1}{2}v^2\right]_{\sqrt{2gL}}^{v} = \left[gx - \frac{g}{2kL}x^2\right]_{0}^{x}$$

$$\Rightarrow \frac{1}{2}\left(v^2 - 2gL\right) = gx - \frac{g}{2kL}x^2 \Rightarrow v = \sqrt{2gL + 2gx - \frac{g}{kL}x^2} = \left(2gL + 2gx - \frac{g}{kL}x^2\right)^{0.5}$$

Maximum value of v occurs when : $\dfrac{dv}{dx} = 0 \Rightarrow \dfrac{dv}{dx} = \dfrac{0.5\left(2g - \dfrac{2g}{kL}x\right)}{\left(2gL + 2gx - \dfrac{g}{kL}x^2\right)^{0.5}} = 0$

$$\Rightarrow \frac{dv}{dx} = 0 \text{ when } 2g - \frac{2g}{kL}x = 0 \quad \text{i.e. when} \quad x = kL$$

$$\Rightarrow V_{max} = \sqrt{2gL + 2gx - \frac{g}{kL}x^2} = \sqrt{2gL + 2gkL - \frac{g}{kL}k^2L^2} \text{ i.e. } V_{max} = \sqrt{2gL + gkL} = \sqrt{gL(2 + k)}$$

29. Particle projected upwards in resistive medium

At time $t = 0$ a particle of mass M kg is projected vertically upwards from a point on the ground. The resistance to the motion is MBv^2 Newtons where v = velocity at time t. Find expressions for:
(a)   The velocity (upwards) at time t
(b)   The vertical distance travelled as a function of velocity, v
(c)   The time taken to reach the highest point
(d)   The vertical distance to the highest point reached

Solution

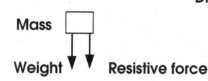

Direction of travel

Fig. 1

Mass

Weight **Resistive force**

(a) Find an expression for velocity (upwards) at time t

Equation of Motion : $M(\text{Acceleration}) = -Mg - MBv^2$

But : Acceleration $= \dfrac{d^2x}{dt^2} = \dfrac{dv}{dt} = v\dfrac{dv}{dx} \Rightarrow M\left(\dfrac{dv}{dt}\right) = -Mg - MBv^2 \Rightarrow \dfrac{dv}{dt} = -g - Bv^2$

$\Rightarrow \displaystyle\int \dfrac{dv}{g + Bv^2} = -\int dt \qquad$ The limits are : $t = 0 \Rightarrow v = u;\ t = t \Rightarrow v = v;$

$\Rightarrow \displaystyle\int_u^v \dfrac{dv}{g + Bv^2} = -\int_0^t dt \Rightarrow \int_v^u \dfrac{dv}{g + Bv^2} = \int_0^t dt$

$\Rightarrow \dfrac{1}{\sqrt{gB}}\left[Tan^{-1}\left(\dfrac{v\sqrt{gB}}{g}\right)\right]_v^u = t \Rightarrow Tan\left(t\sqrt{gB}\right) = \left(\sqrt{\dfrac{B}{g}}\right)(u - v) \Rightarrow v = u - \left(Tan\left(t\sqrt{gB}\right)\sqrt{\dfrac{g}{B}}\right)$ (i)

(b) Find the vertical distance travelled a function of velocity

Equation of Motion : $M(\text{Acceleration}) = -Mg - MBv^2$

But : Acceleration $= \dfrac{d^2x}{dt^2} = \dfrac{dv}{dt} = v\dfrac{dv}{dx} \Rightarrow M\left(v\dfrac{dv}{dx}\right) = -Mg - MBv^2 \Rightarrow v\dfrac{dv}{dx} = -g - Bv^2$

$\Rightarrow \displaystyle\int \dfrac{v\,dv}{g + Bv^2} = -\int dx \qquad$ The limits are : $x = 0 \Rightarrow v = u;\ x = x \Rightarrow v = v;$

$\Rightarrow \displaystyle\int_u^v \dfrac{v\,dv}{g + Bv^2} = -\int_0^x dx \Rightarrow \dfrac{1}{2B}\left[l_n\left(g + Bv^2\right)\right]_u^v = -x \Rightarrow x = \dfrac{1}{2B}l_n\left(\dfrac{g + Bu^2}{g + Bv^2}\right)$ (ii)

(c) The time ($t = T$, say) taken to reach the highest point

At the highest point the velocity $v = 0$. From equation (i):

$v = u - \left(Tan\left(T\sqrt{gB}\right)\sqrt{\dfrac{g}{B}}\right) = 0 \Rightarrow u = Tan\left(T\sqrt{gB}\right)\sqrt{\dfrac{g}{B}} \Rightarrow Tan\left(T\sqrt{gB}\right) = u\sqrt{\dfrac{B}{g}}$

$\Rightarrow T\sqrt{gB} = Tan^{-1}\left(u\sqrt{\dfrac{B}{g}}\right) \Rightarrow T = \dfrac{1}{\sqrt{gB}}Tan^{-1}\left(u\sqrt{\dfrac{B}{g}}\right)$

(d) The vertical distance ($x = H$, say) to the highest point reached

From equation (ii) : $x = \dfrac{1}{2B}l_n\left(\dfrac{g + Bu^2}{g + Bv^2}\right)$ At the highest point, $v = 0 \Rightarrow x = H = \dfrac{1}{2B}l_n\left(\dfrac{g + Bu^2}{g}\right)$

30. Particle projected upwards in resistive medium

A particle of mass $m = 1$ kg is projected vertically upwards from a point O with initial velocity u m/s in a medium whose resistance is kgv^2 where v is the speed of the particle and k is a constant. If H metres is the maximum height reached, find the:

(a)   Particle velocity at a height $H/2$ metres above O on its upward journey
(b)   Velocity at height $H/2$ metres above O on its downward journey
(c)   Ratio of these velocities

(d)   Ratio of the particle's initial and final kinetic energy values (final kinetic energy value is at point where particle has fallen a height H metres)

Solution
The flight profile of the particle is shown in Fig. 1.

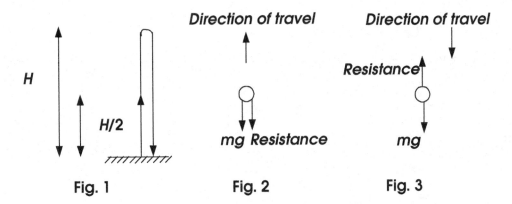

| **Fig. 1** | **Fig. 2** | **Fig. 3** |

(a) Particle velocity at a height $H/2$ metres above O on its upward journey
See Fig. 2 for forces:

Equation of Motion : $m\dfrac{d^2x}{dt^2} = mv\dfrac{dv}{dx} = -mg - kgv^2 = -g[m + kv^2]$

$\Rightarrow v\dfrac{dv}{dx} = -g[1 + kv^2] \Rightarrow \int \dfrac{v}{1 + kv^2}dv = -g\int dx$

The limits are : $v = u$ at $x = 0; v = v$ at $x = h \Rightarrow \displaystyle\int_u^v \dfrac{v}{1 + kv^2}dv = -g\int_0^h dx = -gh$

$\Rightarrow \dfrac{1}{2k}\left[l_n(1 + kv^2)\right]_u^v = -gh \Rightarrow \dfrac{1}{2k}\left[l_n(1 + kv^2)\right]_v^u = gh \Rightarrow \left[l_n(1 + kv^2)\right]_v^u = 2kgh$

$\Rightarrow l_n\left(\dfrac{1 + ku^2}{1 + kv^2}\right) = 2kgh \Rightarrow \dfrac{1 + ku^2}{1 + kv^2} = e^{2kgh}$

When $h = \dfrac{H}{2} \Rightarrow 1 + kv^2 = \dfrac{1 + ku^2}{e^{kgH}} \Rightarrow v^2 = \dfrac{\dfrac{1 + ku^2}{e^{kgH}} - 1}{k}$ (i)

When $h = H \Rightarrow v = 0 \Rightarrow 1 + ku^2 = e^{2kgH}$ (ii)   But, equation (i) then becomes

$v^2 = \dfrac{\dfrac{e^{2kgH}}{e^{kgH}} - 1}{k} = \dfrac{1}{k}\left(e^{kgH} - 1\right) \Rightarrow v = \sqrt{\dfrac{1}{k}\left(e^{kgH} - 1\right)}$ (iii)

(b) Particle velocity at a height $H/2$ metres above O on its downward journey
See Fig. 3 for forces:

Equation of Motion (particle falling): $v\dfrac{dv}{dx} = g - kgv^2 = g\left[1 - kv^2\right]$

$\Rightarrow \displaystyle\int \frac{v}{1-kv^2}\,dv = g\int dx \qquad$ The limits are : $x = 0 \Rightarrow v = 0;\ \ x = x \Rightarrow v = v$

$\Rightarrow \displaystyle\int_0^v \frac{v}{1-kv^2}\,dv = g\int_0^x dx = gx \Rightarrow \frac{1}{-2k}\left[l_n(1-kv^2)\right]_0^v = gx$

$\Rightarrow \left[l_n(1-kv^2)\right]_0^v = -2kgx \Rightarrow l_n(1-kv^2) = -2kgx$

$\Rightarrow 1 - kv^2 = e^{-2kgx} \Rightarrow v^2 = \dfrac{1}{k}\left(1 - e^{-2kgx}\right) = \dfrac{1}{k}\left(1 - \dfrac{1}{e^{2kgx}}\right) = \dfrac{1}{k}\left(\dfrac{e^{2kgx}-1}{e^{2kgx}}\right)$ (iv)

When $\quad x = \dfrac{H}{2} \Rightarrow v^2 = \dfrac{1}{k}\left(\dfrac{e^{kgH}-1}{e^{kgH}}\right) \quad \Rightarrow v = \sqrt{\dfrac{1}{k}\left(\dfrac{e^{kgH}-1}{e^{kgH}}\right)}$ (v)

(c) Ratio of these velocities

From equations (iii) and (v):

Ratio of velocities $= \dfrac{v(up)}{v(down)} = \dfrac{\sqrt{\dfrac{1}{k}\left(e^{kgH}-1\right)}}{\sqrt{\dfrac{1}{k}\left(\dfrac{e^{kgH}-1}{e^{kgH}}\right)}} = \sqrt{e^{kgH}} = e^{0.5kgH}$

(d) Ratio of the particle's initial and final kinetic energy (K.E.) values

Initial K.E. $= \frac{1}{2}mu^2 \quad$ From equation (iv) : Final K.E. $= \dfrac{1}{2}m\dfrac{1}{k}\left(\dfrac{e^{2kgH}-1}{e^{2kgH}}\right)$

But, equation (ii) gives : $1 + ku^2 = e^{2kgH}$

$\Rightarrow$ Final K.E. $= \dfrac{1}{2}m\dfrac{1}{k}\left(\dfrac{1+ku^2-1}{1+ku^2}\right) = \dfrac{1}{2}m\left(\dfrac{u^2}{1+ku^2}\right) = \dfrac{\text{Initial K.E.}}{1+ku^2} \Rightarrow \text{Ratio} = \dfrac{\text{Initial K.E.}}{\text{Final K.E.}} = 1 + ku^2$

Chapter 12   Simple Harmonic Motion

1. Simple Harmonic Motion
Define Simple Harmonic Motion

Solution
A particle moves with simple harmonic motion when it moves so that its acceleration along its path is directed towards a fixed point in that path and varies directly as its distance from this fixed point.

If x is the displacement of a particle from a fixed point O in its path then the magnitude of the acceleration of the particle towards O at this point is $-\omega^2 x$ (the minus sign indicating that the acceleration is in the opposite direction to that in which x increases. Thus, if the particle is at a distance x from O at time t the acceleration in direction of x

increasing $= \dfrac{d^2 x}{d t^2} = -\omega^2 x$)

2. Simple Harmonic Motion
A body moves with Simple Harmonic Motion if its acceleration is proportional to its displacement from a fixed point and is always directed towards that point. Derive expressions for the motion's: Displacement, Amplitude (i.e. maximum displacement), Velocity, Acceleration and Period

Solution

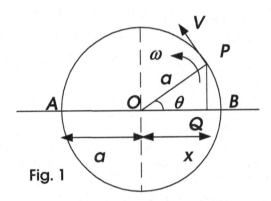

Fig. 1

A particle P rotates about O with uniform angular velocity ω radians/second. $AB =$ diameter of circle formed by motion of P. In time t, the particle turns through an angle θ where: $\theta = \omega t$. See Fig. 1. Particle Q slides along the diameter AB such that $PQ \perp AB$ at all times. Thus, P carries out circular motion with acceleration towards O of $a\omega^2$ in direction PO. Q oscillates about O with simple harmonic motion

(a) Displacement
$x = a \, Cos\, \theta = aCos\, \omega t$ (i)

(b) Amplitude
The maximum value of $x = a$ i.e. a is the amplitude of the motion   (ii)

(c) Velocity

Let v = velocity of Q along AB = component of velocity of P which is parallel to AB

$$\Rightarrow v = \frac{dx}{dt} = -\omega a \, Sin\omega t \quad But: \; Sin\theta = \frac{PQ}{a} = \frac{\pm\sqrt{a^2 - x^2}}{a}$$

$$\Rightarrow v = -\omega a \, Sin\omega t = \omega \, a \frac{\pm\sqrt{a^2 - x^2}}{a} = \pm\omega\sqrt{a^2 - x^2} \quad \text{(iii)} \quad and \quad v^2 = \omega^2(a^2 - x^2)$$

$\Rightarrow v$ has a maximum value when $x = 0$: $v_{max} = \omega a$ and

v has a minimum value when $x = a$: $\qquad v_{min} = 0$

(d) Acceleration

Acceleration of $Q = \dfrac{dv}{dt} = \dfrac{d^2x}{dt^2} = -\omega^2 a \, Cos\omega t = -\omega^2 x$ (iv)

The acceleration has a maximum value when x has a maximum value, i.e. when $x = a$ and the minimum acceleration occurs when $x = 0$. From equation (iv): acceleration is proportional to the displacement from fixed point O and is always directed towards O.

(e) Period

The periodic time is the time taken to make one complete cycle or revolution about O: this is the same as the time for P to perform one complete revolution. Therefore,

periodic time T = (angular) distance / (angular) speed $\quad = \dfrac{2\pi}{\omega}$. The frequency (cycles

per second or hertz, hz) is: $f = \dfrac{1}{T} = \dfrac{\omega}{2\pi} \quad hz$

3. Particle performing simple harmonic motion

A particle of mass 5 kg performs simple harmonic motion with a maximum velocity of 2 m/s and maximum acceleration of 10 m/s². Find the amplitude and frequency of motion.

Solution

From the equations: (i) $m\dfrac{d^2x}{dt^2} = -\omega^2 x$ (ii) $\qquad v^2 = \omega^2(a^2 - x^2)$

Maximum velocity = 2 m/s². This occurs when the distance of the particle from the equilibrium point is 0 (i.e. when $x = 0$ and acceleration = 0)

$\Rightarrow$ Equation (ii) becomes: $v^2 = \omega^2 a^2 \Rightarrow v = \omega a = 2$ m/s (iii)

The maximum acceleration x is 10 m/s. This occurs when $v = 0$, i.e. when x is a maximum $\Rightarrow x$ = amplitude = a

From equation (i):

$m\dfrac{d^2x}{dt^2} = -\omega^2 x$ But (given): $m = 5$, $\dfrac{d^2x}{dt^2} = 10$, $x = a$

$$\Rightarrow 50 = -\omega^2 a \Rightarrow a = \frac{-50}{\omega^2} \Rightarrow |a| = \left|\frac{50}{\omega^2}\right| \quad \text{(iv)}$$

$$\Rightarrow \text{Equation (iii)}: v = \omega a \quad becomes: \; 2 = \omega\frac{50}{\omega^2} = \frac{50}{\omega} \Rightarrow \omega = \frac{50}{2} = 25$$

$$\Rightarrow \text{From eq. (iv)}: \text{amplitude} = |a| = \frac{50}{25^2} = \frac{2}{25} \quad \text{metres, frequency} \quad f = \frac{\omega}{2\pi} = \frac{25}{2\pi} \quad hz$$

4. Differential Methods

A particle of mass m kg suspended from a spring of natural length L metres and stiffness k N/m will extend the spring by D metres (See Figs. 1 and 2). Show that the particle will perform simple harmonic motion if displaced vertically such that the displacement is x metres after t seconds (See Fig. 3).

Solution

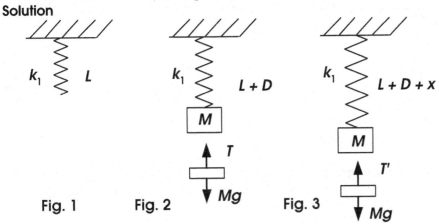

Fig. 1   Fig. 2   Fig. 3

Fig. 1 shows the unloaded spring. Fig. 2 shows the particle suspended in equilibrium:
From Hooke's Law: Tension of the spring, $T = mg = kD$ (i)
where D is the extension of the spring in equilibrium.
When displaced downwards a distance x at time t (See Fig. 3):
Tension of the spring, $T' = k(D + x)$ Equation (ii) and

Equation of motion $= m\dfrac{d^2x}{dt^2} = mg - T' = mg - k(D + x)$ (from equation (ii))

$= kD - k(D + x)$ (from equation (i)) $\Rightarrow m\dfrac{d^2x}{dt^2} = -kx$

Thus the restoring force is proportional to the distance from the equilibrium position $\Rightarrow$ the particle moves with simple harmonic motion about the equilibrium position. Note: the force is always directed to the equilibrium position. In the standard form of the

equation: $\dfrac{d^2x}{dt^2} = -\dfrac{k}{m}x = -\omega^2 x$

5. Simple Harmonic Motion
If simple harmonic motion is taking place with $x = A\,Cos\,(\omega t + \alpha)$ with initial displacement S metres, initial velocity U m/s and ω the usual constant find expressions for: (a) The amplitude of the motion, (b) the phase angle in terms of S, U and ω.

Solution
Using the standard equation: $v^2 = \omega^2 (A^2 - x^2)$ (i) where:
A = Amplitude of motion, ω^2 = constant, x = Displacement at time t
v = Velocity at time t
Given: At $t = 0$, $x = S$, $v = U$
(a) The amplitude of the motion

From equation (i): $U^2 = \omega^2 (A^2 - S^2) \Rightarrow A = \sqrt{\dfrac{U^2}{\omega^2} + S^2}$

(b) The phase angle in terms of S, U and ω.
Also, displacement can be expressed as:
$x = A\,Cos\,(\omega t + \alpha)$ (where α = phase angle)   (ii)

But, at $t = 0$, $x = S$, $v = U$ (given) $\Rightarrow S = A \cos \alpha \Rightarrow \cos\alpha = \dfrac{S}{A} = \dfrac{S}{\sqrt{\dfrac{U^2}{\omega^2} + S^2}}$

Fig. 1

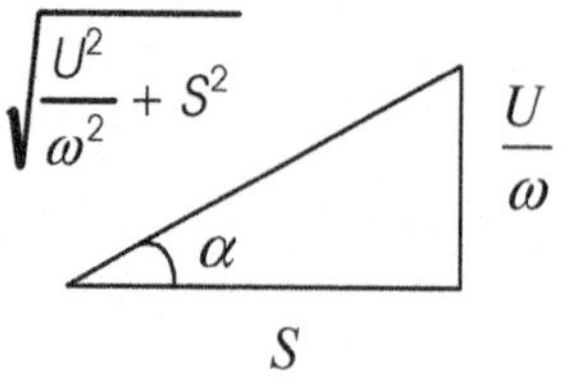

From Fig. 1: $\tan \alpha = \dfrac{U}{\omega S} \Rightarrow \alpha = \tan^{-1}\dfrac{U}{\omega S}$

6. Work done to stretch an elastic string

If a string whose elastic constant (a term which is similar to that of stiffness for a spring) is k N/m (where $k = \dfrac{\lambda}{L}$) is stretched a distance x metres beyond its natural length, L, show that the work done is: $\frac{1}{2} kx^2$

Solution
Let:
L = Natural length in metres
L' = Stretched length in metres
x = Extension (i.e. $L - L'$) in metres
T = Tension of elastic string in Newtons
λ = modulus of elasticity of the string in Newtons

When an elastic string or spring is stretched beyond its natural length, it exerts a restoring force proportion to the amount of the extension beyond its natural length.
i.e. $T = k(L'-L) = kx$

Work done to extend the spring or string from extension x_1 to extension x_2 is

$$\int_{x1}^{x2} T\,dx = \int_{x1}^{x2} kx\,dx = k\left[\dfrac{x^2}{2}\right]_{x1}^{x2} = \dfrac{1}{2}k\left(x_2^2 - x_1^2\right)$$

where $x_1 = 0$, $x_2 = x \Rightarrow$ Work done to stretch string by a distance $x = \frac{1}{2} kx^2$

7. Elastic string

When a string of natural length L metres and elastic constant k N/m is stretched to distances of x and y metres the tensions in the string are T_1 and T_2 respectively. (Note: x, $y > L$). Show that:

(a) $L = \dfrac{yT_1 - xT_2}{T_1 - T_2}$

(b) The work done in stretching the string to a total length of $x + y = \dfrac{1}{2}\dfrac{(xT_1 - yT_2)^2}{(T_1 - T_2)(x - y)}$

Solution

(a) Show: $L = \dfrac{yT_1 - xT_2}{T_1 - T_2}$

Using the approach in the problem above: $T_1 = k(x - L)$, $T_2 = k(y - L)$

$\Rightarrow k = \dfrac{T_1}{x - L} = \dfrac{T_2}{y - L} \Rightarrow T_1 y - T_1 L = T_2 x - T_2 L \Rightarrow L = \dfrac{yT_1 - xT_2}{T_1 - T_2}$

(b) The work done in stretching the string a distance $x + y$

Using the approach in the problem above, the work done to stretch string by a distance $x = \frac{1}{2} kx^2$. In this case the string is stretched to a total length of $x + y$ metres i.e. it is stretched by a distance $x + y - L$ (i)

From (a) above :

$k = \dfrac{T_1}{x - L} = \dfrac{T_1}{x - \left(\dfrac{T_1 y - T_2 x}{T_1 - T_2}\right)} = \dfrac{T_1(T_1 - T_2)}{xT_1 - xT_2 - T_1 y + T_2 x} = \dfrac{T_1(T_1 - T_2)}{T_1(x - y)} = \dfrac{(T_1 - T_2)}{(x - y)}$ (ii)

Using equations (i) and (ii) :

Work done $= \dfrac{1}{2}\dfrac{(T_1 - T_2)}{(x - y)}(x + y - L)^2 = \dfrac{1}{2}\dfrac{(T_1 - T_2)}{(x - y)}\left(x + y - \left(\dfrac{T_1 y - T_2 x}{T_1 - T_2}\right)\right)^2$

$= \dfrac{1}{2}\dfrac{(T_1 - T_2)}{(x - y)}\left(\dfrac{xT_1 + yT_1 - xT_2 - yT_2 - yT_1 + xT_2}{T_1 - T_2}\right)^2 = \dfrac{1}{2}\dfrac{(T_1 - T_2)}{(x - y)}\left(\dfrac{xT_1 - yT_2}{T_1 - T_2}\right)^2 = \dfrac{1}{2}\dfrac{(xT_1 - yT_2)^2}{(T_1 - T_2)(x - y)}$

8. Vibrating plane

A horizontal plane supports a mass of m kg and has a vibration with a period of $\frac{1}{4}$ seconds. Find the maximum amplitude of the plane which will not cause the mass to bounce off the plane (See Fig. 1).

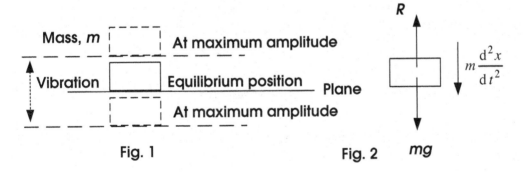

Fig. 1   Fig. 2 mg

Solution

The mass on the plane will bounce off if the reaction, R, between the mass and the plane = 0.

Equation of motion : $m\dfrac{d^2 x}{dt^2} = mg - R$ (See Fig. 2)

For the mass to remain on the plane, (i.e. at the point of bouncing): $R > 0$

$\Rightarrow m\dfrac{d^2 x}{dt^2} < mg \Rightarrow \dfrac{d^2 x}{dt^2} < g$

231

But : $\dfrac{d^2x}{dt^2} = -\omega^2 x$ for simple harmonic motion.

Assume the mass will be at the point of bouncing when x is a maximum i.e. $x =$ the amplitude of the motion $= A$

$$\Rightarrow m\dfrac{d^2x}{dt^2} < mg \Rightarrow m\left|\omega^2 A\right| < mg \Rightarrow \left|\omega^2 A\right| < g \Rightarrow A < \dfrac{g}{\omega^2}$$

But : $g = 9.81\,\text{m/s}^2$ and period of the motion, $T = \dfrac{2\pi}{\omega} = 0.25$ (given)

$$\Rightarrow \omega = 8\pi \Rightarrow A < \dfrac{(9.81)}{(8\pi)^2} < 0.0155 \text{ metres}$$

$\Rightarrow$ maximum amplitude $= 0.0155$ metres for no bouncing of the particle.

9. Vibrating platform

A horizontal platform, on which a particle is resting, oscillates vertically with simple harmonic motion of amplitude 0.2 metres. See Fig. 1. What is the maximum integral number of complete oscillations per minute it can make, if the particle is not to leave the platform.

Solution

The masses on the plane will bounce off if the reaction R between the mass and the plane $= 0$. The forces on the system are then as shown in Fig. 2:

Equation of Motion : $m\dfrac{d^2x}{dt^2} = mg - R$ but $R > 0 \Rightarrow m\dfrac{d^2x}{dt^2} < mg$ (i)

But the platform is oscillating with simple harmonic motion

$$\Rightarrow m\dfrac{d^2x}{dt^2} = m\omega^2 x \text{ (downwards)} \quad \text{(ii)}$$

But at its highest point, the distance, x, between the platform and the point of equilibrium $=$ amplitude $= A$ (say)

$\Rightarrow$ Using equations (i) and (ii) : $m\dfrac{d^2x}{dt^2} = m\omega^2 A < mg \Rightarrow \omega^2 < \dfrac{g}{A} \Rightarrow \omega < \sqrt{\dfrac{g}{A}}$

But : $\omega = 2\pi f \Rightarrow f = \dfrac{\omega}{2\pi} < \dfrac{1}{2\pi}\sqrt{\dfrac{g}{A}}$ oscillations per second

$\Rightarrow f < \dfrac{60}{2\pi}\sqrt{\dfrac{g}{A}}$ oscillations per minute. But : $A = 0.2$ metres (given)

$\Rightarrow f < 66.88$ oscillations per minute

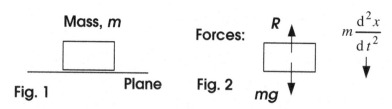

Mass, m

Forces:

R

$m\dfrac{d^2x}{dt^2}$

Plane

Fig. 1

Fig. 2

mg

The particles will not bounce off unless the table vibrates at more than 66 complete oscillations per minute.

10. Mass and Spring

A mass of 10 kg is hung on the end of a vertical spring whose stiffness is $k = 100$ N/m. The mass is then pulled down a distance y metres from its equilibrium position and released from rest. Find the period of vibration.

Solution

At time $t = 0$, $x = y \Rightarrow$ Amplitude $= y$ metres $\Rightarrow$ Period $= T = \dfrac{2\pi}{\omega} = \dfrac{2\pi}{\sqrt{\dfrac{k}{m}}}$

$\Rightarrow \omega^2 = \dfrac{k}{m} = \dfrac{100}{10} = 10 \Rightarrow \omega = \sqrt{10} \Rightarrow T = \dfrac{2\pi}{\sqrt{10}}$ seconds

11. Three Springs

Three springs S_1, S_2 and S_3 have stiffness k_1, k_2, k_3 N/m respectively. If the three springs were replaced by one spring S in each of the two arrangements shown below, find the equivalent stiffness k_{EQ} of this single spring.

Solution
Arrangement 1:
Let extensions of springs be x_1, x_2 and x_3 for springs S_1, S_2, S_3 respectively. Then total extension is x where $x = x_1 + x_2 + x_3$

But Force = stiffness × extension i.e. $F = kx$

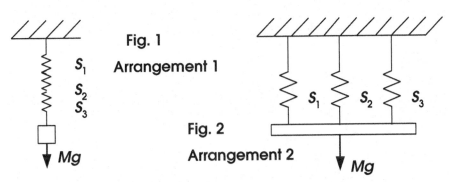

Fig. 1
Arrangement 1

Fig. 2
Arrangement 2

In this case: $Mg = kx \Rightarrow x = \dfrac{Mg}{k}$ But: $x = x_1 + x_2 + x_3$

$\Rightarrow x = \dfrac{Mg}{k_1} + \dfrac{Mg}{k_2} + \dfrac{Mg}{k_3} = \dfrac{Mg}{k_{EQ}} \Rightarrow \dfrac{1}{k_{EQ}} = \left(\dfrac{1}{k_1} + \dfrac{1}{k_2} + \dfrac{1}{k_3} \right)$ N/m

Arrangement 2:
In this case, if a weight of Mg Newtons causes an extension x, then x is the same for each spring i.e. $x = x_1 = x_2 = x_3$.
But the total force $F = F_1 + F_2 + F_3$ (i.e. F = sum of the forces in the springs)
Since: $F = kx = k_1x_1 + k_2x_2 + k_3x_3 = k_1x + k_2x + k_3x = x(k_1 + k_2 + k_3)$
For equivalent single spring: $F = k_{EQ}x$ and $k_{EQ} = (k_1 + k_2 + k_3)$ N/m

12. Two Springs and Mass
Two springs, C and D, of stiffness $k_C = 60$ N/m and $k_D = 40$ N/m respectively are coupled to a mass $m = 0.25$ kg as shown in the diagram. (Assume that the springs are not extended or compressed in equilibrium and that there is no friction involved)

(a) If the mass is displaced by a distance x metres from the equilibrium position, show that it will perform simple harmonic motion.

(b) What is the period of the motion?

(c) If, at time $t = 0$ seconds, the displacement and velocity of the mass are 0.005 m and 0.01 m/s respectively, find the displacement and velocity after 2 seconds.

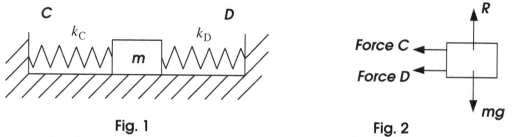

Fig. 1   **Fig. 2**

Solution

(a) Show that simple harmonic motion is performed

Initially the springs are not stretched or compressed. If the mass is displaced by x metres (in the direction of D) then there will be forces acting towards C due to:

The extension of Spring C:

Tension force $= -k_C(x)$ (acting in opposite direction to x)   and

The compression of Spring D:

Compression force $= -k_D(x)$ (acting in opposite direction to x)

The equation of motion is: $m\dfrac{d^2x}{dt^2} = -(k_C + k_D)x$ This is the form of:

Force $= -m\omega^2x$ (i) $\Rightarrow$ the mass performs simple harmonic motion.

(b) The period of the motion

For simple harmonic motion : Acceleration $= -\omega^2 x$

In this case, from equation (i) : Acceleration $= -\left(\dfrac{k_C + k_D}{m}\right)x \Rightarrow \omega = \sqrt{\dfrac{k_C + k_D}{m}}$

Period of motion $= \dfrac{2\pi}{\omega} = 2\pi\sqrt{\dfrac{m}{k_C + k_D}} = 2\pi\sqrt{\dfrac{0.25}{100}} = 0.3142$ seconds

(c) Displacement and velocity after 2 seconds

(Given) : $m = 0.25\,\text{kg}$, $k_C = 60\,\text{N/m}$, $k_D = 40\,\text{N/m} \Rightarrow \omega = \sqrt{\dfrac{100}{0.25}} = 20$

Given : $x = A\cos\omega t + B\sin\omega t$ and $\dfrac{dx}{dt} = -A\omega\sin\omega t + B\omega\cos\omega t$

Putting in the values : At $t = 0$, $\omega = 20$, $x = 0.005$, $\dfrac{dx}{dt} = 0.01 \Rightarrow$

$0.005 = A(1) + B(0) = A$; $0.01 = -20A(0) + 20(B) \Rightarrow A = 0.005$, $B = 0.0005$

At $t = 2$ seconds :

Displacement $= x = 0.005\cos(40\ \text{radians}) + 0.0005\sin(40\ \text{radians}) = -0.00296$ metres

Velocity $= \dfrac{dx}{dt} = -0.1\sin(40\ \text{radians}) + 0.01\cos(40\ \text{radians}) = -0.0812$ m/s

13. Two Springs and Mass

Two springs, C and D, of equal length L metres, and of stiffness k_C and k_D N/m respectively are coupled to a mass m kg as shown in the diagram and attached to two fixed points a distance 3L metres apart. See Fig. 1. (Assume there is no friction involved)
(a)   What is the extension of each spring in equilibrium?
(b)   If spring D breaks, find the period and amplitude of the subsequent motion.
(c)   What is the speed of the mass when the extension = 3/5 th of the amplitude?

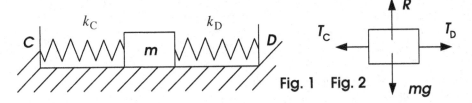

Fig. 1   Fig. 2

Solution
The forces in Newtons acting are shown in Fig. 2: R = Reaction between mass and surface, T = Tension in springs, mg = Weight of mass

(a) What is the extension of each spring in equilibrium?
In equilibrium, $T_C = T_D$
Let extension of Spring $C = x \Rightarrow$ length of Spring $C = L + x$
Let extension of Spring $D = y \Rightarrow$ length of Spring $D = L + y$
$\Rightarrow L + x + L + y = 3L \Rightarrow x + y = L$ (i) $\Rightarrow T_C = -k_C x$ and $T_D = -k_D y = -k_D(L - x)$

But, in equilibrium, $T_C = T_D \Rightarrow k_C x = k_D(L - x) \Rightarrow x = \left(\dfrac{k_D}{k_C + k_D}\right)L,\ y = \left(\dfrac{k_C}{k_C + k_D}\right)L$ metres

(b) If spring D breaks, period and amplitude of subsequent motion
If Spring D breaks then the forces acting on Spring C are (See Fig. 3):

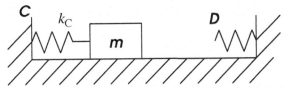   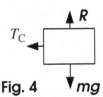

Fig. 3   Fig. 4 mg

$T_C = -k_C x \Rightarrow$ Mass $\times$ Acceleration $= -k_C x$
This is in the format for simple harmonic motion where :
Acceleration $= -\omega^2 x$:
$\Rightarrow$ Acceleration $= -\omega^2 x = -\dfrac{k_C}{m}x \Rightarrow \omega = \sqrt{\dfrac{k_C}{m}} \Rightarrow$ Period $= \dfrac{2\pi}{\omega} = 2\pi\sqrt{\dfrac{m}{k_C}}$

Amplitude = maximum displacement from equilibrium position
Equilibrium position for Spring $C = L$ (i.e. $x = 0$)
$\Rightarrow$ Amplitude = initial position when Spring D is disconnected $= \left(\dfrac{k_D}{k_C + k_D}\right)L$

(c) What is the speed of the mass when the extension = 3/5 of the amplitude?
Using the energy equation: $v^2 = \omega^2(A^2 - x^2)$

If $x = \dfrac{3}{5}A$ then: $v^2 = \omega^2\dfrac{16}{25}A^2 \Rightarrow v = \dfrac{4}{5}\omega A = \dfrac{4}{5}\sqrt{\dfrac{k_C}{m}}\left(\dfrac{k_D}{k_C + k_D}\right)L$ m / s

14. Two Springs and Mass

Two springs, of equal length but of stiffness k_1 N/m and k_2 N/m respectively are connected to a mass m kg and fixed to points A and B as shown in Fig. 1. Find the spring stiffness if the two springs are replaced by an equivalent spring.

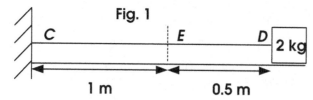

Fig. 1   Fig.2   Fig.3

Solution

Let the mass be displaced a distance x towards B

$\Rightarrow x$ = Extension of spring 1 and compression of spring 2 in metres

$\Rightarrow$ Tensions in Newtons are: T_1 = Tension of spring 1, T_2 = Compressive force of spring 2

$\Rightarrow T_1 = -k_1x, \; T_2 = -k_2x$

$\Rightarrow$ the horizontal forces acting on the mass at displacement x are shown in Figs. 2 and 3 (the vertical forces are not relevant):

Resultant force acting towards point $A = T_1 + T_2 = -k_1x - k_2x = -(k_1 + k_2)x$

If T_{EQ} = tension in equivalent spring and this spring is displaced by x also, then

$T_{EQ} = -k_{EQ}x \Rightarrow T_{EQ} = T_1 + T_2$ Newtons

$\Rightarrow k_{EQ}x = k_1x + k_2x = (k_1 + k_2)x \Rightarrow k_{EQ} = k_1 + k_2$ N/m

15. Mass and Elastic String.

One end of a light flexible elastic string, CD, of natural length 1 metre and elastic constant $k = 10$ N/m is tied to a fixed point, C. A particle of mass 2 kg is attached to the other end, D, and placed on a smooth horizontal table at a distance of 1.5 metres from C and released from rest (See Fig. 1).

(a)   Show that the particle initially moves with simple harmonic motion
(b)   Find the period of motion
(c)   Find the time taken to reach C
(d)   Find the time taken to reach E by considering the energy of the particle

Solution

(a) Show that the particle initially moves with the simple harmonic motion

If x = displacement and t = time:

It is known that: $m\dfrac{d^2x}{dt^2} = -kx$ but $k = 10$ N/m and $m = 2$ kg   (given)

Fig. 1

C E D [2 kg]

1 m   0.5 m

$\Rightarrow 2\dfrac{d^2x}{dt^2} = -10x \Rightarrow \dfrac{d^2x}{dt^2} = -5x$ But the equation

for Simple Harmonic Motion can be expressed in the form

$\dfrac{d^2x}{dt^2} = -\omega^2x \Rightarrow$ the particle is performing simple harmonic motion

(b) Find the period of motion

236

From above : $\omega^2 = 5 \Rightarrow \omega = \sqrt{5} \Rightarrow$ Period $= T = \dfrac{2\pi}{\omega} = \dfrac{2\pi}{\sqrt{5}}$ seconds

(c) Find the time taken to reach C

Consider the distance is travelled in two stages: When the particle is let go it accelerates for 0.5 metres until the string goes slack (at point E, say) , and it then continues on for 1 metres until it reaches C.

Stage 1: Travel from initial position at D to E (where string goes slack):
Use the following general equations:

(i) $x = A \cos \omega t + B \sin \omega t$ (ii) $\dfrac{dx}{dt} = -A\omega \sin \omega t + B\omega \cos \omega t$

At $t = 0$, $x = 0.5$, $\dfrac{dx}{dt} = 0$ (given)

From equation (i): $0.5 = A \cos 0 + B \sin 0 = A \Rightarrow A = 0.5$
From equation (ii): $0 = -A\omega \sin 0 + B\omega \cos 0 \Rightarrow 0 + B \Rightarrow B = 0$

$\Rightarrow x = 0.5 \cos \omega t$ and $\dfrac{dx}{dt} = -0.5\omega \sin \omega t$

But: $\omega = \sqrt{\dfrac{k}{m}} = \sqrt{\dfrac{10}{2}} = \sqrt{5}$ and $T =$ Period $= \dfrac{2\pi}{\sqrt{5}}$ seconds

If $x = 0$ then the particle is at E and $x = 0 = 0.5 \cos \omega t \Rightarrow \cos \omega t = 0$

$\Rightarrow \omega t = \dfrac{\pi}{2} \Rightarrow t = \dfrac{\pi}{2\omega} = \dfrac{\pi}{2 \times \sqrt{5}} = \dfrac{\pi}{2\sqrt{5}}$ seconds

Stage 2: Travel from E to C:

At point E: $\dfrac{dx}{dt} =$ velocity $= -0.5\sqrt{5} \sin \dfrac{\pi}{2} = -0.5\sqrt{5}$ m/s   towards C

Thus the string goes slack after travelling 0.5 metres and its velocity is then $0.5\sqrt{5}$ m/s. As the table is smooth the particle will continue travelling with this velocity until it reaches C.

Time taken to travel the remaining 1 metre to C is: $t = \dfrac{1\,\text{metre}}{0.5\sqrt{5}\ \text{m/s}} = \dfrac{2}{\sqrt{5}}$ seconds $\Rightarrow$ Total

time taken $= \left(\dfrac{\pi}{2\sqrt{5}} + \dfrac{2}{\sqrt{5}} \right)$ seconds.

(d) Find the time taken to travel from D to E by considering the energy of the particle.

The energy components at any point are:
Potential Energy + Kinetic Energy + Energy stored in string.
Assume that Potential Energy = 0 throughout the range of movement of the particle.
Energy of particle at initial position, D:
Velocity = 0 $\Rightarrow$ Kinetic Energy = 0;   Extension of string = 0.5 metres
$\Rightarrow$ Stored Energy $= \frac{1}{2} kx^2 = 0.5\,(10)(0.5)^2 = 1.25$ Joules

Energy of particle at E:
Stored Energy = 0 (as string is slack) But, total energy is same as at initial position $\Rightarrow$
Kinetic Energy $= \frac{1}{2} mv^2 = 1.25$ Joules

$$\Rightarrow v = \frac{\sqrt{5}}{2} \text{m/s} \quad \text{But: } v^2 = \omega^2(A^2 - x^2) \text{ and at } E \colon x = 0$$

Also: A = amplitude = $0.5 \Rightarrow v = \omega A = 0.5\omega \Rightarrow \omega = \sqrt{5} \Rightarrow T = \frac{2\pi}{\sqrt{5}}$ seconds

But travelling from the initial position to E is ¼ of a complete cycle so the time taken is ¼ of $\frac{2\pi}{\sqrt{5}}$ seconds $= \frac{\pi}{2\sqrt{5}}$ seconds, as in (c).

16. Mass on moving table

A mass sits on a rough table which performs simple harmonic motion in a horizontal plane, moving between B and C as shown in Fig. 1. The maximum speed of the table is V m/s and its period of motion is t seconds. Find, in terms of V and t, the minimum coefficient of friction, μ, which will prevent the mass from sliding on the table.

Solution

From the information given, the following apply to the table:

Period $= T = \frac{2\pi}{\omega} = t$ seconds $\Rightarrow \omega = \frac{2\pi}{t}$ (i)

From the standard equation : $v^2 = \omega^2(A^2 - x^2)$,

v is a maximum when : $x = 0 \Rightarrow V = \omega A$

$\Rightarrow A = \frac{V}{\omega} = \frac{Vt}{2\pi}$ metres (where A = amplitude of the motion) Equation (ii)

The maximum acceleration of the table occurs when the table reaches B

and C : maximum acceleration $= -\omega^2 A$ (directed towards the central position)

The range of the motion of the mass and table is shown in Fig. 1. The most likely time for sliding to occur is where the acceleration of the table is at a maximum, i.e. at points B and C. The possible sliding motion is directed away from the central point so the frictional force F will be directed towards the central point. For no sliding to occur the acceleration of the mass must be the same as the acceleration of the table. The forces acting (in Newtons) on the mass at point C are shown in Fig. 2:

The forces acting on the mass at point C are:

Vertical forces: R = Reaction between mass and table, mg = Weight of mass

Horizontal forces: (mass)(acceleration) = $(m)(\omega^2 A)$ = F (i.e. towards central position) where F = Frictional force opposing motion

The equation of motion for no slipping is: $\omega^2 A = \frac{F}{m}$

But : $F = \mu R = \mu mg \Rightarrow$ From equation (iii) : $\omega^2 A = \frac{\mu mg}{m} = \mu g \Rightarrow \mu = \frac{\omega^2 A}{g}$

$\Rightarrow$ Inserting values from equations (i) and (ii) gives minimum value of μ

for no sliding : $\mu = \frac{\omega^2 A}{g} = \frac{\left(\frac{2\pi}{t}\right)^2 \left(\frac{Vt}{2\pi}\right)}{g} = \frac{2\pi V}{gt}$

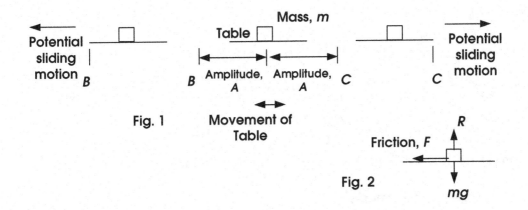

Fig. 1

Fig. 2

17. Mass suspended on spring

A mass of m kg suspended from a spring of length L metres and stiffness k N/m will cause it to extend by X metres. Another mass of M kg is suspended from the same spring and, when put into motion, makes N complete oscillations/minute. Find the value of M and the acceleration of M at $0.5X$ metres below its equilibrium position

Solution

When mass of m is suspended (See Fig. 1): $mg = kX \Rightarrow k = \dfrac{mg}{X}$ N/m

When mass M is making N complete oscillations per minute it is making $\dfrac{N}{60}$ oscillations/s

(a) Value of M

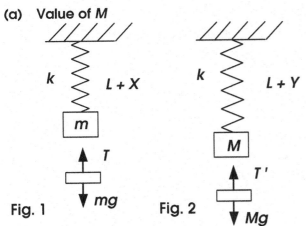

Fig. 1

Fig. 2

$Period = T = \dfrac{60}{N}$ seconds per oscillation $= \dfrac{2\pi}{\omega} \Rightarrow \omega = \dfrac{\pi N}{30}$

But, $\omega = \sqrt{\dfrac{k}{M}} = \sqrt{\dfrac{mg}{XM}} = \dfrac{\pi N}{30} \Rightarrow M = \dfrac{900mg}{X\pi^2 N^2}$ kg

(b) Acceleration of M at $0.5X$ metres below its equilibrium position

When mass m is replaced by mass M the spring, when in equilibrium position, will have an extension Y metres (See Fig. 2):

$$Mg = kY \Rightarrow Y = \frac{Mg}{k} = \frac{MgX}{mg} = \frac{MX}{m}$$

The equation of motion at a distance of $0.5X$ metres below its equilibrium position are :

$$M\frac{d^2x}{dt^2} = Mg - T \Rightarrow \frac{d^2x}{dt^2} = g - \frac{T}{M} \quad \text{But, } T = k(Y + 0.5X) = \frac{mg}{X}\left(\frac{MX}{m} + 0.5X\right)$$

$$\Rightarrow \frac{d^2x}{dt^2} = g - \frac{mg}{MX}\left(\frac{MX}{m} + 0.5X\right) = g - \frac{mMgX}{mMX} - \frac{0.5mgX}{MX} = -\frac{0.5mg}{M} \quad \text{m/s}^2$$

(i.e. acceleration is upwards towards the equilibrium position)

18. Elastic string on smooth inclined plane

A mass of 1 kg lies on a smooth frictionless plane inclined at an angle 30° and is connected to an elastic string of natural length L metres and elastic constant $k = 50$ N/m (See Fig. 1). If the mass is released from rest when the string is not extended, find the amplitude and frequency of the resulting vibration.

Solution

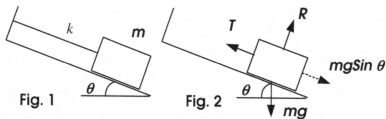

Fig. 1   Fig. 2

(a) Amplitude of the resulting vibration.

The amplitude of the motion will be the equilibrium extension. Assume that when the mass hangs in equilibrium, the string has extension d where

$$T = mgSin\theta = kd \Rightarrow d = \frac{mgSin\theta}{k} \Rightarrow d = \frac{(1)g\frac{1}{2}}{50} \Rightarrow d = \frac{g}{100} \quad \text{metres}$$

(b) Frequency of the resulting vibration.

At extension d the forces acting on the system are shown in Fig. 2.

The restoring force at extension x is kd, directed towards the equilibrium position:

$$\Rightarrow m\frac{d^2x}{dt^2} = -kd \Rightarrow \frac{d^2x}{dt^2} = \left(\frac{-k}{m}\right)d \quad \text{which is in the form} \quad \frac{d^2x}{dt^2} = -\omega^2 d$$

$$\Rightarrow \omega^2 = \frac{k}{m} \Rightarrow \omega = \sqrt{\frac{k}{m}} = \sqrt{\frac{50}{1}} = 5\sqrt{2}$$

But $\omega = 2\pi f \Rightarrow f = \frac{5\sqrt{2}}{2\pi} = \frac{5}{\pi\sqrt{2}} \quad \text{cycles/second(hz)}$

19. Two elastic strings and particle

A heavy particle is hung from points A and B (where $|AB| = D$) by two light elastic strings of natural length $6L$, $5L$ metres and elastic constants 4 and 3 N/m respectively. In equilibrium the two strings make equal angles $\theta = 30°$ with the vertical. Derive an expression for L in terms of D.

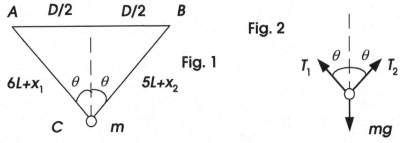

Fig. 1

Fig. 2

Solution

The arrangement of the strings is shown in Fig. 1. The forces in Newtons acting on the particle in equilibrium are shown in Fig. 2:

T = Tension in strings, Mg = Weight of particle

We have : $T_1 \sin\theta = T_2 \sin\theta \Rightarrow T_1 = T_2$

But : $T_1 = 4x_1$ (where x_1, x_2 are extensions of the $6L$ and $5L$ elastic strings

respectively) and : $T_2 = 3x_2 \Rightarrow 4x_1 = 3x_2 \Rightarrow x_1 = \left(\dfrac{3}{4}\right)x_2$

Also : $(6L + x_1)\cos\theta = (5L + x_2)\cos\theta \Rightarrow 6L + x_1 = 5L + x_2$

$\Rightarrow L = x_2 - x_1 = x_2 - \left(\dfrac{3}{4}\right)x_2 = x_2\left(\dfrac{1}{4}\right) \Rightarrow x_2 = 4L \Rightarrow \sin\theta = \dfrac{\frac{D}{2}}{5L + 4L} = \dfrac{\frac{D}{2}}{9L} = \dfrac{D}{18L}$

But (given): $\theta = 30° \Rightarrow \sin 30° = \dfrac{1}{2} = \dfrac{D}{18L} \Rightarrow L = \dfrac{D}{9}$ metres

20. Two vertical springs

A mass m kg is attached to two springs of natural length L_1 and L_2 metres and stiffnesses k_1 and k_2 N/m respectively which are mounted vertically as shown in Fig. 1. The mass causes the springs to be extended by distances x_1 and x_2 metres respectively. The mass is then pulled down a further distance x metres and released. Show that the mass will perform simple harmonic motion and find the period of this motion

Solution

Assume both springs are in tension throughout the motion

Equilibrium conditions:

In equilibrium conditions assume that Spring 1 is stretched by d_1 and Spring 2 is stretched by a distance d_2. See Fig. 1.

For equilibrium (See Fig. 2): $T' = T'' + Mg$

From Hooke's Law: $T' = k_1 d_1$, $T'' = k_2 d_2 \Rightarrow Mg = T' - T'' = k_1 d_1 - k_2 d_2$

Mass displaced from equilibrium position through a distance x:

Equation of motion: $M\dfrac{d^2x}{dt^2} = Mg + T_2 - T_1$ (See Figs. 3 and 4)

From Hooke's Law:

$T_1 = k_1(\text{extension}) = k_1 (d_1 + x)$, $T_2 = k_2 (\text{extension}) = k_2 (d_2 - x)$

But (from above): $Mg + T_2 - T_1 = (k_1 d_1 - k_2 d_2) + k_2 (d_2 - x) - k_1 (d_1 + x)$

$\Rightarrow M\dfrac{d^2x}{dt^2} = (k_1 d_1 - k_2 d_2) + k_2(d_2 - x) - k_1(d_1 + x) = -k_2 x - k_1 x = -(k_1 + k_2)x$

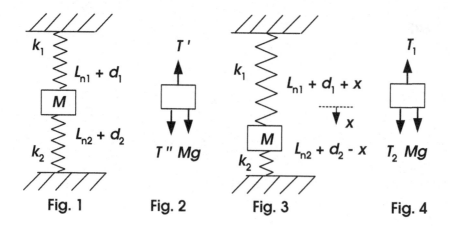

Fig. 1   Fig. 2   Fig. 3   Fig. 4

$$\Rightarrow \frac{d^2x}{dt^2} = \text{acceleration} = a = -\left(\frac{k_1 + k_2}{M}\right)x$$

$\Rightarrow$ Restoring force $\propto$ displacement from the equilibrium position and is directed towards the equilibrium position, in the opposite direction to x

$\Rightarrow$ The mass moves with simple harmonic motion

This is in the format: acceleration $= -\omega^2 x$ where $\omega = \sqrt{\dfrac{k_1 + k_2}{M}}$

$\Rightarrow$ The motion has a period $= \dfrac{2\pi}{\omega} = 2\pi\sqrt{\dfrac{M}{k_1 + k_2}}$

21. Two Elastic Strings

A mass of m kg is suspended from a length, L_1 metres, of elastic string of modulus λ_1 (the other end of which, A, is connected to a fixed point). A second mass, m_2, is suspended by a length, L_2 metres, of elastic string of the same modulus. (See Fig. 1). When the system is in equilibrium find the distance between and A and m_2.

Solution

From Fig. 1: The distance between and A and $m_2 = (L_1 + d_1) + (L_2 + d_2)$

From Fig. 2: $T_1 = m_1 g + T_2 = \dfrac{\lambda_1}{L_1}d_1$ $T_2 = m_2 g = \dfrac{\lambda_1}{L_2}d_2$ $\Rightarrow m_1 g + m_2 g = \dfrac{\lambda_1}{L_1}d_1$

$$\Rightarrow d_2 = \frac{m_2 g L_2}{\lambda_1}, \quad d_1 = \frac{(m_1 + m_2)g L_1}{\lambda_1}$$

$$\Rightarrow \text{Distance} = L_1 + L_2 + \frac{(m_1 + m_2)g L_1}{\lambda_1} + \frac{m_2 g L_2}{\lambda_1} = L_1 + L_2 + \frac{(m_1 + m_2)g L_1 + m_2 g L_2}{\lambda_1}$$

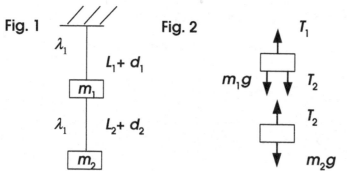

Fig. 1   Fig. 2

242

22. Elastic strings (vertical extension)

A mass m kg is attached to two elastic strings of natural length L_1 and L_2 metres and modulus λ_1 and λ_2 N respectively which are mounted vertically as shown in Fig. 1. The mass causes the strings to be extended by distances d_1 and d_2 metres respectively. The mass is then pulled down a further distance x metres and released. Show that the mass will perform simple harmonic motion and find the period of this motion

Solution

Assume both strings are in tension throughout the motion

Equilibrium conditions:

In equilibrium conditions assume that String 1 is stretched by d_1 and String 2 is stretched by a distance d_2. (See Fig. 1).

For equilibrium (See Fig. 2): $T = T' + Mg$

From Hooke's Law: $T = \dfrac{\lambda_1}{L_{n1}} d_1$, $T' = \dfrac{\lambda_2}{L_{n2}} d_2 \Rightarrow Mg = T - T' = \dfrac{\lambda_1}{L_{n1}} d_1 - \dfrac{\lambda_2}{L_{n2}} d_2$

Mass displaced from equilibrium position through a distance x:

See Figs. 3 and 4. Equation of motion: $M\dfrac{d^2x}{dt^2} = Mg + T_2 - T_1$ (See Figs. 3 and 4)

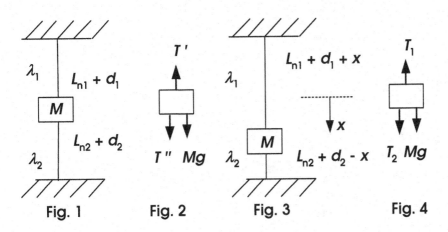

| Fig. 1 | Fig. 2 | Fig. 3 | Fig. 4 |

From Hooke's Law:

$T_1 = \dfrac{\lambda_1}{L_{n1}}(\text{extension}) = \dfrac{\lambda_1}{L_{n1}}(d_1 + x)$, $\quad T_2 = \dfrac{\lambda_2}{L_{n2}}(\text{extension}) = \dfrac{\lambda_2}{L_{n2}}(d_2 - x)$

$Mg + T_2 - T_1 = \left(\dfrac{\lambda_1}{L_{n1}}d_1 - \dfrac{\lambda_2}{L_{n2}}d_2\right) + \dfrac{\lambda_2}{L_{n2}}(d_2 - x) - \dfrac{\lambda_1}{L_{n1}}(d_1 + x) \Rightarrow M\dfrac{d^2x}{dt^2} = -\left(\dfrac{\lambda_1}{L_{n1}} + \dfrac{\lambda_2}{L_{n2}}\right)x$

$\Rightarrow$ Restoring force $\propto$ displacement from the equilibrium position and the restoring force acts towards the equilibrium position and in opposite direction to x). If the Restoring

force $= Ma$ then $Ma = -(\dfrac{\lambda_1}{L_{n1}} + \dfrac{\lambda_2}{L_{n2}})x \Rightarrow$ acceleration $= a = -\dfrac{1}{M}\left(\dfrac{\lambda_1}{L_{n1}} + \dfrac{\lambda_2}{L_{n2}}\right)x$

$\Rightarrow$ The mass moves with simple harmonic motion

This is in the format: acceleration $= -\omega^2 x$ $\quad$ where $\omega = \sqrt{\dfrac{1}{M}\left(\dfrac{\lambda_1}{L_{n1}} + \dfrac{\lambda_2}{L_{n2}}\right)}$

$\Rightarrow$ The motion has a period $= \dfrac{2\pi}{\omega} = \dfrac{2\pi}{\sqrt{\dfrac{1}{M}\left(\dfrac{\lambda_1}{L_{n1}} + \dfrac{\lambda_2}{L_{n2}}\right)}}$

23. Elastic string (vertical extension)

A mass m kg is attached to the mid-point of an elastic string of natural length L metres and modulus λ N respectively which is attached to two points AB in a vertical line a distance $L + d$ metres apart as shown in Fig. 1. In equilibrium conditions the mass causes both halves of the string to be extended by distances d_1 and d_2 metres respectively. The mass is then pulled down a further distance x metres and released. Show that the mass will perform simple harmonic motion and find the period of this motion.

Solution

Assume the string is in tension throughout the motion

Note: The particle divides the string into two halves each of modulus $\square$ and natural length $L/2$ metres

Equilibrium conditions:

In equilibrium conditions assume that String 1 is stretched by d_1 and String 2 is stretched by a distance d_2 where (See Fig. 1): $d = d_1 + d_2$ Equation (i)

For equilibrium (See Fig. 2): $T = T' + Mg$ From Hooke's Law: $T = \dfrac{\lambda}{L/2}\, d_1,\ T' = \dfrac{\lambda}{L/2}\, d_2$

$\Rightarrow Mg = T - T' = \dfrac{2\lambda}{L}\, d_1 - \dfrac{2\lambda}{L}\, d_2 \Rightarrow \dfrac{MgL}{2\lambda} = d_1 - d_2$ Equation (ii)

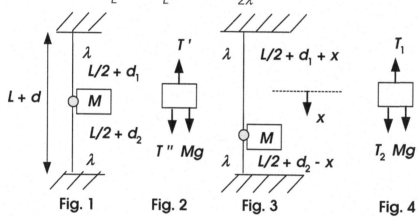

| Fig. 1 | Fig. 2 | Fig. 3 | Fig. 4 |

Adding equations (i) and (ii) gives: $d_1 = \dfrac{MgL + 2\lambda d}{4\lambda}$, $d_2 = \dfrac{2\lambda d - Mgl}{4\lambda}$

Mass displaced from equilibrium position through a distance x:
See Figs. 3 and 4.

Equation of motion : $M\dfrac{d^2 x}{dt^2} = Mg + T_2 - T_1$ (See Figs. 3 and 4)

From Hooke's Law:

$T_1 = \dfrac{2\lambda}{L}(\text{extension}) = \dfrac{2\lambda}{L}(d_1 + x),\quad T_2 = \dfrac{2\lambda}{L}(\text{extension}) = \dfrac{2\lambda}{L}(d_2 - x)$

$Mg + T_2 - T_1 = \left(\dfrac{2\lambda}{L}d_1 - \dfrac{2\lambda}{L}d_2\right) + \dfrac{2\lambda}{L}(d_2 - x) - \dfrac{2\lambda}{L}(d_1 + x) \Rightarrow M\dfrac{d^2 x}{dt^2} = -\left(\dfrac{4\lambda}{L}\right)x$

$\Rightarrow$ Restoring force $= -(\dfrac{4\lambda}{L})\, x$ (towards equilibrium position, in opposite direction to x) $\Rightarrow$

Restoring force $\propto$ displacement from the equilibrium position
$\Rightarrow$ The mass moves with simple harmonic motion

If Restoring force = Ma then $Ma = -\left(\dfrac{4\lambda}{L}\right)x \Rightarrow$ acceleration $= a = -\dfrac{4\lambda}{ML}x$

This is in the format: acceleration $= -\omega^2 x$ where: $\omega = \sqrt{\dfrac{4\lambda}{ML}}$

$\Rightarrow$ The motion has a period $= \dfrac{2\pi}{\omega} = 2\pi\sqrt{\dfrac{ML}{4\lambda}} = \pi\sqrt{\dfrac{ML}{\lambda}}$

24. Elastic string (horizontal extension)

One end of an elastic string of natural length L metres and modulus k N is attached to a point, O, on a smooth horizontal plane. A mass m kg is attached to the other end of the elastic string and mass pulled out to point B, stretching the elastic string by a metres, and then released. Find the period of the subsequent motion.

Solution

The motion is shown in Fig. 1. The mass will travel from B to C and back to B, etc. The motion from C to D, D to C, A to B, B to A will be dependent on the tension exerted by the stretched elastic string and will be simple harmonic motion. During the motion from D to A and A to D the elastic string will be slack and the mass will travel at nil acceleration (i.e. constant velocity) and, therefore, the motion will be periodic but will not be simple harmonic motion.

Simple Harmonic Motion

From Hooke's Law: $T_1 = \dfrac{\lambda}{L}$ (extension) $= -\dfrac{\lambda}{L}a \Rightarrow$ Restoring force $= -\dfrac{\lambda}{L}a$

(towards equilibrium position, O)
$\Rightarrow$ (See Fig. 2) Restoring force $\propto$ displacement from the equilibrium position
$\Rightarrow$ The mass moves with simple harmonic motion
If the Restoring force = (Mass)(acceleration) then

(Mass)(acceleration) $= -\dfrac{\lambda}{L}x \Rightarrow$ acceleration $= -\left(\dfrac{\lambda}{ML}\right)x$

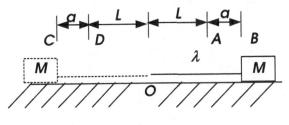

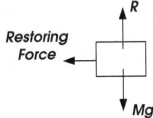

| **Fig. 1** | **Fig. 2** |

This is in the format: acceleration $= T_1 = -\omega^2 x$ where $\omega = \sqrt{\dfrac{\lambda}{ML}}$

$\Rightarrow$ The motion has a period $= \dfrac{2\pi}{\omega} = 2\pi\sqrt{\dfrac{ML}{\lambda}}$

$\Rightarrow$ Time to move from B to $A = \dfrac{1}{4}$ of $\dfrac{2\pi}{\omega} = \dfrac{\pi}{2}\sqrt{\dfrac{ML}{.\lambda}}$ seconds

$\Rightarrow$ Time to move from B to A, D to C, C to D and A to B = four times the time to traverse

one of these distances $= 2\pi\sqrt{\dfrac{ML}{\lambda}}$ seconds

Constant velocity motion

The total energy at B = Potential Energy (i.e. energy in stretched string) + Kinetic Energy.
As the mass is at rest at point B its Kinetic Energy is 0.
The energy required to stretch the elastic string through a distance a metres (from A to B) is: $\frac{1}{2}\left(\frac{\lambda}{L}\right)a^2 \Rightarrow$ Energy of mass at $B = \frac{1}{2}\left(\frac{\lambda}{L}\right)a^2$

As the mass reaches A the elastic string becomes slack so its Potential Energy = 0 but it will have attained a speed v so its Kinetic Energy $= \frac{1}{2}Mv^2$

But the total energy must remain constant $\Rightarrow$ Total Energy at B = Total Energy at A.

$\Rightarrow \frac{1}{2}\left(\frac{\lambda}{L}\right)a^2 = \frac{1}{2}Mv^2 \Rightarrow v = \pm a\sqrt{\frac{\lambda}{ML}}$ ("–" sign indicating direction of travel))

The part of the cycle of motion traversed at constant velocity = $DA + AD = 4L$ metres.
The time required to travel this distance = 4L/speed

$\Rightarrow$ Time to traverse $4L$ metres $= \dfrac{4L}{a\sqrt{\dfrac{\lambda}{ML}}} = \dfrac{4L}{a}\sqrt{\dfrac{ML}{\lambda}}$

Total time to complete one full cycle of motion
Total time = Time spent under Simple Harmonic Motion + Time spent under Constant

Velocity $= 2\pi\sqrt{\dfrac{ML}{\lambda}} + \dfrac{4L}{a}\sqrt{\dfrac{ML}{\lambda}} = \sqrt{\dfrac{ML}{\lambda}}\left(2\pi + \dfrac{4L}{a}\right)$ seconds

25. Two horizontal springs

A mass m kg, sitting on a smooth horizontal plane, is attached to two springs of natural length L_1 and L_2 metres and stiffnesses k_1 and k_2 N/m respectively which are attached to points C and D as shown in Fig. 1. Initially the system is in equilibrium and the springs are stretched by d_1 and d_2 metres respectively. The mass is then displaced through a small distance x metres towards D and released (Assume both springs are in tension throughout the motion). Show that the mass will perform simple harmonic motion and find the period of this motion.

Solution

Fig. 1   **Fig. 2**   **Fig. 3**

Equilibrium conditions:
In equilibrium conditions assume that Spring 1 is stretched by d_1 and Spring 2 is stretched by a distance d_2. (See Fig. 1) Note: Vertical forces: $R = Mg$
For equilibrium: $T' = T''$
From Hooke's Law: $T' = k_1 d_1$, $T' = k_2 d_2 \Rightarrow k_1 d_1 = k_2 d_2$ Equation (i)

Mass displaced from equilibrium position through a distance x:

246

$$\Rightarrow M\frac{d^2x}{dt^2} = T_2 - T_1 \text{ (See Figs. 3 and 4)}$$

From Hooke's Law: $T_1 = k_1(\text{extension}) = k_1(d_1 + x)$, $T_2 = k_2(\text{extension}) = k_2(d_2 - x)$
But (from above): $T_2 - T_1 = k_2(d_2 - x) - k_1(d_1 + x)$

$$\Rightarrow M\frac{d^2x}{dt^2} = k_2(d_2 - x) - k_1(d_1 + x) \quad \text{But}: k_1d_1 = k_2d_2 \Rightarrow M\frac{d^2x}{dt^2} = -k_2x - k_1x = -(k_1 + k_2)x$$

$\Rightarrow$ Restoring force acts towards the equilibrium position and in the opposite direction to
$x \Rightarrow$ Restoring force $\propto$ displacement from the equilibrium position
$\Rightarrow$ The mass moves with simple harmonic motion

$$\Rightarrow \frac{d^2x}{dt^2} = \text{acceleration} = a = -\left(\frac{k_1 + k_2}{M}\right)x$$

This is in the format: acceleration $= a = -\left(\dfrac{k_1 + k_2}{M}\right)x \quad$ where $\omega = \sqrt{\dfrac{k_1 + k_2}{M}}$

$$\Rightarrow \text{The motion has a period} = \frac{2\pi}{\omega} = 2\pi\sqrt{\frac{M}{k_1 + k_2}}$$

Note: If the following are given: Distance $CD = pL_n$, $L_{n1} = qL_n$, $L_{n2} = rL_n$
$\Rightarrow pL_n = qL_n + rL_n + d_1 + d_2 \Rightarrow (p - q - r)L_n = d_1 + d_2 \quad$ Equation (ii)
Using Equations (i) and (ii):

$$k_1d_1 = k_2d_2 \Rightarrow d_1 = \frac{k_2}{k_1}d_2 \Rightarrow \frac{k_2}{k_1}d_2 + d_2 = (p - q - r)L_n$$

$$\Rightarrow d_2 = \left(\frac{k_1}{k_1 + k_2}\right)(p - q - r)L_n \quad \text{and} \quad d_1 = \left(\frac{k_2}{k_1 + k_2}\right)(p - q - r)L_n$$

26. Elastic string (horizonal extension)

A mass m kg is attached to the mid-point of an elastic string of natural length L metres
and modulus λ N respectively which is attached to two points CD in a horizontal line a
distance $L + d$ metres apart as shown in Fig. 1. In equilibrium conditions the mass causes
both halves of the string to be extended by distances d_1 and d_2 metres respectively.
The mass is then displaced a further distance x metres towards D and released. Show
that the mass will perform simple harmonic motion and find the period of this motion.

Solution

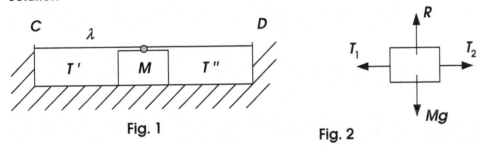

Fig. 1   **Fig. 2**

If the mass is attached to the mid-point of the extended string then both halves of the
string will have natural length $\dfrac{L_n}{2}$ and modulus λ. Assume both halves of the string are in
tension throughout the motion
Equilibrium conditions:
In equilibrium conditions assume that String 1 is stretched by d_1 and String 2 is stretched
by a distance d_2. Note: Vertical forces: $R = Mg$

$d = d_1 + d_2$ Equation (i)

For equilibrium: $T = T'$

From Hooke's Law: $T = \dfrac{\lambda}{L_n/2}\, d_1$, $T' = \dfrac{\lambda}{L_n/2}\, d_2$ $\Rightarrow \dfrac{2\lambda}{L_n}\, d_1 = \dfrac{2\lambda}{L_n}\, d_2 \Rightarrow d_1 = d_2$

$\Rightarrow$ From equation (i): $d_1 = d_2 = \dfrac{d}{2}$

Mass displaced from equilibrium position through a distance x:

See Fig. 2.

Equation of motion: $M\dfrac{d^2x}{dt^2} = T_2 - T_1$ (See Fig. 2) $\qquad$ From Hooke's Law:

$T_1 = \dfrac{2\lambda}{L_n}(\text{extension}) = \dfrac{2\lambda}{L_n}(d_1 + x)$, $\quad T_2 = \dfrac{2\lambda}{L_n}(\text{extension}) = \dfrac{2\lambda}{L_n}(d_2 - x)$

$T_2 - T_1 = \dfrac{2\lambda}{L_n}(d_2 - x) - \dfrac{2\lambda}{L_n}(d_1 + x)$ $\quad$ But: $d_1 = d_2 \Rightarrow M\dfrac{d^2x}{dt^2} = -\left(\dfrac{4\lambda}{L}\right)x$

$\Rightarrow$ Restoring force $= -\left(\dfrac{4\lambda}{L_n}\right)x$ (acting towards equilibrium position and in opposite

direction to x) $\Rightarrow$ Restoring force $\propto$ displacement from the equilibrium position $\Rightarrow$ The mass moves with simple harmonic motion

If the Restoring force $= Ma$ then $Ma = -\left(\dfrac{4\lambda}{L_n}\right)x \Rightarrow$ acceleration $= a = -\left(\dfrac{4\lambda}{ML_n}\right)x$

This is in the format: acceleration $= -\omega^2 x$ $\quad$ where $\omega = \sqrt{\dfrac{4\lambda}{ML_n}}$

$\Rightarrow$ The motion has a period $= \dfrac{2\pi}{\omega} = 2\pi\sqrt{\dfrac{ML_n}{4\lambda}} = \pi\sqrt{\dfrac{ML_n}{\lambda}}$

Printed in the United States
By Bookmasters